Nocturnal Asthma

Mechanisms and Treatment

edited by

Richard J. Martin, M.D.

Professor of Medicine
University of Colorado Health Science Center;
Director, Sleep Research Program
Co-Director, Asthma Program
Staff Physician, Department of Medicine
National Jewish Center for Immunology and Respiratory Medicine
Denver, Colorado

Futura Publishing Company, Inc.
Mount Kisco, NY

Library of Congress Cataloging-in-Publication Data

Nocturnal asthma : mechanisms and treatment / edited by Richard
  J. Martin.
    p.   cm.
  Includes bibliographical references and index.
  ISBN 0-87993-546-4
  1. Asthma. 2. Clinical chronobiology. I. Martin, Richard J.
  (Richard Jay), 1946– .
  [DNLM: 1. Asthma. 2. Circadian Rhythm. 3. Sleep—
physiology. WF 553 N759]
  RC591.N63   1993
  616.2'38—dc20
  DNLM/DLC
  for Library of Congress   92-48725
                                                         CIP

Copyright 1993
Futura Publishing Company, Inc.

Published by
Futura Publishing Company, Inc.
2 Bedford Ridge Road
Mount Kisco, New York 10549

LC #: 92-48725
ISBN #: 0-87993-546-4

Printed in the United States of America.

This book is printed on acid-free paper.

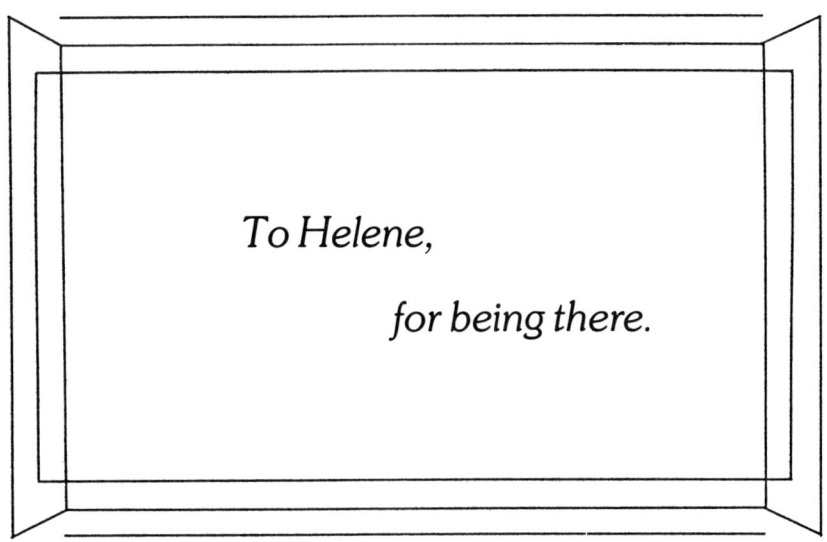

To Helene,

        for being there.

Contributors

Robert D. Ballard, M.D.
Associate Professor of Medicine, University of Colorado Health Sciences Center, Director, Respiratory Therapy, Denver V.A. Medical Center, Denver, Colorado

William R. Beam, M.D.
Instructor of Medicine, University of Colorado Health Sciences Center, Staff Physician, National Jewish Center for Immunology and Respiratory Medicine, Denver, Colorado

William W. Busse, M.D.
Professor of Medicine, Head, Section of Allergy and Clinical Immunology, University of Wisconsin Medical School, Madison, Wisconsin

Gilbert E. D'Alonzo, D.O.
Professor of Medicine, Temple University, Philadelphia, Pennsylvania

Malcolm R. Hill, Pharm. D.
Assistant Professor of Pediatrics, University of Colorado Health Sciences Center, Clinical Research Scientist, Clinical Pharmacology Laboratory, National Jewish Center for Immunology and Respiratory Medicine, Denver, Colorado

Nizar N. Jarjour, M.D.
Assistant Professor of Medicine, Pulmonary and Critical Care Medicine, University of Wisconsin Medical School, Madison, Wisconsin

Richard J. Martin, M.D.
Professor of Medicine, University of Colorado Health Sciences Center; Director of Sleep Research, Co-Director of Asthma Program, National Jewish Center for Immunology and Respiratory Medicine, Denver, Colorado

Ketan K. Sheth, M.D.
Fellow, Allergy/Immunology, Department of Medicine, University of Wisconsin Medical School, Madison, Wisconsin

Michael H. Smolensky, Ph.D.
Professor of Environmental Physiology, School of Public Health; Professor of Pulmonary Medicine, School of Medicine, University of Texas Health Sciences Center-Houston, and Director, Hermann Center for Chronobiology and Chronotherapeutics, Houston, Texas

Stanley J. Szefler, M.D.
Professor of Pediatrics, Professor of Pharmacology, University of Colorado Health Sciences Center, Director of Clinical Pharmacology, National Jewish Center for Immunology and Respiratory Medicine, Denver, Colorado

Sally E. Wenzel, M.D.
Assistant Professor of Medicine, University of Colorado Health Sciences Center, Staff Physician, National Jewish Center for Immunology and Respiratory Medicine, Denver, Colorado

Foreword

As students of the medical sciences, we have been taught much about the pathogenesis of various diseases and their natural histories over long periods of time. Unfortunately, few among us were educated about or observant enough to appreciate the dynamic changes in disease activity that naturally occur over relatively short periods of time. This lack of appreciation of normal biological rhythms and their influence on disease processes is changing. Thanks to the insight and hard work of a core of dedicated and tireless investigators who conduct studies while others sleep, the concepts and applications of medical chronobiology are now being advanced through publication of original investigations as well as summaries of these studies in books such as this.

Asthma is one disease process in which it is important to appreciate normal rhythmic events of a circadian nature if the disease is to receive proper evaluation and treatment. It is abundantly clear that airway patency varies as a circadian rhythm which, in medically unstable patients, can make events occurring during sleep (nocturnal asthma) life-threatening in nature. The statistics reviewed in various chapters of this publication concerning the morbidity and mortality that occur during sleep in subjects with asthma should be firmly anchored within the minds of all who care for patients with this disease.

Nocturnal asthma represents not only an important clinical entity, but also an area of investigation that is providing critical insights in the pathogenesis of this disorder. In this respect, nocturnal asthma represents natural asthma and not disease precipitated within the laboratory by stimuli which may only loosely mimic natural events (allergen challenge). In this respect, the findings of marked changes in airway inflammation that apparently develop and resolve over a 24-hour period of time have made many investigators reassess their views on asthma as well as the factors that produce as well as lead to resolution of inflammatory processes. In this respect, continued study of this natural manifestation of disease activity should provide insight into the inflammatory nature of this disease as well as the factors that normally produce circadian variations in airway function.

This publication is well organized to provide the reader with the necessary background to appreciate the importance of chronobiology in various clinical diseases while concentrating primarily on asthma. Thus, in Chapters 1 and 2, Drs. Smolensky and D'Alonzo present basic concepts of medical chronobiology and review principles of chronopharmacology and therapeutics. This is followed in Chapter 3 by an overview of nocturnal asthma by Dr. Martin and in Chapter 4 by Dr. Ballard's general review of the effects of sleep on various obstructive lung diseases (asthma, chronic obstructive lung disease, cystic fibrosis). Chapter 5 by Drs. Jarjour, Sheth, and Busse and Chapter 6 by Drs. Wenzel and Martin review the inflammatory events associated with nocturnal asthma from the point of view of the various inflammatory cells that may contribute to this process. The final four chapters outline various therapeutic approaches to this problem including use of theophylline and β-adrenergic agents (Drs. D'Alonzo and Smolensky), corticosteroids (Drs. Beam and Martin), and other pharmacologic agents (Drs. Hill and Szefler). In the final chapter, Dr. Martin reviews various nonpharmacological approaches to nocturnal asthma.

In summary, this book provides a current and comprehensive overview of chronobiology, concentrating on nocturnal asthma as the prototypic process in which appreciation of normal biological rhythms can improve patient care and provide insight into the pathogenesis of disease.

GARY L. LARSEN, M.D.
Senior Faculty Member
National Jewish Center for Immunology
  and Respiratory Medicine;
Professor and Head
Section of Pediatric Pulmonary Medicine
University of Colorado School of Medicine
Denver, Colorado

Preface

"I have observed the fit always to happen after sleep in the night where nerves are filled with windy spirits and the heat of the bed has rarified the spirits and humours!" Dr. John Floyer very eloquently described what was occurring with his own asthma during the night in 1698. Not only is he suggesting that the worst time for asthmatic patients occurs during the night, but that the neurogenic system (nerves are filled with windy spirits) and inflammation/mediators (spirits and humors) play an important role in nocturnal asthma.

Indeed, this text is dedicated to a specific disease process, i.e., asthma, but from a relatively new perspective of medicine, that of chronobiological alterations. The vast majority of us were trained from the perspective that the human being is a stable homeostatic system. Thus, what was learned from daytime investigation in regard to both pathophysiology and therapeutics was applied not only to the entire 24-hour cycle, but to all days of the month and all months of the year. This thought process has actually impeded the full understanding of the field of medicine.

This book is directed to clinicians and investigators interested in asthma. It covers general concepts of chronobiology, circadian alterations in physiology and specific cells and mediators of inflammation, as well as chronotherapeutic and nonpharmacological interventions for nocturnal asthma. As Dr. Floyer said, " . . . nerves . . . windy spirits . . . and humours" are much more evident during the sleep-related hours.

RICHARD J. MARTIN, M.D.

Contents

Foreword: Gary L. Larsen, M.D. vii

Preface: Richard J. Martin, M.D. ix

Chapter 1 // Medical Chronobiology: Concepts and Applications
Michael H. Smolensky, Ph.D.
Gilbert E. D'Alonzo, D.O. 1

Chapter 2 // Biological Rhythms and Medications:
Chronopharmacology and Chronotherapeutics
Michael H. Smolensky, Ph.D.
Gilbert E. D'Alonzo, D.O. 25

Chapter 3 // Nocturnal Asthma: An Overview
Richard J. Martin, M.D. 71

Chapter 4 // Effects of Sleep on Respiratory Physiology in
Nocturnal Asthma
Robert D. Ballard, M.D. 117

Chapter 5 // Cellular Mechanisms of Nocturnal Asthma:
The Role of The Eosinophil, Neutrophil,
and Mast Cell
Nizar N. Jarjour, M.D.
Ketan K. Sheth, M.D.
William W. Busse, M.D. 163

Chapter 6 // The Macrophage and Lymphocyte in Nocturnal
Asthma: Potential Roles and Importance
Sally E. Wenzel, M.D.
Richard J. Martin, M.D. 199

Chapter 7   // Chronopharmacology of Theophylline and
               Beta-2 Sympathomimetic Therapies
               Gilbert E. D'Alonzo, D.O.
               Michael H. Smolensky, Ph.D. 221

Chapter 8   // Corticosteroids, Oral and Inhaled, in the
               Treatment of Nocturnal Asthma
               William R. Beam, M.D.
               Richard J. Martin, M.D. 281

Chapter 9   // Other Pharmacological Interventions in
               Nocturnal Asthma
               Malcolm R. Hill, Pharm. D.
               Stanley J. Szefler, M.D. 333

Chapter 10 // Nonpharmacological Interventions for
               Nocturnal Asthma
               Richard J. Martin, M.D. 357

Index 387

1

Medical Chronobiology:

Concepts and Applications

Michael H. Smolensky, Ph.D., and Gilbert E. D'Alonzo, D.O.

Introduction

Biological functions in human beings are organized both in space, in terms of the physical anatomy, and over time, in terms of the sequential stages of growth and development as well as endogenous biological rhythms.[1-3] Biological rhythms are ubiquitous in nature; they are characteristic of most, if not all, bioprocesses in animals, plants, and insects.[4-7] *Chronobiology* is the scientific discipline concerned with the elucidation, mechanisms, and significance of intrinsic biological rhythms of all life forms.

Most clinicians are unfamiliar with the science of chronobiology. The teaching of biology in preparatory and medical school stresses the concept of homeostasis, which purports that the *milieu interne* is maintained in a relatively constant state by a set of specific feedback mechanisms. Chronobiology, on the other hand, stresses the premise that human bioprocesses and functions manifest predictable and recurring variability in time, detectable as biological rhythms, at every level of organization. Homeostatic

Martin RJ (editor): *Nocturnal Asthma: Mechanisms and Treatment,* © Futura Publishing Co., Inc., Mount Kisco, NY, 1993.

mechanisms maintain the moment-to-moment equilibrium of the internal environment, while biological clocks and their associated rhythms ensure that the organism is prepared to cope successfully with predictable-in-time changes in the ambient environment occurring over periods such as 24 hours and the year. These intrinsic biological rhythms and clocks also coordinate biochemical, endocrine, and physiological processes and functions to meet the differential requirements for energy substrates during activity versus sleep over the 24 hours, as well as organize reproductive events and activities during the fertility cycle to ensure conception and the perpetuation of the species.

The concept of chronobiology is not new in medicine. Yet, because it has been perceived as being at odds with the dominating theory of homeostasis, it has not always been accepted as an important field of scientific endeavor by physicians. The conceptual differences between homeostasis and chronobiology are not merely academic but are of practical importance in medicine. The concept of homeostasis infers that the risk and intensity of human disease is of equal probability each hour of the day and night, each day of the menstrual cycle, and each month of the year. Moreover, homeostasis leads one to expect that the major goal of pharmacotherapy should be constancy of blood or tissue concentrations of medications in order to achieve constancy in therapeutic efficacy. Indeed, medications are administered in equal intervals and in equal doses to achieve this goal. In addition, the theory of homeostasis implies that the time, for example, morning versus evening, when medications are administered is inconsequential with regard to their pharmacokinetics and pharmacodynamics. The emerging findings from a great many chronobiological investigations challenge these and other tenets of homeostasis. These call for a reconsideration of several traditional assumptions and practices common to medicine today.

Human Chronobiology:
Concepts and Principles

The focus of this chapter is medical chronobiology, particularly with reference to the topics of this volume—allergy and chest disease. However, before discussing the implications of human biological rhythms to medicine, it is necessary first to define the central concepts, principles, and terms of this new science.

A *biological rhythm* is a self-sustaining oscillation with the period, that is, the amount of time required to complete each repetition, being relatively nonvarying under normal conditions. Biological rhythms are defined by a specific set of characteristics.[1-3,8]

The first one is *period*. The period of a rhythm is the length of time needed to complete its cycle, 24 hours or 28 days, for example. One of the most important bioperiodicities in medicine is the circadian (*circa* = about; *dies* = day, or about 24-hour) rhythm. Biological oscillations that are shorter than 20 hours in duration are referred to as ultradian rhythms. Bioperiodic events of the EEG and ECG are well-recognized examples of ultradian rhythms. Oscillations greater than 28 hours in duration are termed infradian rhythms. Weekly, menstrual, and seasonal cycles in biological function are examples of infradian rhythms. Up to now, the circadian cycle has received the greatest attention in medicine.

Amplitude is a second characteristic of rhythms. Amplitude is the predictable variability in a biological function over time which is directly ascribable to rhythmicity. Some circadian rhythms, such as those in body temperature and heart rate, are of relatively low amplitude. In contrast, others are of quite high amplitude. The circadian fluctuations in blood cortisol, epinephrine, and lymphocyte number are but a few examples of high-amplitude biological rhythms.

Rhythms are characterized, too, by a so-called "time-series average" or *level* around which predictable variability in time is manifested.

Finally, rhythms are characterized by their *phasing*, that is, the occurrence of peak and/or trough values with reference to a given scale of time, such as the 24 hours, month (menstrual cycle), and year.

Collectively, chronobiologists refer to the precisely integrated temporal order of biofunctions and processes as the *biological time structure*. In health, there exists a harmony of function over time, not only with respect to the staging of the numerous intrinsic rhythmic processes of the same period, such as ones of 24 hours, but with reference to rhythms of other (ultradian and infradian) periods as well.[1-3]

Initially, it was hypothesized that rhythms represented a type of direct or conditioned response of organisms to specific cyclical phenomena and events in the ambient environment. However, research findings prove that living beings inherit, by means of genetic transmission, unique clock mechanisms and associated periodicities.[9-11] Endogenous rhythmicity constitutes a fundamental functional characteristic of cells, tissues, and organs. The period and phasing of rhythms are coordinated by pacemaker clocks located at various levels of biological organization, with those in the brain being most important.[12-14]

While it is recognized that rhythmicity itself is an inherited trait, biological clocks are known to be set or reset by certain environmental time cues (termed *synchronizers*). With reference to the circadian time structure, these include the time of sunrise or lights-on in the morning and sunset or lights-off at night in conjunction with our 24-hour cycle of activity and sleep.[15-17]

It is of interest that the circadian rhythms of persons exposed to experimental conditions devoid of time cues continue to persist.[15-18] This has been demonstrated in a series of investigations on volunteers subjected to prolonged (several months in duration) sojourns in the constant environmental conditions of underground caves and special study bunkers. In the absence of external time cues, such as sunrise and sunset or wristwatches and other timepieces, however, the exact period of circadian bioclocks and the rhythms they drive tends to drift or free-run from the expected one of exactly 24.0 hours known for persons dwelling in a normal environment. Under constant conditions, the period of human circadian rhythms commonly lengthens to about 25 hours or longer.[15] For this reason, it is postulated that the inherited period of the human circadian clock is greater than 24 hours.

In environments devoid of time cues, the free-running period of each of the body's circadian rhythms is unique, differing slightly in duration between one another and from precisely 24.0 hours. This means that the normal phase-relationships between circadian bioperiodicities common to the usual conditions of everyday life become radically disordered and desynchronized. This desynchronization of the organism's time structure results in a loss of biological efficiency. Traveling east or west across several time zones by high-speed aircraft results in a similar situation, a temporary desynchronization of the circadian time structure; this constitutes the basis for the myriad of complaints and symptoms referred to as jet lag.[19,20]

Under the conditions of everyday life, the period and phasing of human circadian clocks and rhythms are set, or said to be *synchro-*

nized, primarily by the 24-hour schedule of sleep (in darkness) and activity (in light).[17,21] In persons who adhere to a relatively consistent activity-rest schedule, the time of the peaks and troughs of most circadian rhythms is quite predictable from day to day. Some people, however, do not adhere to consistent sleep-wake routines. People who are employed in industries requiring rotating shift work must periodically alter the timing of their sleep-activity routine. When working days, sleep is taken during the nighttime; in contrast, when working nights, sleep must be taken during the daytime.

The alteration of the sleep-wake synchronizer schedule by rotating shift workers is associated with the phase resetting of their circadian rhythms. The extent of change in the staging of the endogenous circadian rhythms in such workers is equal in magnitude to the change in their sleep-activity routine. In other words, a shift in terms of clock time in a worker's sleep-activity routine of 8 hours induces a corresponding shift in the staging of the worker's circadian time structure of 8 hours. Because of this, the occurrence of the peak and trough of circadian processes in day in comparison to night workers will differ with reference to external clock time.[22,23] For a diurnally active person who awakens around 6:30 AM and retires to bed around 11:00 PM, the peak in the circadian rhythm of plasma cortisol occurs between 6 AM and 8 AM. For a person adjusted to night-shift work and who sleeps during the daytime, the peak plasma cortisol concentration occurs later, when awakening from daytime sleep. Although the occurrence of the peak of the circadian rhythm in plasma cortisol differs with reference to *external* clock time in diurnally versus nocturnally active persons, it is comparable when referenced to *internal* time, such as the peak or trough of an endogenous circadian *marker rhythm*. In this regard, the sleep-wake cycle constitutes a most useful reference for making a judgment about the expected phasing of the multitude of endogenous circadian functions and processes in individuals, especially those compliant to a more or less consistent life routine.[21] This point is crucial for the proper application of chronobiology in medicine.

In summary, the biological time structure of living organisms is an inherited characteristic. Its expression, as in the case of circadian rhythms, however, may be influenced by environmental factors, particularly the schedule of sleep and activity. Other factors such as disease-related phenomena and medications influence the expression of biological rhythms as well.

Circadian Rhythms and the
Manifestation of Disease

Human beings vary greatly in their physiological and biochemical status over the 24 hours due to the staging of numerous circadian rhythms. The periodic alteration of our functional status gives rise to day-night differences in the susceptibility and resistance of patients to the underlying pathophysiology of a disease process. This commonly gives rise to temporal patterns in the signs and symptoms of a disease. Although asthma has received by far the most attention as a circadian rhythm-influenced disorder, it is only one of many diseases which are known to differ in intensity or occurrence in a cyclical fashion over the 24 hours as a result of chronobiological factors. Other diseases that demonstrate circadian variability include:

Allergic rhinitis[24,25] with the major symptoms of sneezing, rhinorrhea, and congestion typically being much worse upon arising from nightly sleep than during the middle of the activity span (Fig. 1).

Asthma[26,27] with the risk of symptoms in most patients being more than 100-fold greater during the span of sleep than activity.

Typical angina[28] with chest pain and electrocardiographic abnormalities being most common during the first 3 to 5 hours of the daily activity span (Fig. 2).

Prinzmetal angina[29–31] with the manifestation of electrocardiographic abnormalities being restricted mainly to the sleep span.

Myocardial infarction[32–34] with the incidence of heart attack, as well as the mortality associated with it, being greatest during the initial hours of the daily activity span (Fig. 2).

Stroke[35] with the highest incidence occurring between the later portion of the sleep span and the initial hours of activity.

Cerebral hemorrhage[3] **with the greatest incidence occurring in the evening about 3 hours prior to bedtime.**

Essential hypertension[36–38] with patients generally manifesting greatest systolic and/or diastolic blood pressure around the middle to later span of daily activity and lowest pressure overnight or a few hours before awakening. In hypertension associated with renal failure,[38,39] the peak and trough of systolic and diastolic blood pressure may be reversed with the highest blood pressure occurring at night during sleep (Fig. 3).

Arthritis[40,41] with the signs and symptoms of rheumatoid

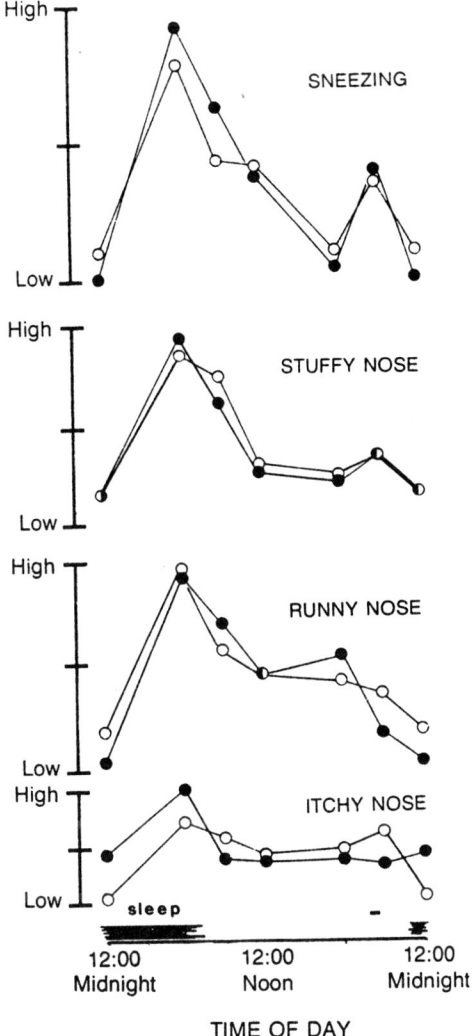

Figure 1. Temporal variation during the 24 hours in the severity of the major symptoms of allergic rhinitis as self-rated by 330 men (●) and 435 women (○) at specific clock hours daily over a 7-day span. Patients were studied without pharmacotherapy of any type. The symptoms of sneezing as well as nasal congestion and rhinorrhea are worse upon awakening after nightly sleep. Bouts of sneezing are also common in the evening. Although the complaint of nasal pruritus is more common in the morning, it is not as variable during the 24 hours as the other self-rated symptoms. Adopted from Reinberg et al.[25]

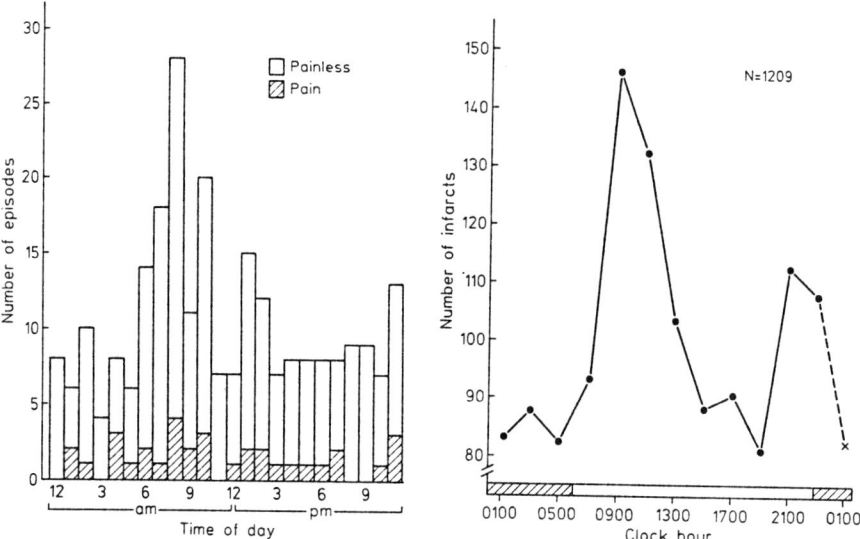

Figure 2. Twenty-four-hour pattern in the occurrence of myocardial ischemia as defined by ST-segment depression in 24 patients with typical angina (left) and occurrence of myocardial infarction in 1,209 presumably diurnally active individuals (right). The risk of angina and of heart attack both is greatest during the initial hours of daily activity. Time shown in military units: 1200 = noontime and 0000 = midnight. Figures reproduced from Rocco et al.[28] and Reinberg and Smolensky.[3]

arthritis being most intense upon awakening and those of osteoarthritis generally being worse around the middle and later portion of daily activity.

Peptic ulcer disease[42,43] with the pain manifesting typically after gastric emptying following daytime meals and occurring prominently during the very early morning hours causing disruption of sleep.

Epilepsy[44-46] with the occurrence of overt seizures often being restricted to particular times of the day or night, with individual differences in patterns between patients.

These are but a few diseases which exhibit rather large-amplitude, day-night patterns in their occurrence or intensity over the 24 hours. It should be mentioned that chronobiological examples pertaining only to the circadian time scale have been emphasized. Important predictable-in-time alterations in physiological and dis-

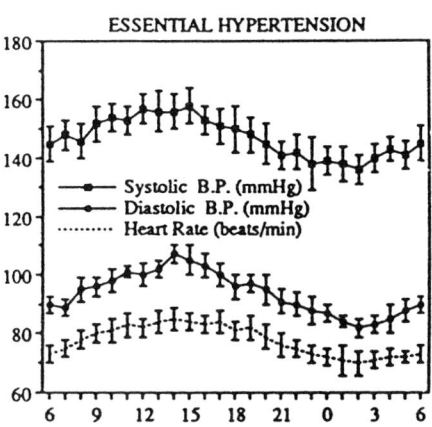

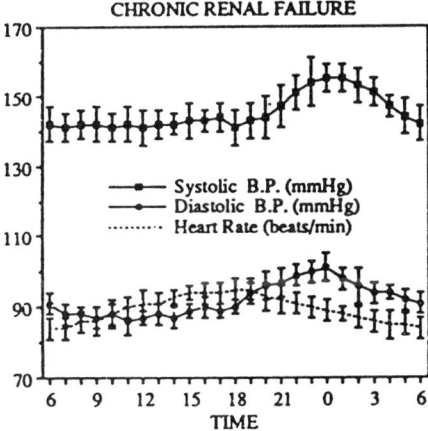

Figure 3. Circadian rhythm in systolic and diastolic blood pressure in groups of 30 patients suffering either from essential hypertension (top) or chronic renal failure (bottom). Diurnally active patients with essential hypertension exhibit highest arterial pressures during the day and lowest ones during the night; patients with chronic renal failure manifest an inverse temporal pattern. The circadian variation in systolic and diastolic pressures is not trivial, averaging around 25 mm Hg. A diagnosis based on single once-daily determination of blood pressure carries the risk of either false-positive or -negative conclusions since unusually high or low systolic and diastolic values may result only from the expression of the circadian rhythm of blood pressure. Time designated in military units: 12 = noontime and 0 = midnight. Figures reproduced from Portaluppi et al.[38]

ease status occur over other time domains as well, such as the week, month (in premenopausal women) and the year.[47–50]

Biological Rhythms and Diagnostic Procedures and Tests

Because the amplitude of circadian rhythms can be rather great, the time when diagnostic assessments are executed may constitute a very significant influence upon clinical findings. This is true for tests utilized in both general medical practice as well as the specialties of allergy and pulmonary medicine.

Hypertension

The early detection and treatment of hypertension is a major priority in general medicine. Typically, blood pressure is assessed in the clinic during the regular daytime office hours to determine if it is within normative values. Generally, these values are based on once-a-day measurements and are without consideration of circadian changes in systolic and diastolic pressures. Blood pressure is not constant during the 24 hours. In diurnally active, normotensive and hypertensive individuals, systolic and diastolic blood pressures generally are higher, by 10–25 mm Hg, in the afternoon in comparison to the sleep-related and morning hours.[51] Due to such circadian change, the clinical assessment of diurnally active, hypertensive patients in the morning might give rise to an underestimate of the severity of the hypertension or even to an occasional false-negative finding (Fig. 3). The circadian rhythm in blood pressure also may affect the evaluation of normotensive persons. Afternoon assessment of persons considered to be normotensive (based on findings of measurements done in the morning) might occasionally produce false-positive findings since this is the time of day when blood pressure is highest. Detection of elevated values in certain individuals when assessed in the afternoon could represent the influence of the circadian time of the measurement rather than the presence of disease.

Some clinicians who are knowledgeable of the circadian pattern in arterial pressure attempt to standardize the clock time of patient appointments when reassessing blood pressure. This may not be an appropriate practice. First, the standardization of clock time for repeat clinic visits, even if the sleep-wake routine of the patient is consistent, enables an assessment of blood pressure at only one specific stage of the circadian cycle. Measurements restricted to a *single* circadian stage are unlikely to be representative of blood pressure at other times during the 24 hours, especially in persons who exhibit high-amplitude rhythmicity in systolic and/or diastolic pressure.[36–39] In certain individuals, the sleep-wake schedule may be variable due to employment in an industry requiring rotating shift or periodic night work. If this is the case, the same clock time of appointment need not correspond to the same biological time of the endogenous blood pressure circadian rhythm. For these reasons, it is often useful to assess blood pressure throughout the entire 24 hours using ambulatory blood pressure monitoring devices. This approach enables a more complete and representative evaluation of the circadian pattern of blood pressure and most importantly provides data over the entire 24 hours upon which to base the differential diagnosis of normotension versus hypertension.

Allergy

Today, the diagnosis of allergy relies heavily on the results of cutaneous allergen testing. The cutaneous response to histamine and the commonly used antigens of house dust and mixed grass pollens, among others, following intradermal injection is affected by the biological time when testing is conducted.[52–54] In diurnally active persons, the cutaneous reactivity to intradermal tests, quantified in terms of the area of erythema and/or induration, is considerably lower during the initial hours of daytime activity than it is prior to bedtime, when it is greatest. On average, the area of response (erythema and/or induration) to injected antigens in the evening is about threefold greater than it is in the morning (Fig. 4). In some cases, patients may be nonresponsive to provocation with an antigen when testing is done in the morning; yet, they may be significantly reactive when testing is conducted 12 hours later.

Ordinarily, intradermal injections utilizing histamine as a control are included in the skin test battery. This allows the reaction of antigens to be compared to that of a standard test substance. Since such controls are incorporated, many clinicians believe that the rhythm in cutaneous reactivity can be disregarded. However, there

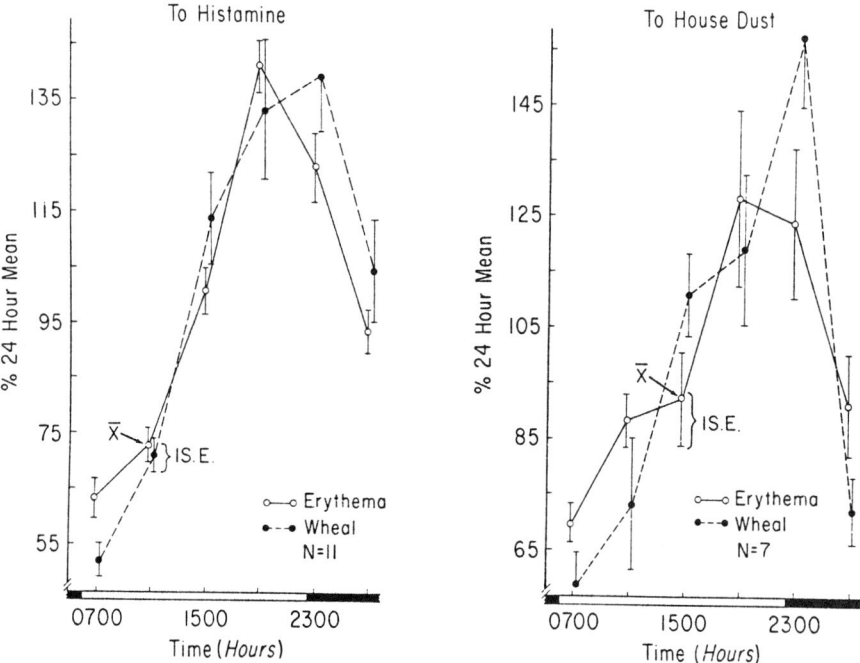

CIRCADIAN RHYTHM IN CUTANEOUS REACTIVITY

Figure 4. Circadian rhythm in cutaneous reactivity (total area of erythema and wheal) to intradermal injection of histamine (left) or house dust extract (right) given in a standardized way to a group of diurnally active, sensitized patients. With reference to the group 24-hour mean (set to 100% in graphs) area of reactivity (determined by tests done every 4 hours for 24 hours), the response to house dust extract varies according to test time by about 60% for erythema and nearly 90% for induration. Reactivity is lowest in the morning and greatest in the evening. The reaction to histamine exhibits a comparable circadian pattern. Time shown in military units: 1200 = noontime and 2400 = midnight. Figures reproduced from Lee et al.[53]

can be substantial differences in the 24-hour mean level and amplitude of the circadian rhythm in cutaneous reactivity between antigens and histamine. For example, in most patients it is rare that there is no cutaneous reaction to histamine when tests are conducted in the morning, a biological time of lowest skin reactivity to both antigens and histamine. In contrast, it is sometimes the case that antigen-sensitized patients fail to manifest a cutaneous reaction to an allergen during morning tests, yet evidence a strong reaction when retested in the afternoon or evening, a time of heightened cutaneous responsiveness. Because of this, the histamine control is not an optimum marker for interpreting the circadian rhythmic influence on allergen skin tests.

Reactive Airway Disease

The diagnosis of reactive airways disease typically involves spirometric studies and may include provocative challenges with methacholine, histamine, and/or specific antigens. With reference to asthma, most patients experience worst respiratory distress overnight, during sleep.[26,27] In-laboratory evaluation, as well as outpatient self-assessment studies, verify that airway patency varies as a circadian rhythm (Fig. 5). Forced expiratory volume determined by spirometry and peak expiratory flow (PEF) determined by peak flow meter are both considerably greater during the afternoon than during the nighttime hours of sleep.[55] Even in patients with very mild asthma, the afternoon peak-to-overnight trough circadian variation in airway patency commonly amounts to 25% of the 24-hour average.[56] In more severely affected patients, the circadian change in spirometry and PEF may be as great as 50–60% of the 24-hour mean level. In general, the more unstable and severe the asthma, the greater the circadian variability and the more reduced the 24-hour mean level of airway patency compared to normal values.[56]

Since the majority of medical clinics are open only during the daytime, spirometric assessments, especially those conducted between 11:00 AM and 5:00 PM, are more likely to underrepresent the severity of the airway obstruction experienced overnight. This is especially true for those patients who exhibit a high-amplitude circadian pattern in airway caliber and who have an extensive medical history of nocturnal asthma. A 5- to 7-day assessment of bronchial patency using a PEF meter, with four to six measurements made during each 24 hours including nocturnal awakenings due to asthma, is usually sufficient to define the circadian pattern of airway function.[57,58]

Asthma is characterized by heightened reactivity of the airways to a variety of chemical and antigenic substances. Typically, airway challenges involving methacholine or histamine are utilized to assess the magnitude of such hyperreactivity. Occasionally, inhala-

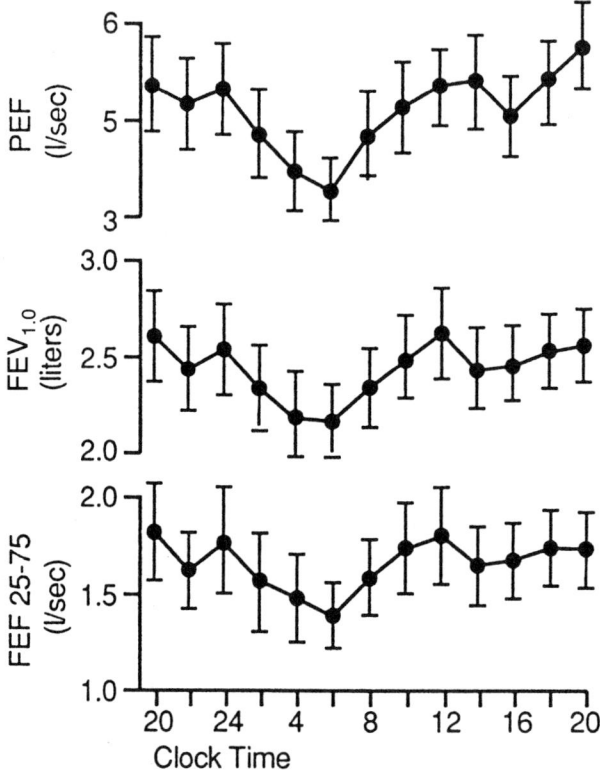

Figure 5. Circadian rhythm of airway patency in 26 diurnally active asthmatics studied while treated with placebo medication. Spirometric assessments done every 2 hours throughout the day and night (when awakened briefly from sleep for assessment) document airway status is poorest overnight and best during the midday and evening. The day-night variation in spirometric indices can be as great as 25–50% of the 24-hour mean level; thus, daytime evaluation of asthma patients is likely to *underestimate* the severity of the airway obstruction and asthmatic condition experienced by the patient during the overnight span. Time given in military units: 12 = noon and 24 = midnight. Figure reproduced from Staudinger and Steinijans.[55]

tional provocations are done with specific antigens. The results of several studies demonstrate that the bronchial reactivity to aerosols of histamine,[59] acetylcholine,[60] and house dust [61] follows a circadian response pattern (Fig. 6). In diurnally active persons, the airways are least vulnerable to provocation with these agents between noon and 5:00 PM and most susceptible between 11:00 PM and 6:00 AM.

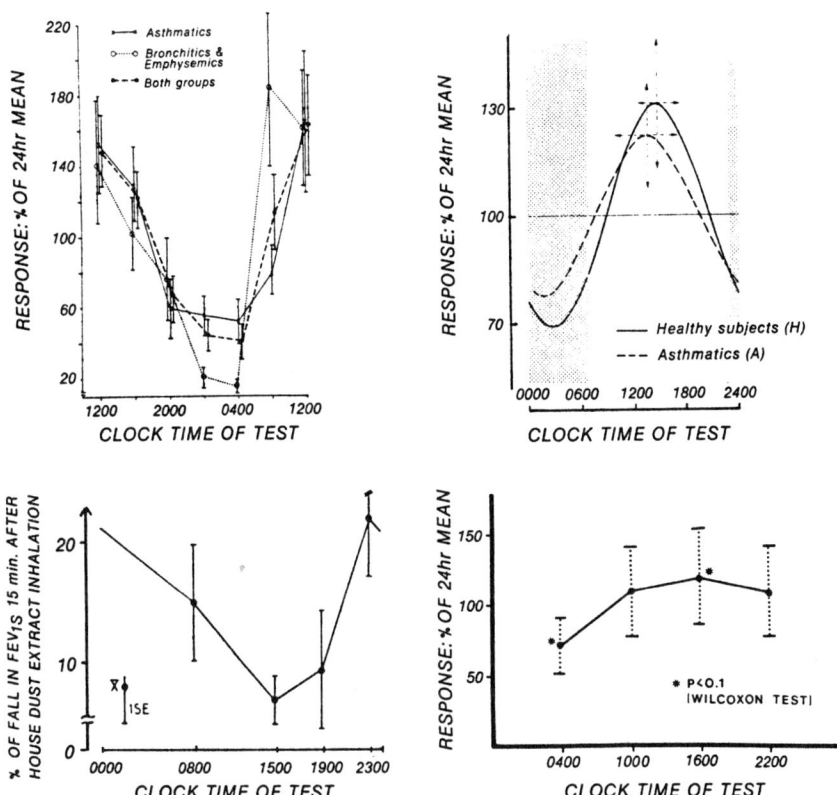

Figure 6. Day-night variation of diurnally active asthmatic or other chronic obstructive pulmonary disease (COPD) patients to provocation with histamine (upper left: % change in PC_{10}), acetylcholine (upper right: % change in PC_{15}), house dust extract (lower left: % change from clock-time control baseline), and cold air (lower right: % change from clock-time control baseline). Except for house dust, the temporal variation is expressed in terms of % deviation from the respective 24-hour group mean (for the several clock-time tests done during the day and night). The airway response to the provocative agents varies significantly over the 24 hours. Results of tests done around midday utilizing histamine, acetylcholine, or house dust tend to *underrepresent* the magnitude of the airway hyperreactivity experienced overnight. The opposite is true for cold air challenge. Figure adapted from DeVries et al.,[59] Reinberg et al.,[60] Gervais et al.,[61] and Sly and Landau.[62]

Although it is doubtful that the diagnosis of asthma would be jeopardized by the choice of test time, the clinical assessment of the severity of the hyperreactivity manifested during the nighttime, when the risk of asthma is greatest, is likely to be underappreciated based on daytime patient studies.

Commonly, chest physicians and allergists conduct spirometric studies utilizing a beta-agonist aerosol to determine the extent to which the airway obstruction is reversible. In patients with stable asthma, the bronchodilator effect of the beta-agonist aerosol, orciprenaline, has been shown to be circadian rhythmic.[63] In stable, day-active asthmatic subjects, the bronchodilator effect of this agent is very strong when inhaled in the morning as well as the evening. In comparison, when inhaled around midday, the effect is rather weak (Fig. 7). These findings imply that the results of airway reversibility studies could be affected by the biological time when they are conducted. Morning trials are likely to show a more dramatic reversibility of the airway obstruction following inhalation of a beta-agonist aerosol, while midday and afternoon trials are likely to show less impressive reversibility.

The circadian time structure is capable of influencing the findings of other commonly used diagnostic procedures and tests.[31,64–69] Current literature suggests that both the glucose tolerance and ACTH-stimulation tests[64,69] can be greatly affected by the time, morning versus afternoon or/and winter versus summer, they are performed. Moreover, the findings of clinical chemistry studies may be affected by the time when blood and other fluids are obtained.[67,68] Time-qualified-for-rhythms reference values are now available for aiding the interpretation of such test findings obtained according to the hour of the day, day of the menstrual cycle, and month of the year.[67,68]

Discussion

Human beings as a species prefer a routine of diurnal activity alternating with nocturnal rest. Because of this, most medical research into the mechanisms of disease are conducted during the daytime hours. As a consequence, our understanding and knowledge of disease pathophysiology is based almost entirely on data derived from daytime studies conducted at the convenience of both patients and investigators. The major exception is asthma. Circadian study of

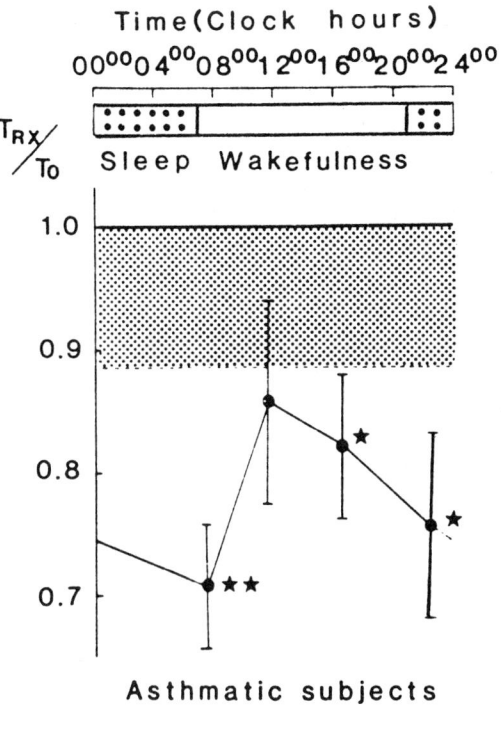

Figure 7. Morning-evening variation in the airway response of diurnally active, stable asthmatic children to the inhalation of the beta-agonist aerosol, orciprenaline. The change (T_{RX}/T_O: the lower the ratio the greater the level of bronchodilation achieved) of airway resistance following treatment (T_{RX}) with reference to the corresponding clock-time pretreatment control baseline (T_O), is greatest in the morning at 6:30 AM (0630) and evening around 11:30 PM (2330) than it is around midday and the afternoon when it is marginal. The findings suggest: (a) Stable asthma patients may not require equal-interval dosings of inhalative beta-agonist medications to maintain the airways patent and (b) the clinical determination of the degree to which airway obstruction is reversible by a beta-agonist bronchodilator agent is strongly influenced by the biological time when assessment is done. Reproduced from Gaultier et al.[63]

this disease and its underlying mechanisms has been pursued since the 1960s. Nonetheless, with deference to the most prevalent medical concept in medicine—homeostasis, the relevance of the *when* (time of day, stage of menstrual cycle, or month of year) patient research is done is rarely explored, or for that matter even considered! With the more complete definition and understanding of the biological time structure plus greater appreciation for rhythmic aspects of the pathophysiology of disease, the necessity for both daytime and nighttime investigations is being increasingly recognized. The implications of biological time structure for medicine, pharmacology, and epidemiology remain to be fully explored.

The field of medical chronobiology, although a relatively young one, is presently developing quite rapidly throughout the world. Initially, chronobiological concepts and findings were slow to be accepted. In recent years the situation has changed. In part, this is because of new developments in medical technology which now enable the continuous 24-hour monitoring of patients both in and outside the hospital environment. Around the clock assessment of esophageal and gastric pH, systolic and diastolic blood pressure, electrocardiography, and body activity—actigraphy—is routinely done in many medical centers as outpatient studies.[37–39,70–74] Such monitorings not only provide physicians with a 24-hour database but they also document the clinical relevance of chronobiology in the diagnosis and treatment of disease. Recent developments in the field of medical chronobiology, coupled with exciting new biomedical technology, have already begun to influence the practice and procedures used in clinical medicine. Continuing advances in medical chronobiology will likely result in new perspectives and understandings of human biology and disease pathophysiology, as well as new concepts in the treatment of illness as discussed in the next chapter.

References

1. Halberg F. Chronobiology. Ann Rev Physiol 1969; 31:675–725.
2. Moore-Ede MC, Sulzman FM, Fuller CA. The Clocks That Time Us. Harvard University Press, Cambridge, MA, 1982.
3. Reinberg A, Smolensky MH. Biological Rhythms and Medicine. Springer-Verlag, New York, 1983.
4. Bünning E. Die physiologische Uhr. Springer-Verlag (2nd Edition), Berlin, 1963.
5. Nelson W, Halberg F. Phase relations of circadian rhythms: Animals.

In: Environmental Biology, Altman PL, Dittmer DS (eds). Fed Amer Soc Exper Biol, Bethesda, MD, 1966, pp 586–596.

6. Brady J. The physiology of insect circadian rhythms. Adv Insect Physiol 10:1–115, in press.

7. Palmer JD. Contributions made to chronobiology by studies of the fiddler crab rhythms. Chronobiol Intl 1991; 8:110–130.

8. Halberg F, Tong YL, Johnson EA. Circadian system phase. An aspect of temporal morphology; procedures and illustrative examples. In: The Cellular Aspects of Biorhythms, H von Mayersbach (ed), Springer-Verlag, New York, 1967, pp 20–48.

9. Sehgal A, Man B, Price JL, Vosshall LB, Young MW. New clock mutations in *Drosophila*. In: Temporal Control of Drug Delivery, Hrushesky WJM, Langer L, Theeuwes F (eds), Ann New York Acad Sci 1991; 618:1–10.

10. Li-Weber M, de Groot MEJ, Schweiger HG. Sequence homology to the *Drosophila per* locus in higher plant nuclear DNA and in *Acetabularia* chloroplast DNA. Mol Gen Genet 1987; 209:1–7.

11. Reinberg A, Touitou Y, Restoin A, Migraine C, Levi F, Montagner H. The genetic background of circadian and ultradian rhythm patterns of 17-hydroxycorticosteroids: a cross-twin study. J Endocr 1985; 105:247–253.

12. Moore RY. The suprachiasmatic nucleus and the mammalian circadian system. In: Trends in Chronobiology, Hekkens WTJM, Kerkhof GA, Rietveld WJ (eds). Pergamon Press, Oxford, 1988, pp 97–105.

13. Halberg F. Temporal coordination of physiologic function. Cold Springs Harbor Sym Quant Biol 1960; 25:289–310.

14. Reiter RJ. Intrinsic rhythms of the pineal gland and associated hormone cycles in body fluids. Ann Rev Chronopharmacol 1988; 4:77–136.

15. Wever RA. The Circadian System of Man. Results of Experiments Under Temporal Isolation. Springer-Verlag, Heidelberg, 1979.

16. Czeisler CA, Allan JS, Strogatz SH, Ronda JM, Sanchez R, Rios CD, Freitag WO, et al. Bright light resets the human circadian pacemaker independent of the timing of the sleep-wake cycle. Science 1986; 233:667–671.

17. Wever RA, Polasek J, Wildgruber CM. Bright light affects human circadian rhythms. Pflügers Arch 1983; 396:85–87.

18. Ghata J, Halberg F, Reinberg A, Siffre M. Rythmes circadiens desynchronisés du cycle social (17-hydroxy-corticostéroides, température rectal, veille-sommeil) chez deux sujets adultes sains. Ann Endocrinol (Paris) 1969; 30:245–260.

19. Comperatore CA, Krueger GP. Circadian rhythm desychronosis, jet lag, shift lag, and coping strategies. In: Occupational Medicine: State of the Art Reviews, Scott AJ (ed), 1990; 5:323–341.

20. Arendt J, Aldhous M, Marks V. Alleviation of jet-lag by melatonin: Preliminary results of controlled double-blind trial. Br Med J 1986; 292:170.

21. Halberg F, Simpson H. Circadian acrophase of human 17- hydroxycorticosteroid excretion referred to midsleep rather than midnight. Human Biol 1967; 39:405–413.

22. Reinberg A (ed). Chronobiological field studies of oil refinery shift workers. Chronobiologia 1979; 6:1–119.
23. Scott JA (ed). Shiftwork Occupational Medicine: State of the Art Reviews 1990; 5:165–428.
24. Nicholson PA, Bogie W. Diurnal variation in the symptoms of hay fever. Implications for pharmaceutical development. Curr Med Res Opin 1973; 1:395–400.
25. Reinberg A, Gervais P, Levi F, Smolensky M, Del Cerro L, Ugolini C. Circadian and circannual rhythms of allergic rhinitis: An epidemiologic study involving chronobiologic methods. J Allergy Clin Immunol 1988; 81:51–62.
26. Dethlefsen U, Repges R. Ein neues Therapieprinzip bei nachtlichem Asthma. Med Klin 1985; 80:44–47.
27. Turner-Warwick M. Epidemiology of nocturnal asthma. Am J Med 1988; 85(Suppl 1B):6–8.
28. Rocco MB, Nabel EG, Selwyn AP. Circadian rhythms and coronary artery disease. Am J Cardiol 1987; 59:13C–17C.
29. Waters DD, Miller DD, Bouchard A, Borsch X, Theroux P. Circadian variation in variant angina. Am J Cardiol 1984; 54:61–64.
30. Kuroiwa A. Symptomatology of variant angina. Jpn Circ J 1978; 42:459–476.
31. Yasue H, Omote S, Takizawa A, Nagao M, Miwa K, Tanaka S. Circadian variation of exercise capacity in patients with Prinzmetal's variant angina: Role of exercise-induced coronary arterial spasm. Circulation 1979; 59:938–948.
32. Master AM, Jaffee HL. Factors in the onset of coronary occlusion and coronary insufficiency. JAMA 1952; 148:794–798.
33. Pepine CJ. Circadian variations in myocardial ischemia. JAMA 1991; 65:386–390.
34. Muller JE, Stone PH, Turin ZG, Rutherford JG, Czeisler CA, Parkers C, Poole WK, et al. The MILIS study group. Circadian variation in the frequency of onset of acute myocardial infarction. N Engl J Med 1985; 313:1315–1322.
35. Marshall J. Diurnal variation in the occurrence of strokes. Stroke 1977; 8:230–231.
36. Lemmer B. Temporal aspects of the effects of cardiovascular active drugs in humans. In: Chronopharmacology: Cellular and Biochemical Interactions, Lemmer B (ed). Marcel Dekker, Inc., New York, 1989, pp 525–541.
37. Meyer-Sabellek W. 24-hour recordings of blood pressure in hypertensives: Technical and methodological background. Chronobiol Intl 8:506–510, 1991.
38. Portaluppi F, Montanari L, Ferlini M, Gilli P. Altered circadian rhythms of blood pressure and heart rate in non-hemodialysis chronic renal failure. Chronobiol Intl 1990; 7:321–327.
39. Portman R, Yetman RJ, West MS. Efficacy of 24-hour ambulatory monitoring in children. J Pediatrics 1991; 118:842–849.
40. Labrecque G, Reinberg AE. Chronopharmacology of non-steroid anti-inflammatory drugs. In: Chronopharmacology: Cellular and Biochemi-

cal Interactions, Lemmer B (ed). Marcel Dekker, Inc., New York, 1989, pp 545–580.

41. Harkness JAL, Richter MB, Panayi GS, Van de Pette K, Unger A, Pownall R, Geddawi M. Circadian variation in disease activity in rheumatoid arthritis. Br Med J 1982; 284:551–554.

42. Moynihan BGA. Duodenal Ulcer. WB Saunders, Philadelphia, 1912.

43. Moore JG, Smolensky MH. Biological rhythms in gastro-intestinal function and processes: Implications for the pathogenesis and treatment of peptic ulcer disease. In: Ulcer Disease. Investigation and Basis for Therapy, Swabb EA, Szabo S (eds). Marcel Dekker, Inc., New York, 1991, pp 55–85.

44. Langdon-Down M, Brain WR. Time of day in relation to convulsions in epilepsy. Lancet 1929; i:1029–1032.

45. Patry FL. The relation of time of day, sleep and other factors to the incidence of epileptic seizures. Am J Psychiatr 1931; 87:789–813.

46. Dreifuss FE, Meinardi H, Stefan H. (eds), Chrono-pharmacology in Therapy of the Epilepsies. Raven Press, New York, 1990.

47. Nicolau GY, Haus E, Popescu M, Sackett-Lundeen L, Petrescu E. Circadian, weekly and seasonal variations in cardiac mortality, blood pressure and catecholamine excretion. Chronobiol Intl 1991; 8:149–159.

48. Dalton KD. The Premenstrual Syndrome. CC Thomas, Springfield IL, 1964.

49. Smolensky MH, Halberg F, Sargent F. Chronobiology of the life sequence. In: Advances in Climatic Physiology, Itoh S, Ogata K, Yoshimura H (eds). Igaku Shoin, Tokyo, 1972, pp 281–318.

50. Gibinski, K. A review of seasonal periodicity in peptic ulcer disease. Chronobiol Intl 1987; 4:91–99.

51. Smolensky MH, Tatar SE, Bergman SA, Losman JG, Barnard CN, Dacso CC, Kraft IA. Circadian rhythmic aspect of human cardiovascular function: A review by chronobiologic statistical methods. Chronobiologia 1976; 3:337–371.

52. Reinberg A, Zagulla-Mally Z, Ghata J, Halberg F. Circadian reactivity rhythms of human skin to house dust, penicillin and histamine. J Allergy Clin Immunol 1969; 44:292–306.

53. Lee RE, Smolensky MH, Leach C, McGovern JP. Circadian rhythm in the cutaneous sensitivity to histamine and selected antigens including phase relationship to urinary cortisol excretion. Ann Allergy 1977; 38:231–236.

54. McGovern JP, Smolensky MH, Reinberg A. Circadian and circamenstrual rhythmicity in cutaneous reactivity to histamine and allergenic extracts. In : Chronobiology in Allergy and Immunology, McGovern JP, Smolensky MH, Reinberg A (eds). CC Thomas, Springfield, IL, 1977, pp 79–116.

55. Staudinger HW, Steinijans VW. Theophylline steady-state pharmacokinetics: Recent concepts and their application in the chronotherapy of bronchial asthma. In: Clinical Chronopharmacology. Clinical Pharmacology, Lemmer B, Huller H (eds). 1990; 6:136–147.

56. Smolensky MH, Barnes PJ, Reinberg A, McGovern JP. Chronobiology

and asthma. I. Day-night differences in bronchial patency and dyspnea and circadian rhythm dependencies. J Asthma 1986; 23:321–343.

57. Hetzel MR, Clark TJH. Comparison of normal and asthmatic circadian rhythms in peak expiratory flow rate. Thorax 1980; 35:732–738.

58. Random B, Smolensky MH, Hsi B, Albright D, Burge S. Field survey of circadian rhythm in PEF of electronics workers suffering from colophony-induced asthma. Chronobiol Intl 1987; 4:263–271.

59. DeVries K, Goei JT, Booy-Noord H, Orie NGM. Changes during 24 hours in lung function and histamine hyperreactivity of the bronchial tree in asthmatic and bronchitic patients. Intl Arch Allerg Appl Immunol 1962; 20:93–101.

60. Reinberg A, Gervais P, Morin M, Abulker C. Rythme circadien humain du seuil de la reponse bronchique a l'acetylcholine. CR Acad Sci (Paris) 1971; 272:1879–1881.

61. Gervais P, Reinberg A, Gervais C, Smolensky MH, DeFrance O. Twenty-four-hour rhythm in the bronchial hyperreactivity to house dust in asthmatics. J Allergy Clin Immunol 1977; 59:207–213.

62. Sly PD, Landau LI. Diurnal variation in bronchial responsiveness in asthmatic children. Pediatr Pulmonol 1986; 2:207–213.

63. Gaultier C, Reinberg A, Motohashi Y. Circadian rhythm in total pulmonary resistance of asthmatic children. Effects of a beta-agonist agent. Chronobiol Intl 1988; 5:285–290.

64. Zimmet PZ, Wall JR, Rome R, Stimmler L, Jarrett RJ. Diurnal variation in glucose tolerance: Associated changes in plasma insulin, growth hormone and nonesterified fatty acids. Br Med J 1974; ii:485–488.

65. Nugent CA, Nichols T, Tyler FH. Diagnosis of Cushing's syndrome: Single dose dexamethasone suppression test. Arch Intern Med 1965; 116:172–176.

66. Carroll BJ, Feinberg M, Greden JF, Tarika J, Albala AA, Haskett RF, James NM, et al. A specific laboratory test for the diagnosis of melancholia. Arch Gen Psychiatry 1981; 38:15–22.

67. Haus E, Nicolau GY, Lakatua D, Sackett-Lundeen L. Reference values for chronopharmacology. Ann Rev Chronopharmacol 1988; 4:333–424.

68. Haus E, Touitou Y. Time-qualified reference values for laboratory medicine. Chronobiol Intl 10: in press, 1993.

69. Touitou Y. Time-dependent effects of drugs used as diagnostic agents on the hypothalamo-pituitary adronal axis In: Chronopharmacology: Cellular and Biochemical Interactions, Lemmer B (ed). Marcel Dekker, New York, 1989, pp 597–614.

70. Tan WC, Martin RJ, Pandey R, Ballard R. Effects of spontaneous and simulated gastroesophageal reflux on sleeping asthmatics. Am Rev Respir Dis 1990; 141:1394–1399.

71. Aubert-Tulkens G, Culee C, Hartman-Van Rijckevorsel K, Rodenstein DO. Ambulatory evaluation of sleep disturbance and therapeutic effects in sleep apnea syndrome by wrist activity monitoring. Am Rev Respir Dis 1987; 136:851–856.

72. Brown A, Smolensky MH, D'Alonzo GE, Redmond D. Actigraphy: A

means of assessing circadian patterns in human activity. Chronobiol Intl 1990; 7:125–133.

73. Hausmann D, Lichtlen PR, Nikutta P, Wenzlaff P, Daniel WG. Circadian variation of mycocardial ischemia in patients with stable coronary artery disease. Chronobiol Intl 1991; 8:385–398.

74. Lemmer B. (ed). Recent advances in the chronopharmacology of the cardiovascular system. Part 2. Blood pressure regulation, hypertension and drug treatment. Chronobiol Intl 1991; 8:441–538.

2

Biological Rhythms and Medications:

Chronopharmacology and Chronotherapeutics

Michael H. Smolensky, Ph.D.,
and Gilbert E. D'Alonzo, D.O.

Introduction

The fundamental strategy underlying the conventional pharmacotherapy of most diseases involves treatment schedules and drug delivery technologies that meet the homeostatic goal of constancy of blood and tissue drug levels. Sustained-release dosage forms make possible twice-a-day and even once-a-day treatment regimens so that treatments can be dosed at equal intervals and in equal doses. Decisions relating to the scheduling of medications emphasize patient compliance. It is recommended that they be timed to coincide with regularly occurring events or activities of the patient, such as the commencement or termination of daily activity, or the consumption of breakfast and/or supper. According to the prevailing medical dictum of homeostasis, constancy of drug levels ensures constancy in

Martin RJ (editor): *Nocturnal Asthma: Mechanisms and Treatment,* © Futura Publishing Co., Inc., Mount Kisco, NY, 1993.

drug efficacy and safety. Homeostatic theory also leads one to expect that the pharmacokinetics and dynamics of therapies are comparable regardless of their timing during the 24 hours, menstrual cycle, or year. New findings from the field of medical chronobiology challenge these assumptions and expectations. As discussed herein, the kinetics and dynamics of medications may not be the same when they are timed in the morning versus the evening because of chronobiological determinants. Moreover, the documentation of biological rhythms in the pathophysiology of disease processes implies that the requirement for pharmacotherapy during the day and night may not be the same.

Chronopharmacology: Concepts and Principles

Chronopharmacology is the study of biological rhythms and medications.[1–5] This science is concerned primarily with the elucidation of administration-time dependencies in the pharmacokinetics and effects of medications, and the mechanisms underlying such phenomena. The findings of numerous investigations verify that the kinetics and dynamics of medications can be profoundly affected by the time of their administration. A secondary but, nonetheless, significant concern of chronopharmacology is the manner in which medications and their timings affect the body's biological clocks and rhythms. Since the human time structure constitutes a key aspect of biological efficiency, any medication-induced disruption or desynchronization of it could be deleterious for an already compromised patient.[1,3,6]

Research in the field of chronopharmacology has resulted in several new concepts of importance to medicine and clinical pharmacology.[1–5] *Chronokinetics* refers to administration-time (biological rhythm-dependent) differences in the extent and rate of drug absorption, distribution, metabolism, and elimination. The chronokinetics of a medication can be described by conventional indices used in pharmacology, for example, the T_{max} (the lag time following dosing to peak blood level), C_{max} (the concentration of drug at T_{max}), AUC (the area under the time-concentration curve), V_d (the volume of distribution for the drug), and K_e (drug elimination rate), among others.[7]

Chronopharmacokinetic phenomena result from biological

rhythms in visceral organ function especially the gut, liver, and kidney.[8–10] For example, the circadian rhythm in gastric acid secretion can exert a significant influence on the absorption kinetics of certain dosage forms. In diurnally active persons, gastric acid secretion is high during the evening and the initial hours of sleep, and low during the early and middle portion of the daily activity span.[11] The time of drug ingestion is likely to significantly affect the absorption kinetics of those dosage forms whose dissolution rate is pH-dependent. The circadian rhythm of gastric emptying also can affect the absorption kinetics of medications. The rate of gastric emptying is considerably greater in the morning after arising from nighttime sleep than in the evening and overnight when it is slowest.[12] High-amplitude circadian rhythms in hepatic and renal blood flow,[13,14] hepatic enzyme activity and concentration,[15,16] and renal function[9] further contribute to administration time-dependent variation in the pharmacokinetics of medications. Although chronokinetic phenomena are not explainable by body posture or meal timing and composition, it must be acknowledged that these and other factors are capable of affecting the kinetics of medications.[17,18]

Figures 1 to 4 present some examples of chronokinetic phenomena of selected sustained-release and parenteral medications. Such administration-time dependencies are known for certain theophylline,[19] benzodiazepine,[20] beta-receptor antagonist and agonist,[21–23] and nonsteroidal antiinflammatory drugs (NSAIDs),[24] to mention but a few. Several recently published papers and books review the vast number of medications that exhibit chronokinetic phenomena when administered orally, parenterally, or percutaneously.[1–3,5,25–27]

Chronesthesy is defined as biological rhythm-dependent changes in the biosystem's susceptibility to medications, taking into account chronobiological phenomena occurring at the cellular level.[1–5] Chronesthesies result from biological rhythms in drug receptor number or conformation, rate-limiting steps in metabolic pathways, and/or the free-to-bound fraction of medications. According to conventional homeostatic considerations, the effect of a medication is hypothesized to be primarily dependent on its concentration in target tissues. However, due to circadian and other rhythms in cells, tissues, and organs, time-dependent variation in the effects of medications is possible even when a constant drug level is maintained. The concept of chronesthesy accounts for the observation that biosystems can be relatively unresponsive to a medication

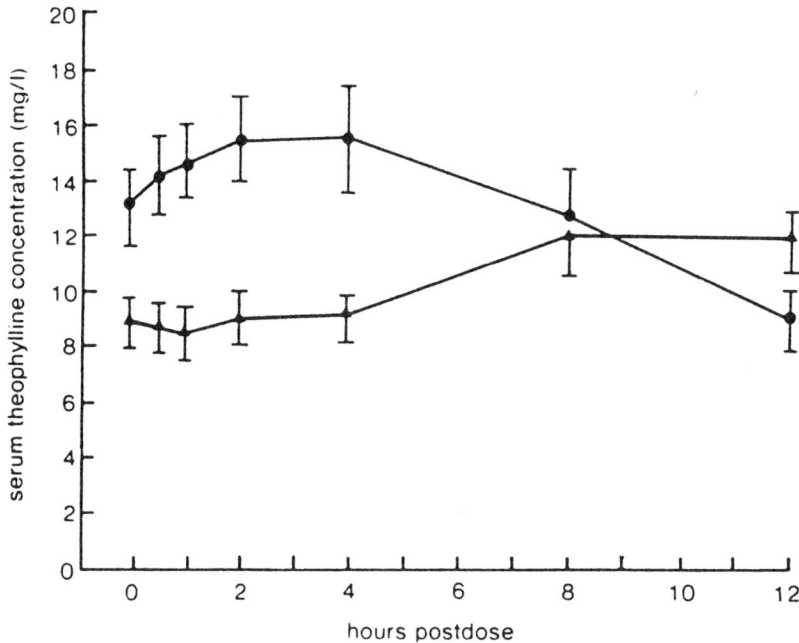

Figure 1. Dosing-time-dependent differences under steady-state conditions in the pharmacokinetics of theophylline in 13 asthmatic children, 7 to 17 years of age treated with TheoDur® (Schering) at equal intervals and in equal doses.[28] The 8 AM ingestion was scheduled before breakfast; the 8 PM one followed supper by two hours. Theophylline kinetics from TheoDur® after the morning and evening treatments were dramatically different. After the morning dosing, the serum theophylline concentration (STC) peaked within 4 to 6 hours and the trough level occurred just prior to the evening treatment, as expected. After the evening drug ingestion, the STC trough occurred at the beginning (rather than end) of the nighttime dosing interval, and the STC peak occurred at the end (rather than beginning) of it. Moreover, the peak-to-trough difference in the STC was considerable, averaging 7 mg/L. These findings were confirmed by the same authors in a second study on a different but comparable group of young asthmatic children.[19] Reproduced with permission, Scott et al.[28]

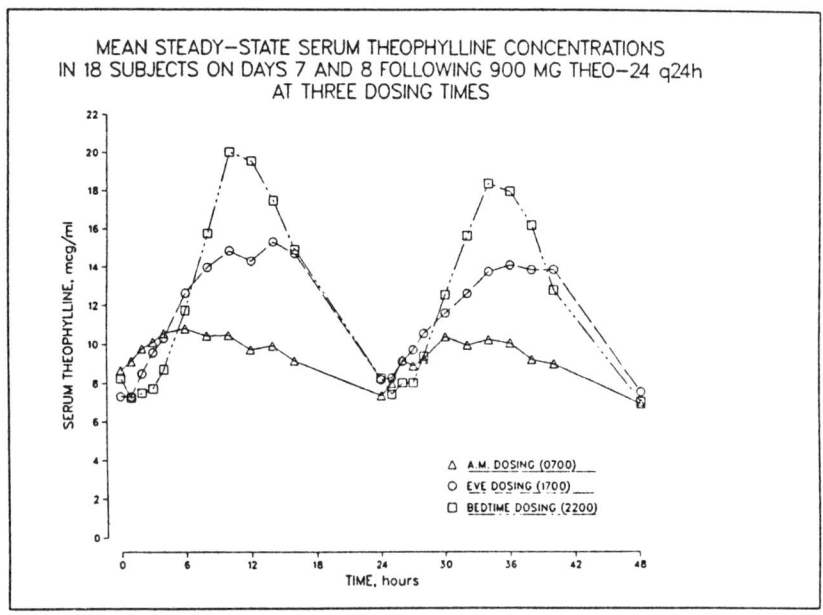

MEAN STEADY–STATE SERUM THEOPHYLLINE CONCENTRATIONS
IN 18 SUBJECTS ON DAYS 7 AND 8 FOLLOWING 900 MG THEO–24 q24h
AT THREE DOSING TIMES

Figure 2. Treatment-time-dependent difference in the AUC (bioavailability) of theophylline from a once-a-day formulation (Theo-24®, Searle) in a group of 18 healthy adult subjects. Under steady-state conditions (7 to 8 days of treatment) and with control of meal content and timings, the AUC for a 900 mg dose of Theo-24® was nearly twofold greater following dosing at 5 PM (1700) and nearly threefold greater following dosing at 10 PM (2200) as compared to dosing at 7 AM (0700). Shown is the mean serum theophylline concentration every 3 hours over two consecutive 24-hour dosing intervals for each treatment time (i.e., 7 PM, 5 PM or 10 PM). From Smolensky et al., with permission.[29]

when administered at one time, yet highly responsive when administered in the same dose at another.

Chronesthesies are known for many classes of medications as illustrated in Figures 5 to 9. Analgesics,[32] anticoagulants,[33] beta-receptor antagonists and agonists,[21,34–37] corticosteroids,[38,39] NSAIDs[24] and theophyllines,[19,40–42] among others,[43–48] exhibit chronesthesies. Chronesthesies give rise also to biological rhythms in the dose-response relationship of medications (Figs. 10–12) as demonstrated for theophylline,[40,41] beta-agonists,[36] H_1-receptor antagonist,[48] and other medications.[49]

(text continues on p. 39)

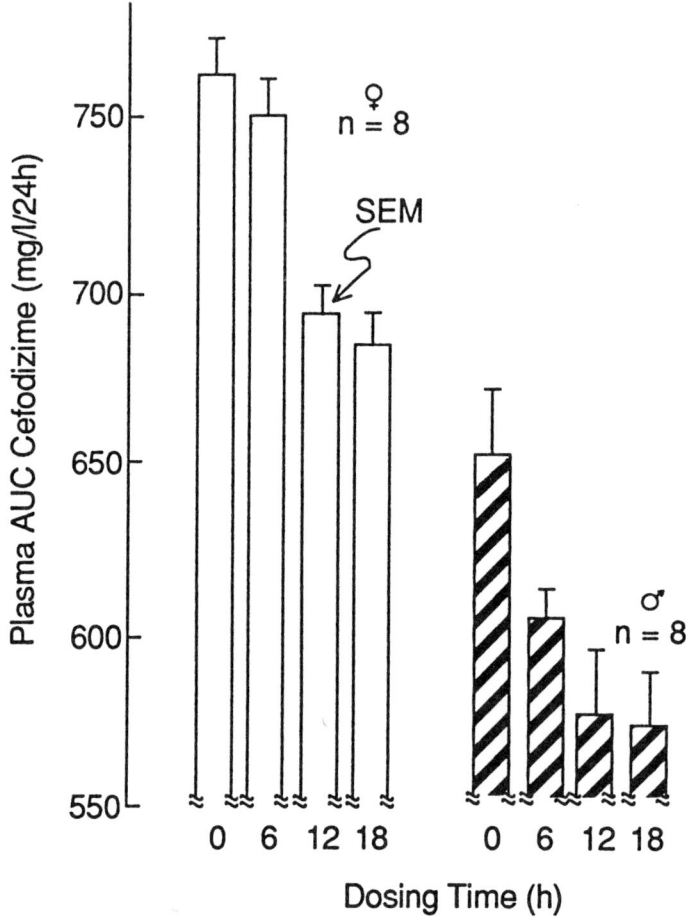

Figure 3. Differences in the plasma bioavailability (AUC) of a cephalosporin antibiotic, cefodizime (Hoechst, Italy) administered by intravenous infusion (2 gm) at midnight, 6 AM, noon, and 6 PM in a series of drug dosing trials. Groups of eight men and eight women in good health were investigated under carefully controlled and standardized study conditions. Gender and dosing-time effects in AUC were detected ($P < 0.01$). The time of greatest and lowest AUC for the diurnally active subjects, regardless of gender, was achieved after the midnight and 6 PM cefodizime infusions, respectively. At each time of treatment the AUC for males exceeded that of females. Reproduced from Jonkman et al., with permission.[30]

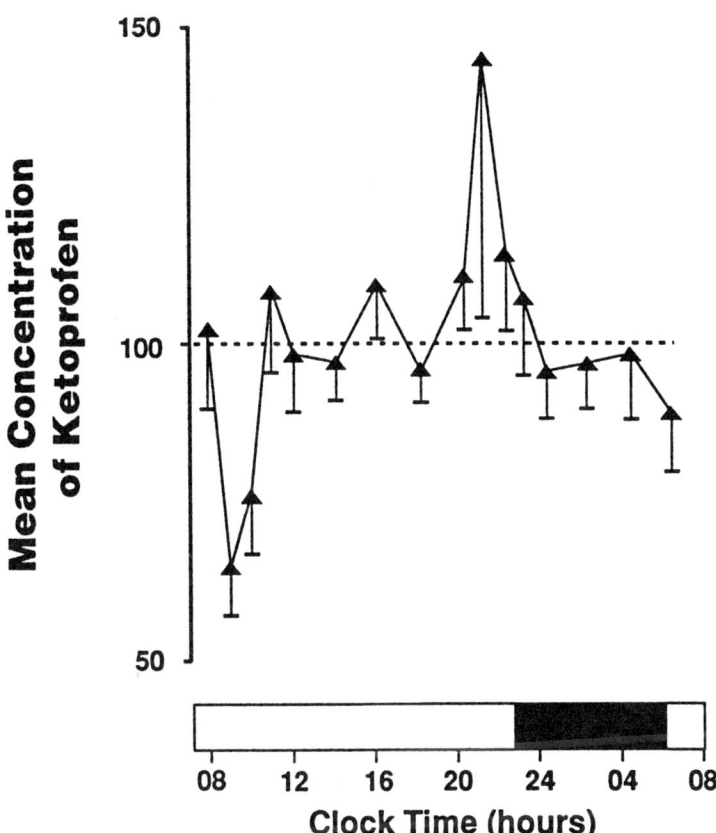

Figure 4. Circadian rhythm in the mean plasma concentration of ketoprofen infused intravenously at a constant rate in a dosage of 5 mg/kg/24 hr. Blood samples were obtained from eight diurnally active sciatica patients every two hours during the last 24 hours of the 48-hour drug infusion study. The hourly concentration of ketoprofen, expressed as values relative to the 24-hour mean for the group of eight patients, varied dramatically. The highest drug level ($\bar{X}$ and SEM: 3.95 ± 0.5 mg/L) around 9 PM and lowest drug level around 9 AM (1.99 ± 0.5 mg/L) differed significantly (P<0.01) even though ketoprofen was infused continuously at a constant rate. Reproduced from Decousus et al., with permission.[31]

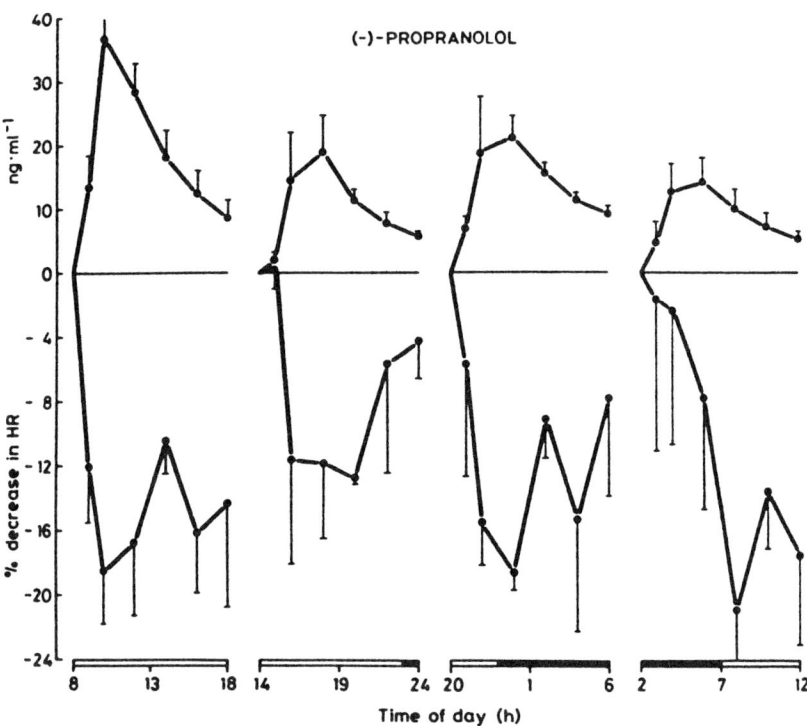

Figure 5. Circadian time-dependency in the plasma concentration of $(-)$-propranolol and in heart-rate decrease (determined in sitting position) in relation to clock-time control values in healthy subjects after oral intake of 80 mg of racemic propranolol either at 8 (8 AM), 14 (2 PM), 20 (8 PM) or 2 (2 AM) with a 1-week washout between each study. The upper portion of the figure reveals significant administration-time-dependent variation in the pharmacokinetics (AUC, C_{MAX}, and elimination half-life) of propranolol. The circadian rhythm in the heart rate lowering effect of propranolol did not directly coincide with the circadian rhythm in its pharmacokinetics. The strong heart rate lowering effect observed following drug dosing at 8 AM coincided with the dosing time of greatest AUC. However, an equally strong heart rate lowering effect followed dosing at 2 AM when the AUC for propranolol was lowest. This dichotomy between drug AUC (and level) and pharmacodynamic effect arises from the circadian chronesthesy of the cardiovascular system for propranolol. Reproduced from Langner et al., with permission.[22]

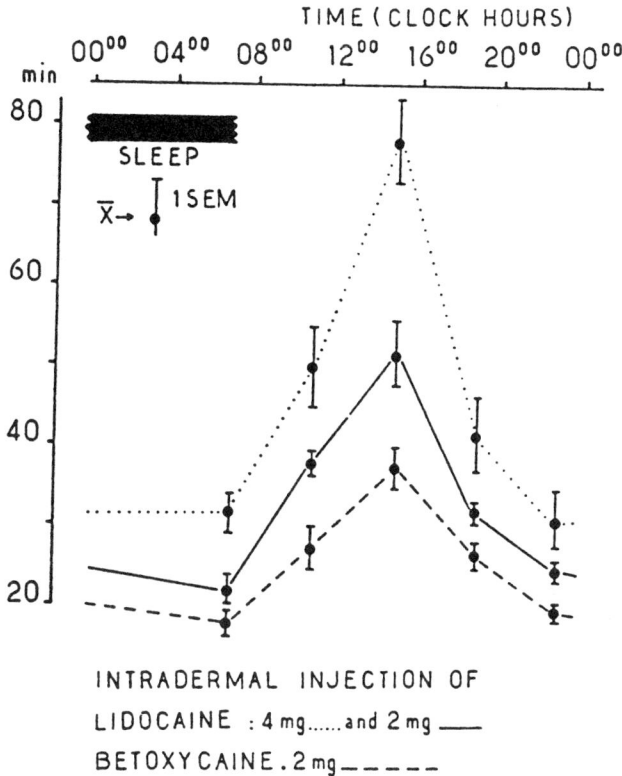

Figure 6. Circadian changes in the duration of local analgesia produced by lidocaine (2 and 4 mg) and by betoxycaine (0.2 mg) in six healthy diurnally active adults. Cutaneous analgesia was induced by intradermal injection (0.1 ml) of each medication into the flexor surface of both forearms at fixed clock hours: 0700 (7 AM), 1100 (11 AM), 1500 (3 PM), 1900 (7 PM) and 2300 (11 PM). The duration of analgesia was approximately twice as long when the medications were injected at 3 PM, than at 7 AM or 11 PM. The circadian pattern in lidocaine effect was dose independent. Reproduced from Reinberg A. and Reinberg MA, with permission.[32]

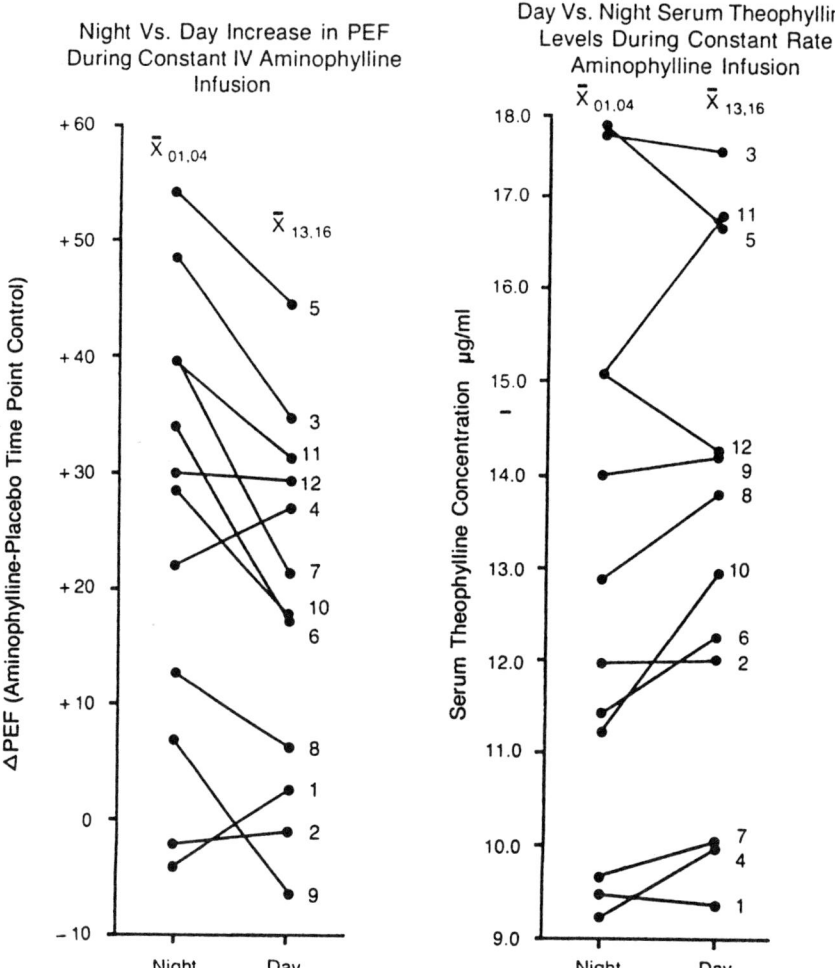

Figure 7. Airway response to an IV constant rate, aminophylline infusion dosed to achieve therapeutic concentrations of theophylline (9–18 μg/ml) in a group of 12 young asthma patients (6 to 17 yrs.). Although no day (1 and 4 PM)-night (1 and 4 AM) difference in serum theophylline concentration was observed, statistically significant variation in the effect (chronesthesy) upon the airways, quantified by PEF, was detected. With reference to placebo-control baselines, the magnitude of the theophylline-induced bronchodilator effect (Δ PEF) was highest at night when PEF was poorest. Reproduced from Smolensky et al., with permission.[40]

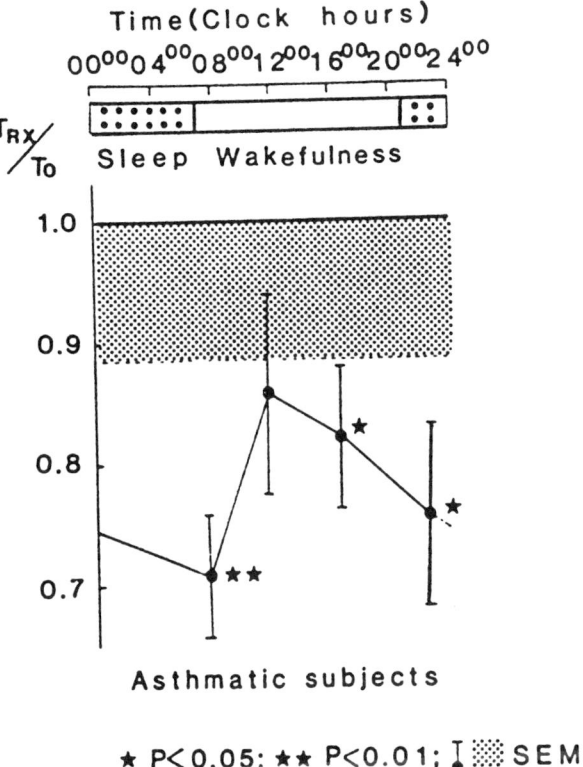

Figure 8. Diurnal variation in the effect of metaproterenol aerosol given in identical doses at different times of day on the airways of stable asthmatic children. The reduction in airway resistance following inhalation of the beta-agonist bronchodilator (T_{RX}) versus the baseline control level before treatment (T_O), expressed as T_{RX}/T_O (the lower the ratio, the better the bronchodilator effect), differed according to its timing. The bronchodilator effect was very strong at 0630 (6:30 AM) and before bedtime, yet minimal around midday. The time-dependent variation in drug effect constitutes a chronesthesy for metaproterenol. Reproduced with permission.[35]

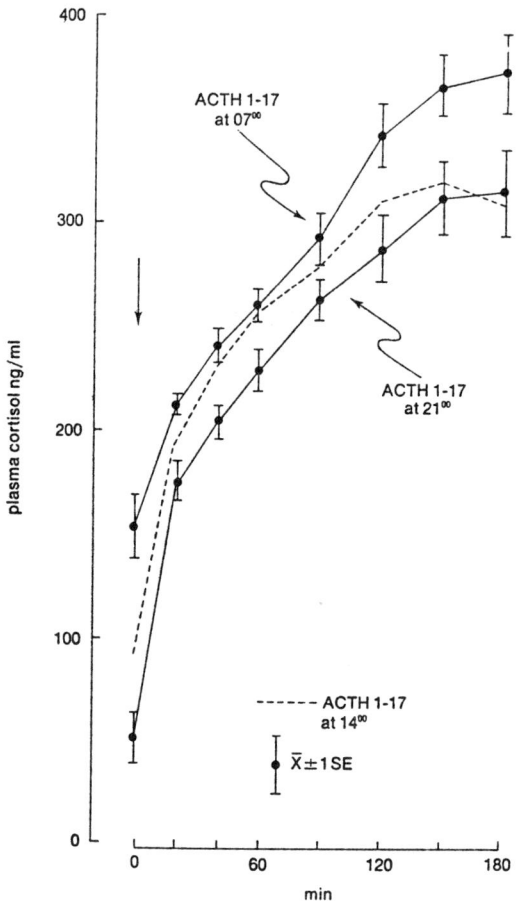

Figure 9A. Circadian rhythm in the effect of an intramuscular injection of 100 µg ACTH (ACTH 1-17; Synchrodyn®, Hoechst, Italy) on the plasma cortisol (time-concentration) response curve and urinary excretion (Fig. 9B) of 17-OHCS (by-product of cortisol metabolism) in eight diurnally active, healthy young men. The plasma response cortisol curve shown above was greatest when ACTH 1-17 was administered at 0700 (7 AM) and lowest at 2100 (9 PM).

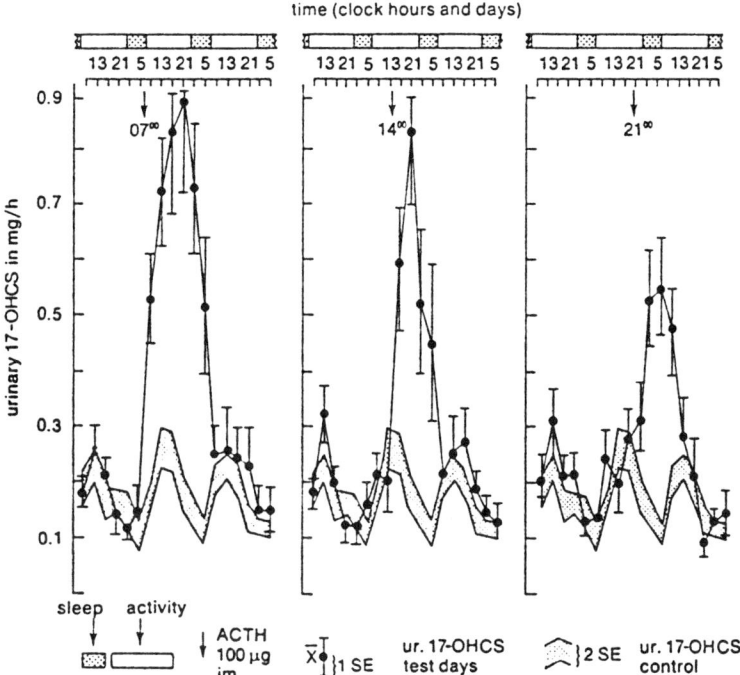

Figure 9B. The greatest change in the area under the time-concentration curve for the urinary excretion of 17-OHCS followed treatment with ACTH 1-17 at 7 AM (0700); the smallest change followed treatment at 9 PM (2100). (Shaded area: circadian reference rhythm of 17-OHCS for the group documented during three 72-hour baseline studies on each subject. Shaded area represents ± 1 SE of the group mean according to clock time. Solid line: Display of X̄ ± 1 SE of the 17-OHCS level at the designated clock times for the 24 hours before ($R_x - 1$ day), during (R_x day) and after ($R_x + 1$ day) ACTH injection for each treatment time.) Reproduced from Reinberg.[44]

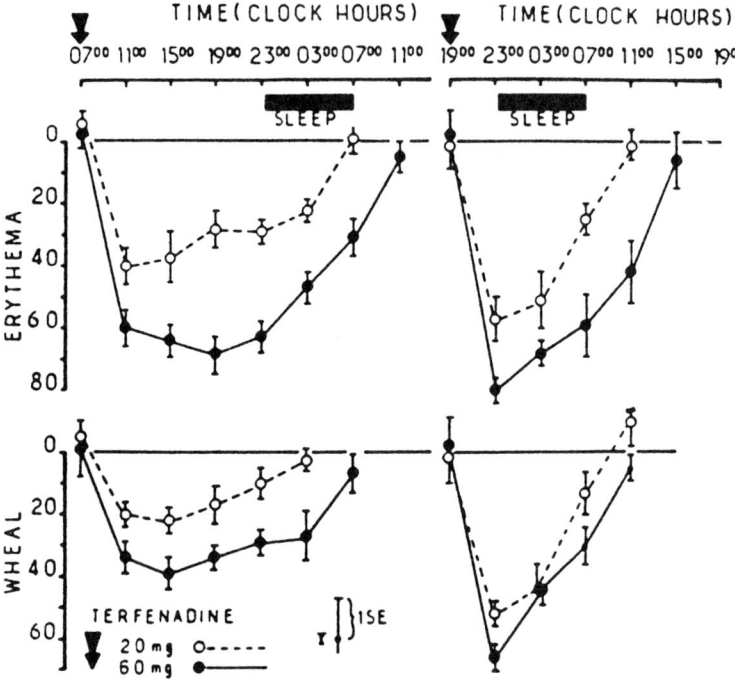

Figure 10. Dosing-time-dependent differences in terfenadine-induced inhibition of the cutaneous reaction to intradermally injected histamine (2 μg/0.1 ml). A double-blind and randomized procedure was used to assess the effects of placebo and terfenadine (20 and 60 mg doses) in a series of trials done on 10 healthy diurnally active subjects at 7-day intervals. The cutaneous responses to histamine following terfenadine dosing was expressed relative to the corresponding circadian time point reference values obtained during placebo treatment. Terfenadine in a dose-dependent manner exhibited greater duration of effect when ingested in the morning at 0700 (7 AM) than at 1900 (7 PM). In contrast, terfenadine exhibited a more potent and immediate antihistaminic effect when dosed at 1900 (7 PM) than at 0700 (7 AM). Reproduced from Reinberg et al., with permission.[48]

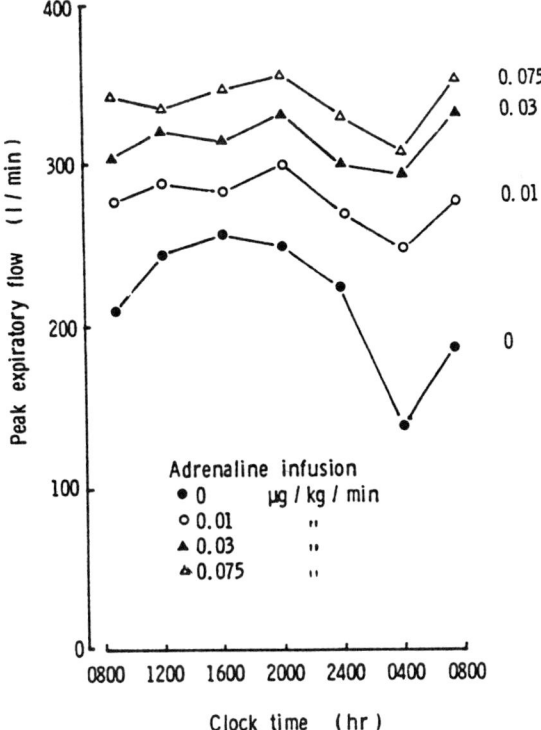

Figure 11. Circadian variation in the effect of graded doses of adrenaline given as acute IV infusions at selected clock times to asthmatic subjects. In terms of the absolute PEF level attained, adrenaline was more effective, in a dose-dependent manner, when given during the day than at night. In terms of the percent improvement in PEF from baseline (placebo = 0 µg/kg/min adrenaline), adrenaline was better, in a dose-dependent manner, when administered at 4 AM (0400). Reproduced from Barnes et al., with permission.[36]

Chronotoxicity refers specifically to biological rhythm-dependent variation in the undesired or adverse effects of medications as a function of their administration time. Very significant circadian chronotoxicities are known for a variety of medications, particularly antitumor (Fig. 13) and antibiotic ones.[1–5,50–53]

Chronotolerance is the opposite of chronotoxicity; it is defined as the biological rhythm-dependent difference in the tolerance of cells, tissues, and organs to medications and their active metabolites.[1–5] Chronotoxicities and chronotolerances, once determined, provide a means of enhancing the safety profile of pharmacotherapy. Medications can be administered once or several times a day in unequal

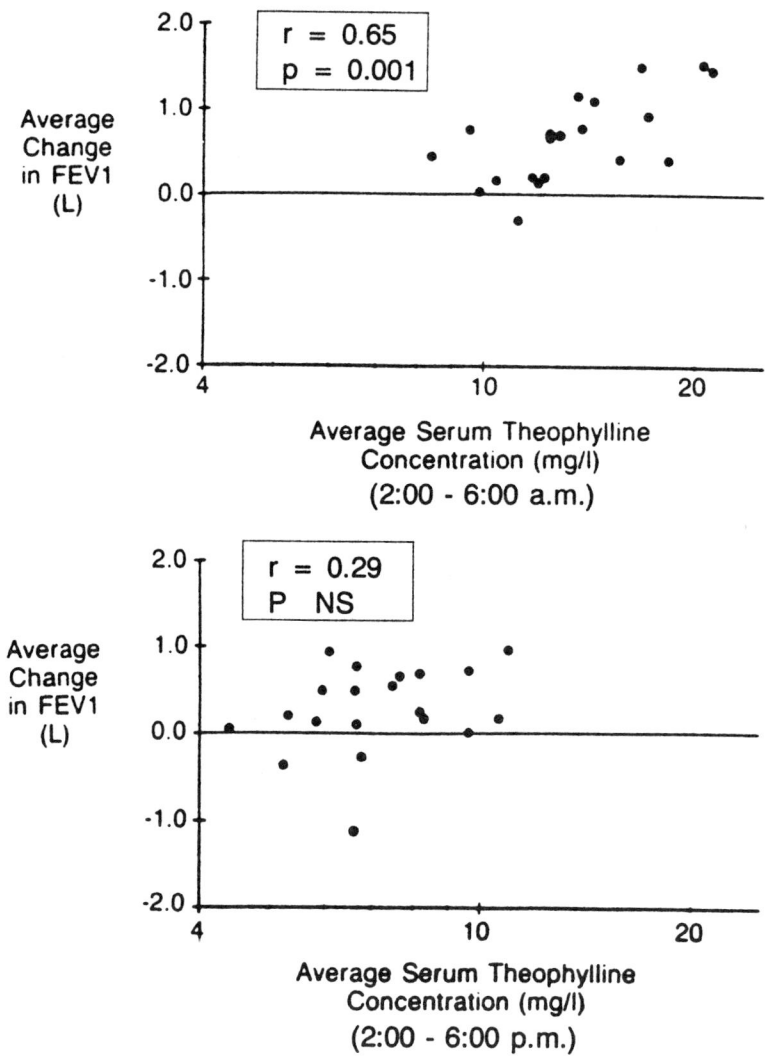

Figure 12A. Day-night difference in the serum theophylline-bronchodilator effect (FEV_1) relationship in adult asthmatics. Under steady-state conditions for once-daily theophylline treatment ingested at 8 PM, a serum theophylline-airway response relationship was demonstrable during the nighttime (between 2 and 6 AM), only (top); no such relationship was detectable during the daytime (between 2 and 6 PM) (bottom).

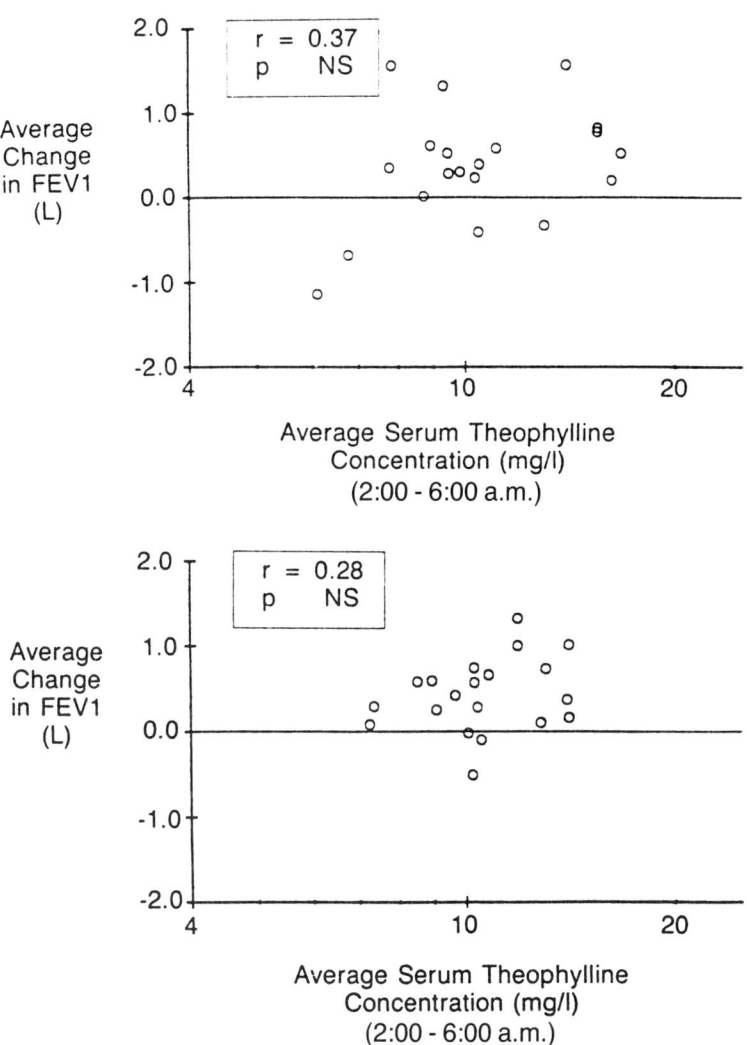

Figure 12B. Equal-interval, twice-daily dosing (8 AM and 8 PM) of a sustained-release theophylline formulation did not result in a demonstrable serum theophylline-bronchodilator effect relationship, during the night (top) or day (bottom). Reproduced from D'Alonzo et al., with permission.[41]

doses according to predetermined chronotolerances so as to reduce or avert drug-induced side effects. Such chronopharmacological considerations are especially relevant for those medications that have high potential for inducing toxicity and for which the therapeutic range is relatively narrow. This is the case for cancer and tablet corticosteroid medications, for example.[50-55]

Basic Concepts of Chronotherapeutics

The pharmacotherapy of most human diseases has long involved treatment schedules and drug delivery technologies that accomplish the goal of constancy of blood and tissue drug levels. It is assumed that the need of patients for medication is the same over time, and that constancy of drug level ensures consistency of drug effect. Knowledge of day-night and other temporal patterns in the pathophysiology of disease and/or the occurrence and intensity of symptoms, as discussed in the previous chapter, coupled with evidence for biological rhythm-dependencies in the kinetics and effects of medications, argue for a new approach to the pharmacotherapy of illness.[1-5]

Chronotherapeutics is the administration of medications according to biological rhythm considerations. Chronotherapeutics takes into account: (a) biological rhythm-related variation in the pathophysiology of disease processes and the manifestation and intensity of symptoms, (b) chronokinetics of the specific dosage forms of medications, (c) chronesthesy of medications and their active metabolites, and (d) individual differences between patients for the timing of treatment.[1-5]

Chronotherapeutics is especially relevant under the following conditions:

When the risk and/or intensity of diseases and their symptoms are known to present as a recurrent and predictable pattern over time during the 24 hours, menstrual cycle, etc. This is quite often the case for allergic rhinitis, regular and variant-type angina, rheumatoid and osteoarthritis, asthma, myocardial infarction, postsurgical pain, and peptic ulcer disease, for example.[3,21,43,56-68]

When it is known that the toxic effects of a medication vary predictably according to biological rhythms. This is true for most, if not all, antitumor medications.[50-53]

When the kinetics and/or dynamics of medications are known to be biological rhythm-dependent and thus influenced by their administration time. This is the case for antihistamines, antihypertensives, bronchodilators, heparin, and nonsteroidal antiinflammatory drugs, among others.[13,19,24,33,34,48,69]

When the intent of treatment is substitution therapy with synthetic analogues to simulate the temporal pattern of hormone levels of normal, healthy individuals. So far this entails the treatment of adrenocortical insufficiency with synthetic corticosteroids and hypothalamic amenorrhea with LHRH analogues.[70,71] A chronotherapeutic approach is seemingly pertinent to optimizing the treatment of insulin-dependent diabetes, although the advantage of such awaits clinical substantiation.[72,73]

When the desired effect of a given medication can be achieved only by dosing in a rhythm-modulated or pulsatile manner. This is the case for LHRH using ambulatory, programmed-in-time infusion pumps to treat hypothalamic amenorrhea.[71]

In the United States and other countries, physicians are already prescribing chronotherapeutic interventions, although perhaps without recognizing that they are doing so. Several examples are cited below:

Evening, once-daily chronotherapy of peptic ulcer disease by means of H_2-receptor antagonist medication.[74,75]

Evening, once-daily chronotherapy of asthma with *specially formulated* theophylline,[19,41,42,76–79] and oral and aerosol beta$_2$-agonist[19,80–84] medications as discussed in other chapters.

The treatment of the morning symptoms of rheumatoid arthritis by once-daily evening sustained-release NSAIDs, and of osteoarthritis, which tends to be most severe in the afternoon and evening, by once-daily noontime or evening sustained-release NSAID dosings.[24,85–87]

Unequal, day-night dosing of first generation, or evening-only dosing of sustained-effect, second generation[69,88,89] H_1-receptor antagonists to optimally manage the symptoms of allergic rhinitis, which tend to be worse in the morning upon arising in many patients (Fig. 14).

Use of ambulatory, programmable-in-time, drug-delivery infusion pumps or other procedures to clock antitumor medications to

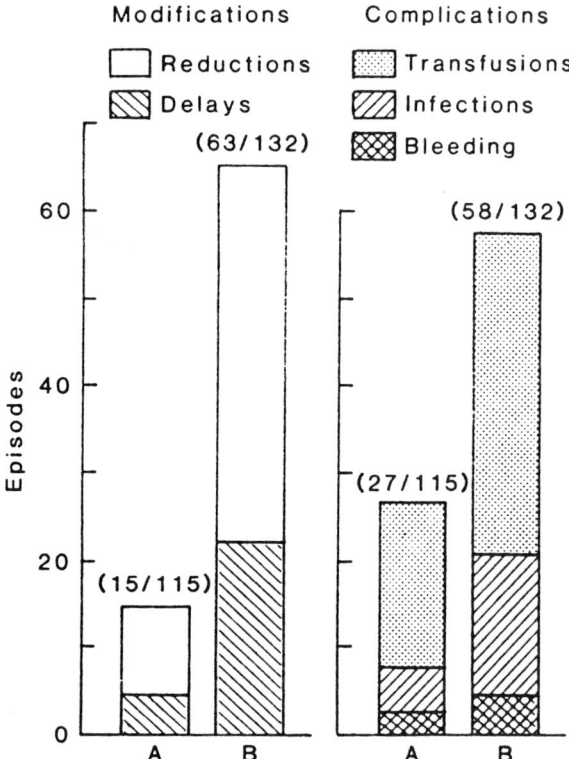

Figure 13. Circadian chronotoxicity of the antitumor medications of doxorubicin and cisplatin. Preclinical studies on rodents determined doxorubicin to be less toxic when administered at the commencement of daily activity, and for cisplatin when administered toward the end of the activity span. In human trials a total of 247 courses of doxorubicin-cisplatin therapy was administered; 115 courses according to schedule A—doxorubicin at 6 AM and cisplatin at 6 PM (the schedule predicted from preclinical studies to be least toxic to patients)—and 132 courses according to schedule B—cisplatin at 6 AM and doxorubicin at 6 PM. Dose reductions due to toxicity were three times more frequent when the medications were administered according to schedule B (P<0.01). Despite the schedule B dose reductions and treatment delays, 44% of schedule B, as opposed to only 23% of schedule A, treatments were associated with episodes of bleeding, infection and/or transfusion requirement (P<0.01). Reproduced from Hrushesky, with permission.[53]

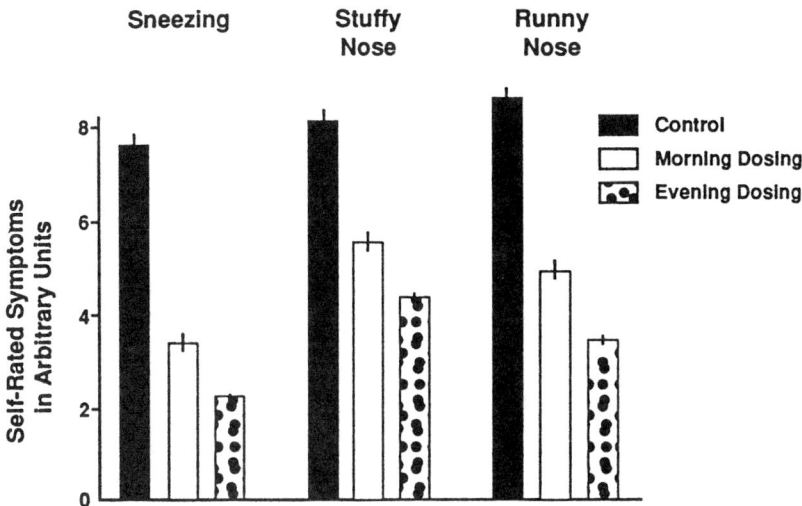

Figure 14. Comparative effectiveness, according to treatment time, of a sustained-release, second-generation H_1-antagonist in 98 diurnally active patients suffering from prominent morning symptoms of allergic rhinitis. Self-ratings of symptoms were done 5 to 6 times daily during the waking hours for 6 to 7 days per treatment schedule. In comparison to placebo-control conditions (blackened vertical bars), both the once-daily evening (stippled vertical bars) and morning (uncolored vertical bars) antihistamine treatment regimens significantly lessened the severity of the symptoms. However, the evening once-a-day treatment schedule was most effective, especially for patients who suffer from prominent morning symptoms of allergic rhinitis. Reproduced from Reinberg, with permission.[69]

biological rhythms to minimize their toxicity and optimize efficacy (Figs. 13,15).[52,53,90,91]

Chronoradiotherapy of solid tumors[92] using the circadian rhythm of tumor temperature as a marker for the timing of treatment (Fig. 16).

Use of ambulatory, programmable-in-time infusion pumps to administer LHRH at 90-minute intervals to optimize the treatment of hypothalamic amenorrhea.[71]

Use of programmable-in-time infusion pumps to deliver sympathomimetic, tocolytic medication according to a circadian schedule to control or avert the uterine contractions of premature labor.[93]

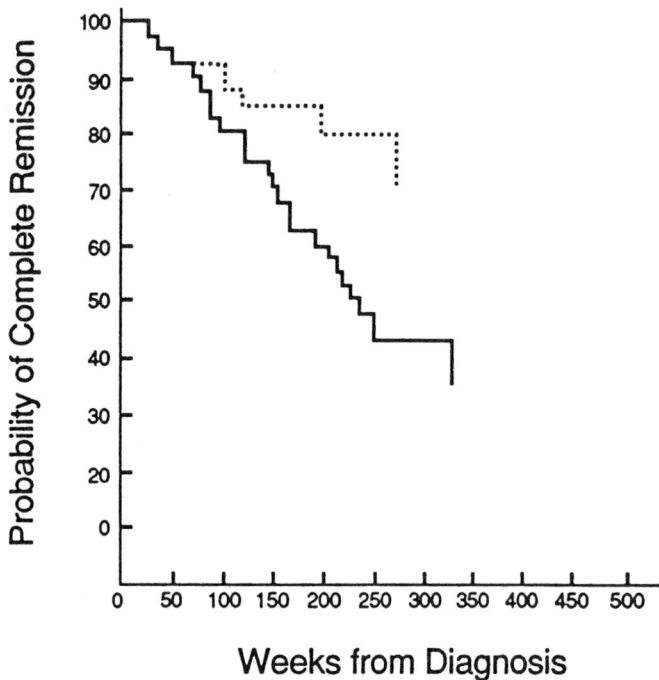

Figure 15. Circadian timing—morning versus evening—of the antitumor medications, 6-mercaptopurine (6-MP) daily, methotrexate (MTX) weekly, and vincristine and prednisolone monthly, for childhood acute lymphoblastic leukemia resulted in a siqnificant difference in the 5-year disease-free interval. Selection of the timing of chemotherapy was decided by the children's parents. The cohort of 36 children regularly treated with 6-MP and MTX in the morning before 10 AM (solid line) exhibited a relatively poor 5-year disease-free interval; the cohort (N = 82) so treated after 5 PM (dotted line) fared significantly better. Reproduced from Rivard et al., with permission.[91]

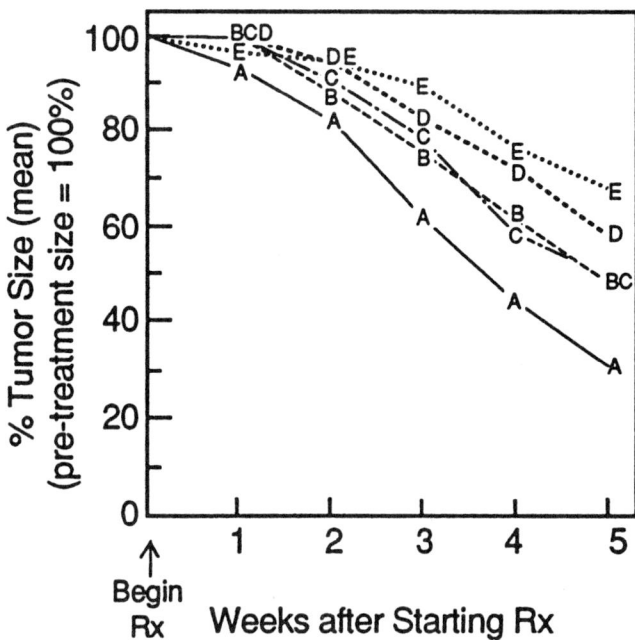

Figure 16. Circadian timing of radiotherapy for solid tumors of the oral cavity. The 24-hour pattern of tumor temperature of each patient was initially assessed. Different cohorts of patients were treated with radiotherapy for 5 weeks consistently at one of the following times: at the circadian temperature peak (Group A), 4 or 8 hours before (Groups B & D) or 4 or 8 hours afterwards (Groups C & E). Regression in tumor size was much more rapid in patients treated at the peak time of tumor temperature (Group A); regression was slowest in those treated near the trough of the 24-hour pattern in tumor temperature (Groups D & E). Adapted from Gupta and Deka, with permission.[92]

Various new chronotherapeutic interventions are under development and trial presently. Among these are so-called chronobiotics—special medications that possess the capability of facilitating the resetting and readjustment of selected bioclocks and associated bioperiodicities.[6,94] Chronobiotics are of particular importance for managing the desynchronization of the circadian time structure resulting from the alteration of the sleep-wake cycle when working rotating shift schedules or when rapidly displaced by jet aircraft across several time zones.

Chronotherapy of Antiasthma Medications

Early medical writings noted that asthma worsens or occurs only at night. However, only recently has the day-night pattern in asthma been convincingly demonstrated through epidemiological studies. The now often-cited investigation of Margret Turner-Warwick[59] in England during the 1980s on more than 7,700 patients documented that sleep disturbance due to asthma is a common problem for a great many asthmatics. The sleep of nearly 65% of those surveyed was disturbed at least three nights per week, while the sleep of approximately 40% was disturbed nightly due to asthma. Overall, almost 75% of the patients studied experienced at least one night of sleep disruption per week due to nocturnal asthma.

Although the major intent of this epidemiological investigation was to determine the incidence of nocturnal asthma, the findings of the study have equally significant implications about the efficacy of the medications used to treat the disease. When the study was conducted in England, the prevailing pharmacotherapies of asthma were equal-interval beta$_2$-agonist, vagolytic and antiinflammatory (corticosteroid or cromolyn) aerosols, equal-interval, low-dose oral theophylline, and tablet corticosteroids. A major conclusion of the study by Turner-Warwick[59] was "the drugs as currently available and as currently used do not seem to control nocturnal symptoms in many patients as well as is often supposed." In spite of the fact the majority of the study participants were under the routine care of their primary physician and were believed to use their medications as prescribed (at equal intervals and in equal doses according to homeostatic conventions), asthma occurred nocturnally—in many patients several nights per week.

With the documentation that asthma is a nocturnal problem for so great a number of patients, there has been renewed academic and industrial interest in exploring new concepts and approaches to optimize the efficacy of the antiasthma medications presently marketed, as well as those currently under development. In this regard the chronotherapy of bronchodilator and antiinflammatory medications has been and continues to be exploited.[19,54,55,77] A subsequent chapter in this volume addresses in depth the chrono-pharmacology and chronotherapy of theophylline and beta$_2$-agonists. Herein, the chronopharmacology and chronotherapy of tablet corticotherapy is explored to illustrate the application of findings in these fields to the clinical management of airway inflammation.

Chronopharmacology of Tablet Corticosteroids

The morning daily or alternate-day timing of methylprednisolone during the 1960s is considered to be the very first chronotherapy introduced into clinical medicine.[95] Ironically, this advance took place in the United States where the concept of chronotherapeutics today is neither well recognized nor well understood by the majority of physicians.

The management of asthma today emphasizes the control of airway inflammation and hyperreactivity. The current trend is to employ topical aerosol corticosteroid medications to achieve these clinical goals. Aerosol, rather than tablet, pharmacotherapy seems to be preferred because inhaled corticoids are said to be targeted directly to lung tissue. The perception is the dosing of corticoids by inhalation minimizes or averts the adverse effects of tablet corticosteroids, such as suppression of the hypothalamic-pituitary-adrenal axis (HPAA), growth inhibition in children, cataracts, hypertension, osteopenia, and others.

Over the past 30 years, it has been demonstrated convincingly that the circadian timing of tablet corticosteroids, such as prednisolone and methylprednisolone, is crucial for the optimization of their efficacy in asthma, and for minimizing, or even averting, adverse effects.[54,55] Ingestion of tablet corticosteroids late in the day or in the evening results in very great risk of drug-induced HPAA suppression. On the other hand, when doses of up to 10 to 15 mg of certain corticosteroids are ingested in the morning, at the commencement of the daily activity period, the risk of HPAA suppression may be modulated or completely avoided.[54,55] This fact relating to the chronotolerance of patients to synthetic corticosteroids demands major consideration in the design of dosing schedules for both tablet and aerosol forms of these medications. The optimization of tablet corticosteroids for asthma also necessitates consideration of the circadian rhythm in inflammatory processes as demonstrated by investigations on laboratory rodents and human beings.[24,96] Generally, the inflammation and associated hyperreactivity of the airways of asthma patients is worse during the overnight sleep period than during diurnal activity.[96-99] The nocturnal exacerbation of inflammation is common to other diseases besides asthma, such as allergic rhinitis and rheumatoid arthritis.[24] Some clinicians feel it best for patients who suffer from nocturnal symptoms of asthma to ingest tablet corticosteroids in the evening, before they retire to sleep at night. However, the optimization of tablet corticotherapy for the control of airway inflammation is not achievable by dosing at night. Oral corticosteroid medications do not attain their maximum effect rapidly. Moreover, the time during the 24 hours when tablet corticosteroids are ingested is a critical determinant of their effectiveness in controlling airway inflammation. Evening dosing for a prolonged span might even be harmful since it tends to potentiate adrenal suppression.[54,55,100,101]

The development of a chronotherapy of tablet corticosteroids requires an understanding of their circadian chronokinetics. Surprisingly, relatively few studies have been conducted on the chronopharmacokinetics of corticosteroid tablets.[102-105] The study by English and colleagues[105] on prednisolone is the most germane. In their investigation, a small group of diurnally active volunteers were studied for the pharmacokinetics of the drug following acute, single oral dosings (2 mg/kg) at specific times of the day and night. Trials were conducted under very controlled conditions at 7-day intervals when prednisolone was dosed in a randomized order at either 6 AM, noon, 6 PM, or midnight. Statistically significant differences were detected in several of the pharmacokinetic parameters. Greatest C_{max} and AUC and shortest elimination half-life followed after the drug ingestion at noontime. The T_{max} and the apparent volume of distribution (V_d), were greatest with the 6 PM dosing. Comparable pharmacokinetic findings were detected as well in the free prednisolone blood concentrations.

Circadian variation in corticosteroid binding globulin (CBG), also referred to as transcortin, was documented by Angeli and colleagues[106] in 1978. CBG is a serum glycoprotein that binds with cortisol and synthetic corticosteroids in the blood. Prednisolone, in particular, has a rather high affinity for CBG, similar to that of cortisol. In normal persons without a medical history of corticosteroid use, the circadian rhythm in CBG for cortisol and prednisolone is easily documented. In diurnally active individuals, transcortin binding to cortisol is nearly twice as great between 8 AM and 4 PM than it is between midnight and 4 AM. The circadian pattern of transcortin binding with prednisolone is different. Binding capacity for prednisolone is about 50% greater between 8 PM and 4 AM than it is 12 hours later. These findings suggest that the acute administration of synthetic corticosteroids such as prednisolone during the first 4 to 8 hours of the daily activity period is likely to result in a greater unbound serum concentration of drug than a late afternoon or evening one. Study of patients treated with 25 to 50 mg of prednisolone for at least 6 months reveals a reduced transcortin binding capacity for both cortisol and prednisolone, and absence of circadian variation.[106] While it is presumed the alteration in CBG capacity and temporal patterning results from corticosteroid treatment itself, it remains to be determined to what extent it is affected by the pathophysiology of the disease, nutritional state, and/or drug dosage and scheduling. Whatever the explanation, the increase in the unbound fraction of both natural and synthetic corticosteroids results in an enhancement of their pharmacodynamic activity.

It is of major clinical significance that corticosteroids exhibit administration-time dependencies both in their desired and adverse effects. In a study conducted by Reinberg and colleagues[38] during the 1970s, it was demonstrated that the effect of methylprednisolone on airway patency depends on the time of its administration. A group of young asthmatic boys was studied for their response to a single 40-mg dose of methylprednisolone sodium succinate administered as a subcutaneous injection. Individual drug dosing studies were conducted at 7 AM, 3 PM, 7 PM, and 3 AM on different days, each separated by at least a 1-week interval. Airway status was monitored by PEF five to six times daily before, during and after each timed injection. The magnitude of improvement in the 24-hour mean level of PEF varied according to the timing of the methylprednisolone treatment. The best effect upon airway function resulted from the drug being injected at 3 PM or 7 PM when the improvement in the 24-hour PEF averaged 17% and 13%, respectively, with reference to the placebo-control baseline values. The effect was minimal, only amounting to a 4% improvement in PEF, when the drug was injected at 3 AM. The effect of the 7 AM injection was satisfactory in inducing a 10% increase in the 24-hour mean PEF, although it was somewhat less than the effect produced by the 3 PM and 7 PM injections.

Numerous studies have addressed the matter of the chronotolerance of patients to tablet corticosteroids,[54,55] mostly in terms of adrenal suppression. The occurrence or extent of this adverse effect of synthetic corticosteroids is dependent not only upon the dosage administered but also on its circadian timing. This was demonstrated in the 1960s by Ceresa and colleagues.[39] As shown in Figure 17, when these investigators infused methylprednisolone at a rate of 660 μg/hour for 8 hours at the commencement of diurnal activity, no suppression of the HPAA was induced. This was not the case when the corticosteroid was infused in the evening or late at night, even if the amount of drug dosed was reduced to one-half that given in the morning trial. The results of the study conducted by Ceresa and coworkers demonstrate that at least one adverse effect of synthetic

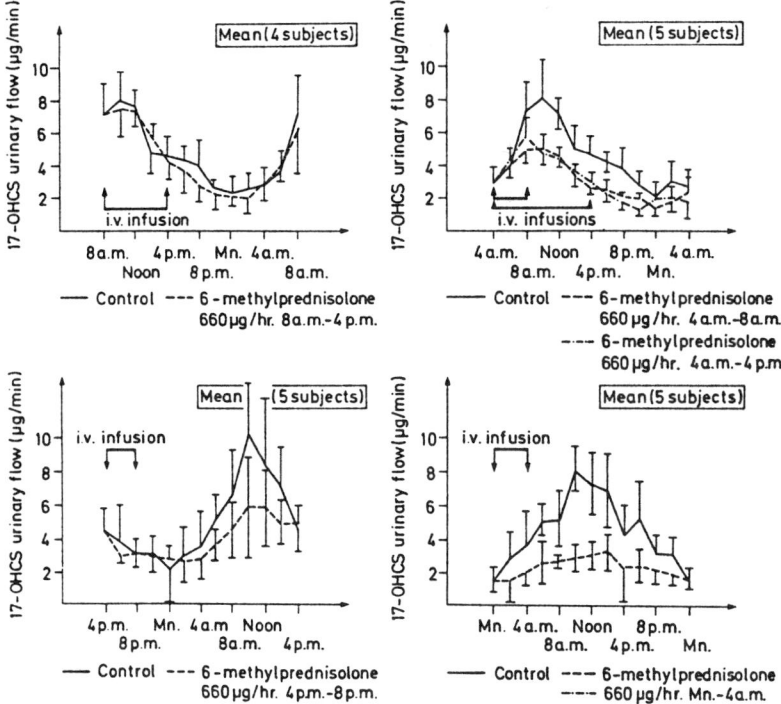

Figure 17. The infusion of 6-methylprednisolone (660 μg/h) during different 4-, 8- or 12-hour spans of the day or night has a differential effect on adrenal secretory activity as gauged by changes in the urinary concentration of 17-OHCS (the metabolic by-product of cortisol). When the corticosteroid was infused around the commencement of the daily activity span, *no* HPAA suppression resulted; the 17-OHCS levels and circadian pattern following saline and drug infusion were comparable. In contrast, moderate to severe HPAA suppression resulted when methylprednisolone was infused late in the afternoon and evening or after midnight. Reproduced from Ceresa et al., with permission.[39]

corticosteroids, HPAA suppression, can be minimized or even averted by a chronotherapeutic approach.

Chronotherapy of Tablet Corticosteroids

The chronotherapy of tablet corticosteroids for asthma has been investigated for many years.[54,95] One approach consists of administering the corticosteroids once daily as a single treatment coincident with the commencement of the activity span, in the morning, daily, or on alternate days, to diurnally active persons.[54,55,95,107] If control of the patient's asthma is not achieved, an increase in the once-daily morning dose is often instituted. Another approach has been devised and exploited by physicians residing in certain European countries.[108,109] This second option consists of dividing the total daily corticosteroid dose such that approximately two-thirds is ingested in the morning upon awakening, with the remaining one-third being taken later in the day around 2:00 to 3:00 PM— within 8 to 9 hours after the commencement of the patient's daily activity span. Both dosing regimens take advantage of the circadian chronotolerance of the HPAA to the adverse effects of tablet corticosteroids, as well as the circadian chronesthesy of the airways to their antiinflammatory effect.[54,55,107–109]

The splitting of the daily dosage of tablet corticosteroids into unequal morning-afternoon administrations seems to be beneficial for several reasons. First, in diurnally active patients, the morning and early afternoon dosing times of these tablet medications coincide with their heightened effect on airway inflammation and patency, and result in better control of asthma. Second, the timing of the larger corticosteroid dose in the morning around 8 AM and the second smaller one in the afternoon around 3 PM minimizes the risk of drug-associated side effects since this dosing schedule mimics the natural circadian pattern of endogenous cortisol secretion. Finally, this type of treatment schedule constitutes a type of rhythm-adjusted substitution therapy for patients suffering from severe HPAA inhibition induced by chronic high-dose, equal-interval corticotherapy administered without regard to chronobiological considerations. In patients with suppressed adrenocortical function, a desynchronization of certain circadian functions and processes may result from the alteration or disruption of the circadian rhythm of adrenocortical hormone secretion. Under ordinary circumstances, the day-night variation in adrenocortical hormones acts as a synchronizer of numerous cellular and metabolic 24-hour rhythms. The once daily morning and unequal morning-afternoon corticosteroid tablet dosing schedules serve to help resynchronize the circadian time structure of patients suffering from HPAA suppression as demonstrated by Reinberg and colleagues[70] in studies on patients suffering from adrenal insufficiency (Fig. 18). In patients with normal adrenocortical function, such dosing schedules reinforce the circadian time structure of the HPAA and associated bioperiodicities.

In Europe, one pharmaceutical company has devised a chronocorticotherapy based on a regimen of unequal morning-afternoon dosings of synthetic corticosteroids. This product (Dutimelan 8-15; Hoechst, Italy) has proven successful in improving airway function and controlling nighttime asthma without inducing HPAA suppression, as illustrated in Figure 19.[108] Studies also confirm that the chronotherapy of prednisolone for many asthma patients may be achieved by a once-daily morning tablet dosing schedule.[107] In patients who suffer from very severe nocturnal asthma, the addition of a second small corticosteroid dose in the afternoon around 3 PM may prove beneficial.[108,109] Studies of both corticosteroid treatment regimens confirm they are effective in controlling asthma with few or minimal side effects, even in patients managed for as long as 11 years in such a manner.[107]

Chronotolerance to Aerosol Corticosteroid Medication

The foregoing discussion regarding the chronopharmacology of tablet corticosteroids is relevant to aerosol forms as well. Patients treated with topical aerosol corticosteroids may be concomitantly subjected to systemic doses of these medications. This occurs from the absorption of drug from the lungs when the aerosol is deposited in the airways and/or the gastrointestinal tract when the corticosteroid aerosol delivered to the oral cavity is inadvertently swallowed. Adverse drug effects may result from the systemic absorption of aerosol corticosteroid medications, especially when they are administered in high doses.

Although many investigators[110–118] have reported aerosol corti-

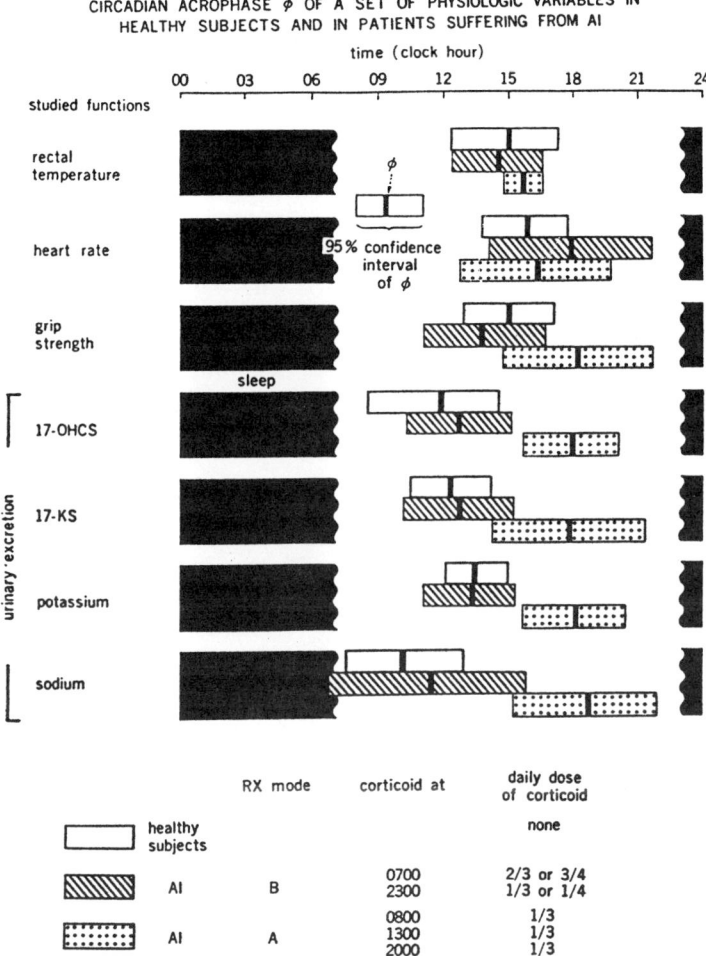

Figure 18. The charting of the circadian acrophase ϕ (peak time of rhythm) of selected physiological variables in healthy subjects (HS) and patients with adrenocortical insufficiency (AI) treated by means of either: (1) a circadian substitution chronocorticotherapy of cortisol (two-thirds or three-quarters of the daily dose at 7 AM and a smaller second one at 11 PM), or (2) a homeostatic, three times a day (at 8 AM, 1 PM, and 8 PM) equal-dose treatment schedule. Patients treated by the homeostatic cortisol schedule exhibited an abnormal circadian time structure in comparison to that of the referenced healthy control group. In contrast, those patients treated with the cortisol chronotherapy exhibited a normal circadian time structure comparable to that of the controls. A chronocorticotherapy of cortisol for adrenocortical insufficiency is considered a better means of treatment since it helps restore and preserve the normal circadian time structure. Reproduced from Reinberg et al., with permission.[70]

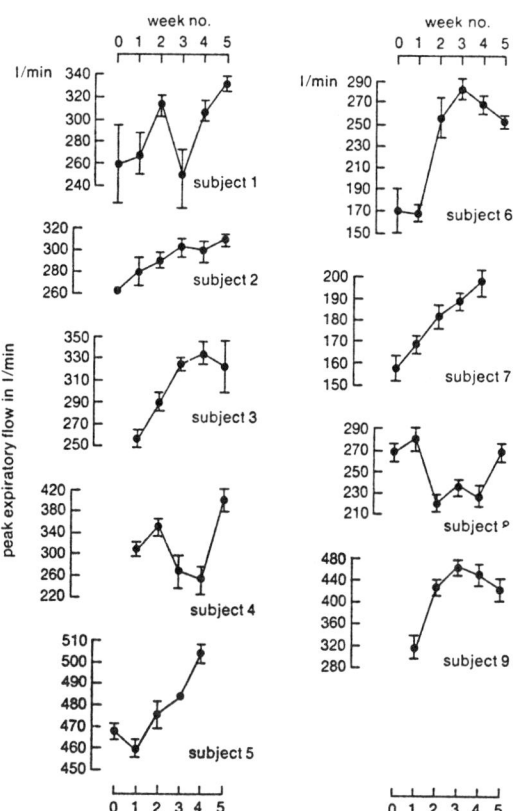

Figure 19A: Week-to-week changes in peak expiratory flow (PEF) in nine corticosteroid-dependent adult asthmatic patients before (week 0) and during one-month (5 weeks) treatment with Dutimelan 8-15® (Hoechst, Italy). This twice-a-day corticosteroid medication consists of one tablet composed of 4 mg prednisolone acetate and 2 mg prednisolone alcohol for ingestion at 8 AM and second tablet consisting of 2 mg prednisolone alcohol and 8 mg cortisone acetate for ingestion at 3 PM. In all but patients 4 and 8, who experienced severe respiratory infection after week two of the study, a progressive improvement in the 24-hour mean (based on five measurements daily during the waking span) PEF resulted while under treatment. In certain patients the improvement in airways status was remarkable.

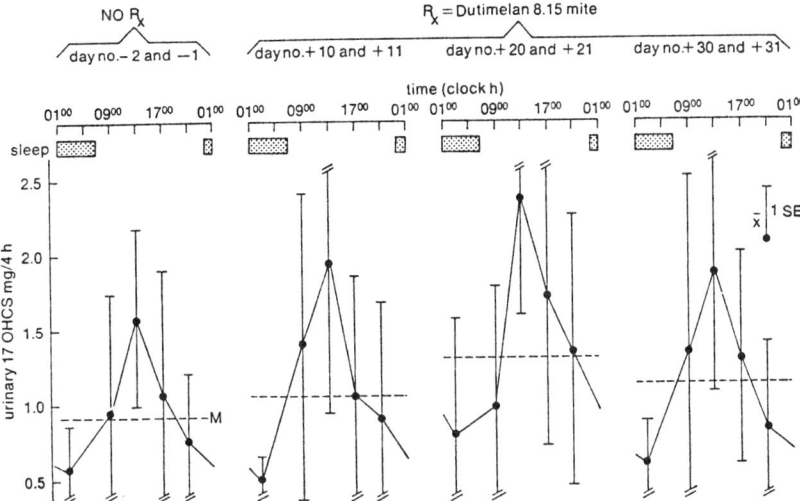

Figure 19B: Circadian rhythm in urinary 17-OHCS excretion before (no R_x, days -2 and -1) and during (days 10 to 11, 20 to 21 and 30 to 31) the one-month Dutimelan 8-15® treatment in the same nine asthmatic patients. This circadian rhythm in 17-OHCS excretion persisted; only minor and nonstatistically significant alterations occurred without evidence of HPAA suppression. This chronocorticotherapy achieved control of asthma without HPAA suppression. Reproduced from Reinberg et al., with permission.[108]

costeroids to be effective and safe for asthma patients, a considerable number of publications[119–155] have noted side effects similar to those seen in patients chronically treated with tablet corticotherapy. In both children and adults, altered adrenal function can occur as a consequence of high-dose corticoid aerosol pharmacotherapy.[122–131] In general, use of beclomethasone dipropionate and budesonide in low dosage (300 to 400 μg/kg/day) is associated with minimal risk of drug-induced HPAA suppression. As the dose is escalated the risk of HPAA suppression is increased, especially in children.[120–129] Other serious adverse effects of aerosol corticosteroids have been reported recently. These include osteopenia or altered bone metabolism,[132–139] posterior subcapsular cataract,[140–144] growth suppression in children,[145–150] and behavioral or mood alterations,[151–153] among others.[154,155]

The risk of side effects from topical aerosol corticosteroids appears to vary among the different products. They seem to be somewhat more common with beclomethasone and triamcinolone than budesonide. According to published studies, one major difference among the formulations is the rate of their hepatic metabolism;[156,157] however, the timing of the daily dosings during diurnal activity also may be contributory.[54,55] In diurnally active patients, the systemic absorption of aerosol-delivered corticosteroids when dosed late in the afternoon and evening is likely to exert greater suppressive effect on the HPAA than when they are dosed in the morning, as demonstrated by studies conducted with corticosteroid tablets and parenteral solutions. It is not known if the other reported side effects of aerosol corticoids also are administration-time dependent. The chronotolerance of patients to synthetic corticosteroids, at least in terms of HPAA suppression, whether administered by ingestion, infusion, or inhalation, is an especially important clinical concern for asthmatic patients of small stature, such as children, and the elderly.

Despite the fact that aerosol corticosteroid medications are considered, and indeed marketed, as topical agents, they in fact can produce undesired systemic effects. Because of this, salient adverse effects—cataracts, osteoporosis, growth retardation, hypertension, etc.—must be suspected as possible complications of inhalant corticosteroid therapy.[158] Future generation topical aerosol corticosteroids having reduced risk of side effects are under development. Although the chronotherapy of aerosol corticosteroids is presently under investigation, the data are yet insufficient to recommend how the daily dose should be apportioned during the 24 hours to optimize these dosage forms.[159]

Discussion

Medical chronobiology entails an understanding of how biological rhythms influence the diagnosis, manifestation, and treatment of disease. The biology of human beings is not constant over the 24 hours, month, and year. Instead, it varies quite predictably in time due to the phasing of the great multitude of endogenous bioperiodicities. Traditionally, medicine has been concerned with the questions of *why* (based on diagnosis), and *what* (type and form of) medication to prescribe, as well as with *how much* (dose or dosage) to administer. Today, we must ask additional questions pertaining to *when* to conduct diagnostic tests, *when* the symptoms and diseases manifest or worsen and, finally, *when* to medicate. The optimization of asthma medications necessitates that the biological time structure be taken into account not only in terms of the chronokinetics and chronesthesy of each particular medication, but in regard to the day-night variation in the pathophysiology of the disease and its symptoms.

The past decade has witnessed very rapid advances in the field of medical chronobiology, particularly in understanding the chronopharmacology and chronotherapeutics of asthma medications. Clinical investigations have substantiated that high-amplitude circadian rhythms contribute to the pathophysiology of asthma. Nonetheless, the goal of pharmacotherapy for asthma (and other human diseases) has been constancy in drug concentration to satisfy the hypothesis that constancy in drug level ensures constancy in effect. There is mounting evidence that the need for medication by most asthma patients is not the same throughout the 24 hours as might be predicted by the concept of homeostasis. In the majority of patients the risk of asthma and the intensity of its symptoms are greater during the nighttime than during diurnal activity. Therefore, it is crucial that the dosing of antiasthma medications take this key fact into account. Chronotherapeutic interventions endeavor to exploit the biological time structure of asthma patients to achieve an optimization of bronchodilator and antiinflammatory medications. Chapter 7 reviews the chronopharmacology of sustained release theophylline and beta$_2$-agonist medications as well as the progress realized thus far in establishing their chronotherapy for optimizing the management of asthma.

References

1. Reinberg A, Labrecque G, Smolensky MH. Chronobiologie et Chronotherapeutique. Heure Optimale d'Administration des Medicaments. Flammarion, Paris, 1991.
2. Lemmer B. Chronopharmacology: Cellular and Biochemical Interactions. Marcel Dekker, Inc., New York, 1989.
3. Reinberg A, Smolensky MH. Biological Rhythms and Medicine. Springer-Verlag, New York, 1983.
4. Reinberg A, Smolensky MH, Labrecque G. New aspects in chronopharmacology. Annu Rev Chronopharmacol 1986;2:3–26.
5. Reinberg A. Concepts in chronopharmacology. Annu Rev Pharmacol Toxicol 1992;32:51–66.
6. Reinberg A, Smolensky MH, Labrecque G. The hunting of a wonder pill for resetting all biological clocks. Annu Rev Chronopharmacol 1988;4:171–204.

7. Gibaldi M, Perrier D. Pharmacokinetics. Marcel Dekker, Inc., New York, 1982.
8. Vener KJ, Moore JG. Chronobiologic properties of the alimentary canal affecting xenobiotic absorption. Annu Rev Chronopharmacol 1988;4:257–281.
9. Waterhouse JM, Minors DS. Temporal aspects of renal drug elimination. In: Chronopharmacology: Cellular and Biochemical Interactions, Lemmer B (ed). Marcel Dekker, Inc., New York, 1989. pp 35–50.
10. Bruguerolle B. Temporal aspects of drug absorption and drug distribution. In: Chronopharmacology: Cellular and Biochemical Interactions, Lemmer B (ed). Marcel Dekker, Inc., New York, 1989, pp 3–13.
11. Moore JG, Englert E Jr. Circadian rhythm of gastric acid secretion in man. Nature 1969;226:1261–1262.
12. Goo RH, Moore JG, Greenberg E, Alazaki NP. Circadian variation in gastric emptying of meals in man. Gastroenterology 1987;93:515–519.
13. Lemmer B, Nold G. Circadian changes in estimated hepatic blood flow in healthy subjects. Br J Clin Pharmacol 1991;32:627–629.
14. Labrecque G, Dore F, Belanger PM, Lalande M. Circadian variation in the blood flow to different organs in the rat. Annu Rev Chronopharmacol 1988;5:445–488.
15. Feuers RJ, Scheving LE. The chronobiology of hepatic enzymes. Annu Rev Chronopharmacol 1988;4:209–256.
16. Belanger PM. Chronobiologic variation in the hepatic elimination of drugs and toxic chemical agents. Annu Rev Chronopharmacol 1988;4:1–46.
17. Jonkman JHG. Food interactions with once-a-day preparations. A review. Chronobiol Int 1987;4:459–466.
18. Warren JB, Turner C, Barnes PJ. Posture and theophylline kinetics. Br J Clin Pharmacol 1985;19:707–709.
19. Smolensky MH. Chronopharmacology of theophylline and beta-sympathomimetics. In: Chronopharmacology: Cellular and Biochemical Interactions, Lemmer B (ed). Marcel Dekker, Inc., New York, 1989, pp 65–113.
20. Nakano S. Temporal aspects of the pharmacokinetics and effects of psychotropic drugs in humans. In: Chronopharmacology: Cellular and Biochemical Interactions, Lemmer B (ed). Marcel Dekker, Inc., New York, 1989, pp 267–280.
21. Lemmer B (guest ed). Recent advances in the chronopharmacology of the cardiovascular system, Parts I & II. Chronobiol Int 1991;8(5/6):307–538.
22. Langner B, Lemmer B. Circadian changes in the pharmacokinetics and cardiovascular effects of oral propranolol in healthy subjects. Eur J Clin Pharmacol 1988;33:619–624.
23. Koeter GH, Postma DS, Keyzer JJ, Meurs H. Effects of oral slow-release terbutaline on early morning dyspnoea. Eur J Clin Pharmacol 1985;28:159–162.
24. Labrecque G, Reinberg AE. Chronopharmacology of non-steroid anti-inflammatory drugs. In: Chronopharmacology: Cellular and Biochemi-

cal Interactions, Lemmer B (ed). Marcel Dekker, Inc., New York, 1989, pp 545–580.
25. Reinberg A, Smolensky MH. Circadian changes in drug disposition in man. Clin Pharmacokinet 1982;7:401–420.
26. Hrushesky WJM, Langer R, Theeuwes F. (eds). Temporal control of drug delivery. Ann New York Acad Sci 1991;618: 1–641.
27. Bruguerolle B. Données recentes en chronopharmacocinétique. Path Biol 1987;35:925–934.
28. Scott PH, Tabachnik E, MacLeod S, Correia J, Newth C, Levison H. Sustained-release theophylline for childhood asthma: evidence for circadian variation of theophylline pharmacokinetics. J Pediatr 1981;99:476–479.
29. Smolensky MH, Scott PH, Harrist RB, Hiatt PH, Wong TK, Baenziger JC, Klank BJ, et al. Administration-time-dependency of the pharmacokinetic behavior and therapeutic effect of a once-a-day theophylline in asthmatic children. Chronobiol Int'l 1987;4:435–447.
30. Jonkman JHG, Reinberg A, Oosterhuis B, deNoord OE, Kerkhof FA, Motohashi Y, Levi F, et al. Dosing time and sex-related differences in the pharmacokinetics of cefodizime and in the circadian cortisol rhythm. Chronobiologia 1988;15:89–102.
31. Decousus H, Ollagnier M, Cherrah Y, Perpoint B, Queneau P. Chronokinetics of ketoprofen infused intravenously at a constant rate. Ann Rev Chronopharmacol 1986;3:321–322.
32. Reinberg A, Reinberg M. Circadian changes of the duration of action of local anaesthetic agents. Naunyn Schmiedebergs Arch Pharmacol 1977;297:149–157.
33. Decousus H, Croze M, Levi F, Jaubert J, Perpoint B, Reinberg A, Queneau P. Circadian changes in anticoagulant effect of heparin infused at a constant rate. Br Med J 1985;290:341–344.
34. Lemmer B. Temporal aspects of the effects of cardiovascular active drugs in humans. In: Chronopharmacology: Cellular and Biochemical Interactions, Lemmer B (ed). Marcel Dekker, Inc., New York, 1989, pp 525–541.
35. Gaultier C, Reinberg A, Motohashi Y. Circadian rhythm in total pulmonary resistance of asthmatic children. Effects of a beta-agonist agent. Chronobiol Int'l 1988;5:285–290.
36. Barnes P, Fitzgerald GA, Dollery CT. Circadian variation in adrenergic responses in asthmatic subjects. Clin Sci 1982;62:349–354.
37. Beaty R, Harman E, Molino L, Hendeles L. The dose-response of albuterol MDI during acute attacks of nocturnal asthma. Am Rev Respir Dis 1992;145(4):A66.
38. Reinberg A, Halberg F, Falliers C. Circadian timing of methylprednisolone effects in asthmatic boys. Chronobiologia 1974;1:333–347.
39. Ceresa F, Angeli A, Boccuzzi A, Molino G. Once-a-day neurally stimulated and basal ACTH secretion phases in man and their responses to corticoid inhibition. J Clin Endocrinol Metab 1969;29:1074–1082.
40. Smolensky MH, Scott PH, McGovern JP, Albright D. Circadian differ-

ences in theophylline effect during constant rate infusion with aminophylline. Ann Rev Chronopharmacol 1986; 3:139–142.

41. D'Alonzo GE, Smolensky MH, Feldman S, Gianotti LA, Emerson MB, Staudinger H, Steinijans VM. Twenty-four-hour lung function in adult patients with asthma: Chronoptimized theophylline therapy once daily in the evening versus conventional twice-daily dosing. Am Rev Respir Dis 1990;142:84–90.

42. Reinberg A, Pauchett F, Ruff F, Gervais A, Smolensky MH, Levi F, Gervais P, et al. Comparison of once-daily dosing for nocturnal asthma. Chronobiol Int'l 1987;4:409–420.

43. Yasue H, Omote S, Takizawa A, Nagao M, Miwa K, Tanaka S. Circadian variation of exercise capacity in patients with Prinzmetal's variant angina: role of exercise-induced coronary arterial spasm. Circulation 1979;59:938–948.

44. Reinberg A, Guillemant S, Ghata NJ, Guillemant J, Touitou Y, Dupont W, Lagoguey M, et al. Clinical chronopharmacology of ACTH 1-17. I. Effects on plasma cortisol and urinary 17-hydrocorticosteroids. Chronobiologia 1980;7:513–523.

45. Lemmer B, Scheidel B, Blume H, Becker HJ. Clinical chronopharmacology of oral sustained-release isosorbide-5-mononitrate in healthy subjects. Eur J Clin Pharmacol 1991;40:71–75.

46. White C, Smolensky MH, Sanders SW, Buchi KN, Moore JG. Day-night and individual differences in response to constant-rate ranitidine infusion. Chronobiol Int'l 1991;8:56–66.

47. Sanders SW, Ballersteros MA, Hogan DL, Koss MA, Isenberg JI. Effect of basal gastric acid secretion on the pharmacodynamics of ranitidine. Chronobiol Int'l 1991;8:186–193.

48. Reinberg A, Levi F, Guillet P, Burke JT, Nicolai A. A chronopharmacologic study of antihistamines with special reference to terfenadine. Eur J Clin Pharmacol 1978;14:245–252.

49. Reinberg A, Levi F. Dose-response relationships in chronopharmacology. Ann Rev Chronopharmacol 1990;6:25–46.

50. Scheving LE. Chronotoxicology in general and experimental chronotherapeutics of cancer. In: Scheving LE, Halberg F (eds). Chronobiology: Principles and Applications to Shifts in Schedules, NATO Advanced Institute Series, Sijthoff and Noordhoff, The Netherlands, 1980, pp 455–479.

51. Mormont C, Boughatts N, Levi F. Mechanisms of circadian rhythms in the toxicity and efficacy of anticancer drugs: Relevance for the development of new analogs. In: Lemmer B (ed). Chronopharmacology. Cellular and Biochemical Interactions, Marcel Dekker, New York, 1989, pp 395–437.

52. Hrushesky WJM, Roemeling RV, Sothern R. Preclinical and clinical cancer chronotherapy. In: Arendt J, Minors DS, Waterhouse JW (eds). Biological Rhythms in Clinical Practice, Wright, London, 1989, pp 225–252.

53. Hrushesky WJM. Circadian timing of cancer chemotherapy. Science 1985;228:73–75.

54. Reinberg A. Chronopharmacology of corticosteroids and ACTH. In:

Lemmer B (ed). Chronopharmacology. Cellular and Biochemical Interactions, Marcel Dekker, New York, 1989, pp 137–178.

55. Reinberg A, Smolensky MH, D'Alonzo GE, McGovern JP. Chronobiology and asthma. III. Timing corticotherapy to biological rhythms to optimize treatment goals. J Asthma 1988;25:219–248.

56. Nicholson PA, Bogie W. Diurnal variation in the symptoms of hay fever. Implications for pharmaceutical development. Curr Med Res Opin 1973;1:395–400.

57. Reinberg A, Gervais P, Levi F, Smolensky M, Del Cerro L, Ugolini C. Circadian and circannual rhythms of allergic rhinitis: An epidemiologic study involving chronobiologic methods. J Allergy Clin Immunol 1988;81:51–62.

58. Dethlefsen U, Repges R. Ein neues Therapieprinzip bei nachtlichem Asthma. Med Klin 1985;80:44–47.

59. Turner-Warwick M. Epidemiology of nocturnal asthma. Amer J Med 1988;85(Suppl. 1B):6–8.

60. Rocco MB, Nabel EG, Selwyn AP. Circadian rhythms and coronary artery disease. Amer J Cardiol 1987;59:13C–17C.

61. Waters DD, Miller DD, Bouchard A, Borsch X, Theroux P. Circadian variation in variant angina. Amer J Cardiol 1984;54:61–64.

62. Kuroiwa A. Symptomatology of variant angina. Jpn Circ J 1978;42:459–476.

63. Master AM, Jaffee HL. Factors in the onset of coronary occlusion and coronary insufficiency. JAMA 1952;148:794–798.

64. Muller JE, Stone PH, Turin ZG, Rutherford JG, Czeisler CA, Parkers C, Poole WK, et al. The MILIS study group. Circadian variation in the frequency of onset of acute myocardial infarction. N Engl J Med 1985;313:1315–1322.

65. Auvil-Novak SE, Novak RD, Smolensky MH, Morris MM, Kwan JW. Temporal variation in the self-administration of morphine sulfate via patient controlled analgesia in postoperative gynecologic cancer patients. Ann Rev Chronopharmacol 1990;7:253–256.

66. Harkness JAL, Richter MB, Panayi GS, Van de Pette K, Unger A, Pownall R, Geddawi M. Circadian variation in disease activity in rheumatoid arthritis. Br Med J 1982;284:551–554.

67. Moynihan BGA. Duodenal Ulcer, W. B. Saunders, Philadelphia, 1912.

68. Moore JG, Smolensky MH. Biological rhythms in gastro-intestinal function and processes: Implications for the pathogenesis and treatment of peptic ulcer disease. In: Swabb EA, Szabo S (eds). Ulcer Disease. Investigation and Basis for Therapy, Marcel Dekker, Inc., New York, 1991, pp 55–85.

69. Reinberg A. Chronopharmacology of H_1-antihistamines. In: Lemmer B (ed). Chronopharmacology. Cellular and Biochemical Interactions, Marcel Dekker, New York, 1989, pp 115–135.

70. Reinberg A, Ghata J, Halberg F, Apfelbaum M, Gervais P, Albulker C, Dupont J. Distribution temporelle du traitement de l'insuffisance cortico-surrénalienne. Essai de chronothérapeutique. Ann Endocrinol 1971;32:566–573.

71. Gompel A, De Plunkett T, Mauvais-Jarvais P. Induction de l'ovulation par pompe à la LH-RH. Gynecologie 1986;37:309.

72. Mejean L, Bicakova-Rocher A, Kolopp M, Vilaume C, Levi F, Debry G, Reinberg A, et al. Circadian and ultradian rhythms in blood glucose and plasma insulin of healthy adults. Chronobiol Int'l 1988;5:227–236.

73. Kollop MA, Bicakova-Rocher A, Reinberg A, Drouin P, Mejean L, Levi F, Debry G. Ultradian, circadian and circannual rhythms of blood glucose and injected insulins documented in six self-controlled diabetics. Chronobiol Int'l 1986;3:265–280.

74. Colin-Jones DG, Ireland A, Gear P, Golding PL, Ramage JK, Williams JG, Leicester RJ, et al. Reducing overnight secretion of acid to heal duodenal ulcers. Comparison of standard divided dose of ranitidine with a single dose administered at night. Amer J Med 1984;77(suppl 5B):116–122.

75. Merki H, Witzel L, Harre K, Scheurle E, Neumann J, Rohmel J. Single dose treatment with H_2 receptor antagonists: Is bedtime administration too late? Gut 1987;28:451–454.

76. Neuenkirchen H, Wilkens JH, Oellerich M, Sybrecht GW. Nocturnal asthma: effect of once per evening dose of sustained release theophylline. Eur J Respir Dis 1985;66:196–204.

77. Smolensky MH, McGovern JP, Scott PH, Reinberg A. Chronobiology and asthma. II. Body-time-dependent differences in the kinetics and effects of bronchodilator medications. J Asthma 1987;24:90–134.

78. Wilkens JH, Wilkens H, Heins M, Kurtin L, Oellerich M, Sybrecht GW. Treatment of nocturnal asthma: the role of sustained-release theophylline and oral beta-$_2$-mimetics. Chronobiol Int'l 1987;4:386–396.

79. Goldenheim PD, Conrad EA, Schein LK. Treatment of asthma by a controlled-release theophylline tablet formulation: a review of the North American experience with nocturnal dosing. Chronobiol Int'l 1987;4:397–408.

80. Postma DS, Koeter GH, v.d. Mark TW, Reig RP, Sluiter HJ. The effect of oral slow-release terbutaline on the circadian variation in spirometry and arterial blood gas levels in patients with chronic airflow obstruction. Chest 1985;87:653–657.

81. Persson G, Gnosspelius Y, Anehus S. Comparison between a new once-daily, bronchodilating drug, bambuterol, and terbutaline sustained-release, twice daily. Eur Respir J 1988;1:223–226.

82. Creemers JD. A multicentric comparative study of salbutamol controlled release (Volmax) and sustained-release theophylline (Theo-Dur) in the control of nocturnal asthma. Eur Respir J 1988;1(Suppl 2):333S. Abstract.

83. D'Alonzo GE, Rennard SI, Ratner PR, Findlay SR, Henochowicz SI, Nathan RA, Paull BR, et al. Twice-daily inhaled salmeterol as maintenance therapy for asthma. Am Rev Respir Dis 1992;145(4):A65.

84. Jenkins MM, Price K, Pounsford JC, Stack ID, Luce P, Demars DD, Alexander WJ. Efficacy and safety of salmeterol in elderly patients with asthma. Amer Rev Respir Dis 1992;145(4):A65.

85. Pownall R, Pickvance NJ. Does treatment timing matter? A double blind crossover study of ibuprofen 2400 mg per day in different dosage

schedules in treatment of chronic low back pain. Brit J Clin Pract 1985;39:7:267–275.

86. Lévi F, LeLouran C, Reinberg A. Chronotherapy of osteoarthritic patients: optimization of indomethacin sustained release (ISR). Ann Rev Chronopharmacol 1984;1:345–348.

87. Reinberg A, Manfredi R, Kahn MF, Chaouat D, Chaout Y, Delcambre B, Le Goff P, et al. Chronothérapie due ténoxicam. Therapie 1991;46:101–108.

88. Leickly F, Higgins D, Ownby D. A comparative cost-effectiveness study of two treatment modalities for ragweed hay fever. J Allergy Clin Immunol 1987;79:189.

89. Reinberg A, Gervais P, Ugolini C, Del Cerro L, Bicakova-Rocher A, Nicolai A. A multicentric chronotherapeutic study of mequitazine in allergic rhinitis. Ann Rev Chronopharmacol 1986;3:441–444.

90. Levi F, Misset JL, Brienza S, Adam R, Metzer G, Itzakhi M, Caussanel JP, et al. A chronopharmacologic phase II clinical trial with 5-fluoruracil, folinic acid, and oxaliplantin using an ambulatory multichannel pump. Cancer 1992;69:893–900.

91. Rivard G, Infante-Rivard C, Hoyeux C, Champagne J. Maintenance chemotherapy for childhood acute lymphoblastic leukemia: better in the evening. Lancet 1985;ii:1264–1266.

92. Gupta BD, Deka AC. Application of chronobiology to radiotherapy of tumor of the oral cavity. Chronobiologia 1975;2(Suppl. 1):125.

93. Lam F, Gill P, Smith M, Kitzmiller JL, Katz M. Use of the subcutaneous terbutaline pump for long-term tocolysis. Obstet Gynecol 1988;72:810– 813.

94. Arendt J, Aldhous M, Marks V. Alleviation of jet-lag by melatonin: Preliminary results of controlled double-blind trial. Brit Med J 1986;292:170.

95. Harter JG, Reddy WJ, Thorn GW. Studies on an intermittent corticosteroid dosage regimen. N Engl J Med 1963;296:591–595.

96. Martin RJ, Cicutto LC, Smith HR, Ballard RD, Szefler SJ. Airways inflammation in nocturnal asthma. Am Rev Respir Dis 1991;143:351– 357.

97. DeVries K, Goei JT, Booy-Noord H, Orie NGM. Changes during 24 hours in lung function and histamine hyperreactivity of the bronchial tree in asthmatic and bronchitic patients. Arch Allergy Appl Immunol 1962;20:93–101.

98. Reinberg A, Gervais P, Morin P, Albulker C. Rhythme circadian humain du seuil de la reponse bronchique a l'acetylcholine. CR Acad Sci 1971;272:1879–1881.

99. Gervais P, Reinberg A, Gervais C, Smolensky MH, DeFrance O. Twenty-four-hour rhythm in the bronchial hyperreactivity to house dust in asthmatics. J Allergy Clin Immunol 1977;59:207–213.

100. Hayek A, Crawford J, Bode HH. Single dose dexamethasone in treatment of congenital adrenal hyperplasia. Metabolism 1971;20:897–901.

101. Moeller H. Chronopharmacology of hydrocortisone and 9-alpha-fluorohydrocortisone in the treatment of congenital adrenal hyperplasia. Eur J Pediatr 1985;144:370–373.

102. Morselli PL, Marc V, Garattini S, Zaccala M. Metabolism of exogenous cortisol in humans—diurnal variations in plasma disappearance rate. Biochem Pharmacol 1970;19:1643–1647.
103. English J, Marks V. Diurnal variations in methylprednisolone metabolism in the rat. IRCS Med Sci 1981;9:721.
104. McAllister WAC, Mitchell DM, Collins JV. Prednisolone pharmacokinetics compared between night and day in asthmatic and normal subjects. Br J Clin Pharmacol 1981;11:303–304.
105. English J, Dunne M, Marks V. Diurnal variation in prednisolone kinetics. Clin Pharmacol Ther 1983;33:381–385.
106. Angeli A, Frajria R, De Paoli R, Fonzo D, Ceresa F. Diurnal variation of prednisolone binding to serum corticosteroid binding globulin in man. Clin Pharamcol Ther 1978;23:47–53.
107. Reinberg A, Touitou Y, Bothol M, Gervais P. Preservation of the adrenal function with oral morning dosing of corticoids in long term (3–11 years) treated cortico-dependent asthmatics. Ann Rev Chronopharmacol 1990;7:327–330.
108. Reinberg A, Guillet P, Gervais P, Ghata J, Vignaud D, Abulker C. One month chronocorticotherapy (Dutimelan® 8-15 mite). Control of the asthmatic condition without adrenal suppression and circadian alteration. Chronobiologia 1977;4:295–312.
109. Reinberg A, Gervais P, Chaussade M, Fabroulet G, Duburque B. Circadian changes in effectiveness of corticosteroids in eight patients with allergic asthma. J Allergy Clin Immunol 1983;71:425–433.
110. Godfrey S, Konig P. Beclomethasone aerosol in childhood asthma. Arch Dis Child 1974;48:665–670.
111. Costello J, Clark TJH. Response of patients receiving high dose beclomethasone dipropionate. Thorax 1974;29:571–573.
112. Kerrebijn K. Beclomethasone dipropionate in long term treatment of asthma in children. J Pediatr 1976;89:821–826.
113. Francis RS. Long-term beclomethasone dipropionate aerosol therapy in juvenile asthma. Thorax 1976;31:309–314.
114. Klein R, Waldman D, Kershnar H, Berger W, Coulson A, Katz RM, Rachelefsky GS, et al. Treatment of chronic childhood asthma with beclomethasone dipropionate aerosol: I. A double blind crossover trial in non steroid-dependent patients. Pediatrics 1977;60:7–13.
115. Goldstein DE, Konig P. Effect of inhaled beclo methasone dipropionate on hypothalamic-pituitary-adrenal axis function in children with asthma. Pediatrics 1983;72:60–64.
116. Williams H, Reed GF, Vernier-Jones ER, Hughes IA. Effect of inhaled beclomethasone dipropionate on saliva cortisol concentration. Arch Dis Child 1984;59:553–556.
117. Freigang B, Ashford DR. Adrenal cortical function after long-term beclomethasone aerosol therapy in early childhood. Ann Allergy 1990;64:342–344.
118. Haahtela T, Markku J, Kava T, Kiviranta K, Koskinen S, Lehtonen K, Nikander K, et al. Comparison of a beta-two-agonist, terbutaline, with an inhaled corticosteroid, budesonide, in newly detected asthma. N Engl J Med 1991;325:388–392.

119. Siegel SC, Heimlich EM, Richards W, Kelley VC. Adrenal function in allergy. IV. Effect of dexamethasone aerosols in asthmatic children. Pediatrics 1964;33:245–250.
120. Linder WR. Adrenal suppression by aerosol steroid inhalation. Arch Intern Med 1964;113:655–656.
121. Choo-Kang YFJ, Cooper EJ, Tribe AE, Grant IWB. Beclomethasone dipropionate by inhalation in the treatment of airways obstruction. Brit J Dis Chest 1972;66:101–106.
122. Gaddie J, Reid I, Skinner C, Petrie GR, Sinclair DJM, Palmer KNV. Aerosol beclomethasone dipropionate: a dose response study in chronic bronchial asthma. Lancet 1973;ii:280–281.
123. Wyatt, R, Waschek J, Weinberger M, Sherman B. Effects of inhaled beclomethasone dipropionate and alternate-day prednisone on pituitary-adrenal function in children with chronic asthma. N Engl J Med 1978;299:1387–1392.
124. Vaz R, Senior B, Morris M, Binkiewicz A. Adrenal effects of beclomethasone inhalation therapy in asthmatic children. J Pediatr 1982;100:660–662.
125. Smith M, Hodson M. Effects of long term high dose beclomethasone dipropionate on adrenal function. Thorax 1983;38:676–681.
126. Law CM, Marchant JL, Honour JW, Preece MA, Warner JO. Nocturnal adrenal suppression in asthmatic children taking inhaled beclomethasone dipropionate. Lancet 1986;i:942–944.
127. Prahl P, Jensen T, Bjerregarrd-Andersen H. Adrenocortical function in children on high dose steroid aerosol therapy. Allergy 1987;42:541–544.
128. Bisgaard H, Nielsen MD, Anderson B. Adrenal function in children with bronchial asthma treated with beclomethasone dipropionate or budesonide. J Allergy Clin Immunol 1988;81:1088–1095.
129. Priftis K, Milner AD, Conway E, Honour JW. Adrenal function in asthma. Arch Dis Child 1990;65:838–840.
130. Tabachnik E, Zadik Z. Diurnal cortisol secretion during therapy with inhaled beclomethasone dipropionate in children with asthma. J Pediatr 1991;118:294–297.
131. Toogood JH, Lefcoe NM, Haines DSM, Jennings B, Errington N, Baksh L, Chung L. A graded dose assessment of the efficacy of beclomethasone dipropionate aerosol for severe chronic asthma. J Allergy Clin Immunol 1977;59:298–308.
132. Ali NJ, Ward MJ. High dose inhaled beclomethasone dipropionate increases bone resorption in normal men. Br J Clin Pharmacol 1988;26:634P–635P.
133. Ali NJ, Capewell S, Ward MJ. Bone turnover during high dose inhaled corticosteroid therapy. Thorax 1991;46:160–164.
134. Ali NJ, Morrision D, Capewell S, Ward MJ. Beclomethasone and osteocalcin. Br Med J 1991;302:1080.
135. Luengo M, Picado C, Piera C, Guanabens N, Montserrat JM, Rivera J, Setoain J. Intestinal calcium absorption and parathyroid hormone secretion in asthmatic patients on prolonged oral or inhaled steroid treatment. Eur Respir J 1991;4:441–444.

136. Padfield PL, Teelucksingh S, Gough KJ, Host P, Tibi L. Beclometha-sone and osteocalcin. Lancet 1991;339:369–370.
137. Teelucksingh S, Padfield PL. Inhaled corticosteroids, bone formation and osteocalcin. Lancet 1991;338:60–61.
138. Toogood JH, Jennings B, Hodsman A, Bakerville J, Fraher LJ. Effects of dose and dosing schedule of inhaled budesonide on bone turnover. J Allergy Clin Immunol 1991;88:572–580.
139. Toogood JH, Hodsman AB, Fraher LJ, Jennings BH. Effect of inhaled budesonide (BUD) and prednisolone (PRED) on bone turnover. Chest 1990;98:109S.
140. Allen MB, Ray S, Dhillon B, Cullen R, Leitch AG. Do steroid aerosols cause cataracts in asthmatic patients? Thorax 1988;43:845P–846P.
141. Allen MB, Ray SG, Leitch AG, Dhillon B, Cullen B. Steroid aerosols and cataract formation. Br Med J 1989;299:432–433.
142. Dyson C, McCormick DS, Toogood J. Prevalence of posterior subcapsu-lar cataracts in an asthmatic population undergoing long-term sur-veillance for adverse effects of inhaled steroid therapy. J Allergy Clin Immunol 1991;87:258.
143. Fraunfelder FT, Meyer SM. Posterior subcapsular cataracts associated with nasal or inhalation corticosteroids. Am J Ophthalmol 1990;109:489–490.
144. Karim AK, Thompson GM, Jacob TJ. Steroid aerosols and cataract formation. Br Med J 1989;299:918.
145. Littlewood JM, Johnson AW, Edwards PA, Littlewood AE. Growth retardation in asthmatic children treated with inhaled beclometha-sone dipropionate. Lancet 1988;i:115–116.
146. Storr JNP. Growth of asthmatic children. Br Med J 1991;303:719.
147. Wales JKH, Barnes ND, Swift PGF. Growth retardation in children on steroids for asthma. Lancet 1991;338:1535.
148. Zimmerman B, Tremblay D, Holland FJ, Brjnach L, Szlapetis G. Cortisol levels and growth in asthmatics less than age 5 treated with high dose nebulized steroid (budesonide 2mg/day). J Allergy Clin Immunol 1991;87(part 1):253.
149. Reed CE. Aerosol steroids as primary treatment of mild asthma. N Engl J Med 1991;325:425–426.
150. Wolthers OD, Pedersen S. Growth of asthmatic children during treatment with budesonide: a double blind trial. Br Med J 1991;303:163–165.
151. Lewis LD, Cochrane GM. Psychosis in a child inhaling budesonide. Lancet 1983;ii:634.
152. Connett G, Lenney W. Inhaled budesonide and behavioral distur-bances. Lancet 1991;338:634–635.
153. Phelan MC. Beclomethasone mania. Br J Psychiatry 1989;155:871–872.
154. Capwell S, Reynolds S, Shuttleworth D, Edwards C, Finlay AY. Purpura and dermal thinning associated with high dose inhaled corticosteroids. Br Med J 1990;300:1548–1551.
155. Kruszynska YT, Greenstone M, Home PD, Cooke NJ. Effect of high

dose inhaled beclomethasone dipropionate on carbohydrate and lipid metabolism in normal subjects. Thorax 1987;42:881–884.

156. Ryrfeldt A, Andersson P, Edsbacker S, Tonnesson M, Davies D, Pauwels R. Pharmacokinetics and metabolism of budesonide, a selective glucocorticoid. Eur J Respir Dis 1982;63(Suppl.):86–95.

157. Johansson SA, Andersson KE, Brattsand B, Gruvstad E, Hedner P. Topical and systemic glucocorticoid potencies of budesonide, beclomethasone dipropionate and prednisolone in man. Eur J Respir Dis 1982;63(Suppl.):74–82.

158. Kamada AK, Parks DP, Szefler SJ. Inhaled glucocorticoid therapy in children: How much is safe? Ped Pulmonol 1992;12:71–72.

159. Toogood JH, Baskerville JC, Jennings B, Lefcoe NM, Johansson SA. Influence of dosing frequency and schedule on the response of chronic asthmatics to the aerosol steroid, budesonide. J Allergy Clin Immunol 1982;70:288–298.

3

Nocturnal Asthma:
An Overview

Richard J. Martin, M.D.

Introduction

"The sleeping patient is still a patient. His disease not only goes on while he sleeps, but indeed may progress in an entirely different fashion from its progression during the waking state."[1] This prophetic statement was put forth by Eugene Robin over 30 years ago. Today we can see how accurate that thought was, as any disease process of any organ system has the potential to worsen during sleep.

The area of medicine related to alterations during sleep is still in its relative infancy, but information has been rapidly forthcoming in the last decade (see Chapters 1 and 2). Examples of acrophases (peak periods) of certain chronobiological rhythms would be surgical deaths at 1:00 AM, myocardial infarctions and diabetic ketoacidosis at 4:00 AM, menses onset at 5:00 AM, and mental and physical ability during the afternoon. Patients with asthma are particularly at risk for alterations in lung function during sleep. Understanding the alterations that occur with any disease during sleep will allow the appropriate studies to be initiated by the physician and, based on these studies, appropriate therapeutic intervention applied.

Martin RJ (editor): *Nocturnal Asthma: Mechanisms and Treatment,* © Futura Publishing Co., Inc., Mount Kisco, NY, 1993.

Sleep related asthma, which will be referred to as nocturnal asthma, although it can be associated with sleep at any time of the day, is an intriguing problem that can be difficult to treat. The available information on this process, which is increasing yearly, will be discussed but, as will become evident, an extensive fund of knowledge is not available. The reader will be referred to various chapters in this book for greater detail and information regarding specific topics.

Since asthma in general is relatively easy to treat and severe problems appear overall to occur uncommonly, two striking statistics that are not generally appreciated by the medical community reveal the severity of nocturnal asthma. Cochrane and Clark found that 68% of the adult asthmatic deaths occurred between midnight and 8:00 AM[2] and Hetzel et al. found that 8 of 10 asthmatic ventilatory arrests occurred between midnight and 6:00 AM.[3] Additionally, asthma mortality is rising worldwide,[4] with most of the deaths occurring at night.[5]

Background

Nocturnal symptoms arising from bronchial asthma were realized as far back as 1698 by Floyer[6] and in 1882 by Salter.[7] Very insightful and interesting statements were made by both physicians. Dr. Floyer said, "I have observed the fit always to happen after sleep in the night where nerves are filled with windy spirits and the heat of the bed has rarified the spirits and humors!" Dr. Salter felt that ". . . this fact is that sleep favours asthma—that spasm of the bronchial tubes is more prone to occur during the insensibility and lethargy of sleep than during the waking hours." Many investigators feel that worsening of symptoms or decrements in airway function frequently occur in the asthmatic population during sleep. To determine the frequency of nocturnal asthma in a nonhospital-based population, Turner-Warwick[8] conducted a survey of 7,729 asthmatic patients in the United Kingdom. Ninety-four percent responded that they awoke at least one night a month with symptoms of asthma. Seventy-four percent awoke at least one night a week, 64% at least three nights a week, and 39% awoke every night (Table 1). The occurrence of these symptoms is also reflected in mortality statistics. For all age groups, 83 of 168 (53%) asthma mortality cases over a 1-year period occurred at night.[5] For the population of 168 patients,

Table 1

Frequency of Nocturnal Asthma

Frequency	Percent
Every night	39
At least 3 nights/week	64
At least 1 night/week	74

Adapted from: Turner-Warwick M. Epidemiology of nocturnal asthma. Am J Med 1988; 85(1B):6–8.

79% had premortem complaints of asthma affecting their sleep, and this occurred every night in 61 patients (42%). Bagg and Hughes[9] also found morning decreases in peak expiratory flow rates in 30 of 40 stable asthmatics. In a "wash-out" phase of a large pharmacological study, 1,525 out of 1,631 dyspneic episodes in 3,129 patients occurred between 10:00 PM and 7:00 AM with the peak at 4:00 AM (Fig. 1).[10] Thus, the incidence of this potentially serious problem is higher than previously realized.

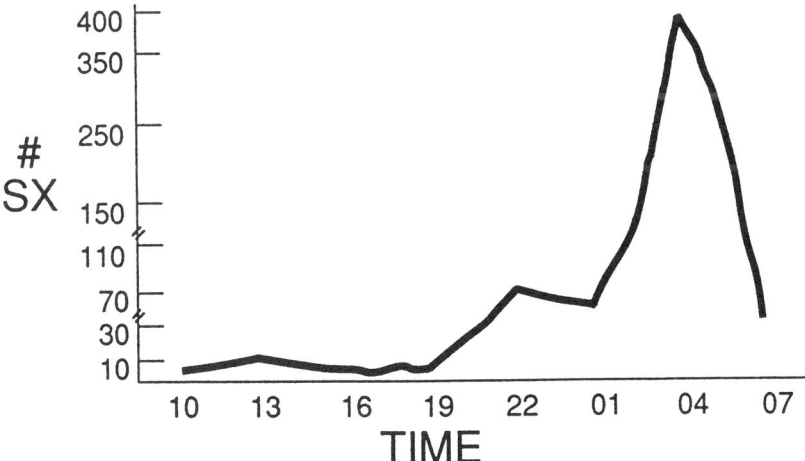

Figure 1. This figure demonstrates the marked frequency of nocturnal symptoms (Sx) in 3,129 mainly asthmatic patients. Redrawn from data in reference 10.

Mortality, Arousal Abnormalities, Respiratory Arrests

In 1971, 1,198 deaths were attributed to asthma in England and Wales, 149 of which occurred in the greater London hospitals where Cochrane and Clark made their survey.[2] Forty-seven (32%) were in the age range of 35 to 64 on which the study was focused. Nine patients were excluded because other conditions may have been the cause of death. Also excluded were 15 because death occurred prior to hospital admission, 2 because patients were not admitted for acute asthma, and 2 because case records were unobtainable. Although the population size was drastically reduced to 19 patients, important observations were made nevertheless.

Thirteen of the 19 patients (68%) died between midnight and 8:00 AM because of asthma. The remainder died at even intervals throughout the rest of the day. Of interest, only one patient was admitted to an intensive care unit and that was for a period of only 2 days. This patient died 6 days later on a general ward. Treatment would be important in considering the possible contributing factors to nocturnal death. Apparently a primary consideration as to the cause of death in this study was not a lack of bronchodilator therapy, but rather the administration of sedatives. Fifteen of the 19 patients received sedation, and 11 patients were given sedatives within 12 hours of death. In four patients, deterioration of their clinical condition rapidly followed sedative administration. However, in comparing this group to 19 surviving asthmatic patients, no difference could be found in the treatment programs. This raises the possibility of a patient subgroup that does not respond appropriately to sleep or sedation; that is, their arousal to important stimuli is blunted. This blunted arousal may be more important in the older asthmatic age group, as they have a markedly higher proportion of asthma deaths.[5]

The possibility of a blunted arousal mechanism was studied in two adolescent males who had a history of severe nocturnal asthma that led to respiratory arrest.[11] These patients and control subjects were studied during sleep with leads for the measurement of sleep staging and a tight-fitting face mask attached to a pneumotachygraph and an occlusion device for the measurement of arousal response time. Occlusion of the external airway was performed at end-expiration and released when alpha waves (arousal pattern) appeared on the electroencephalogram. Arousal from stage 2 sleep was the same between the controls and two patients (8.6 ± 4.9 SD seconds for the control vs. 8.4 ± 3.5 and 9.8 ± 1.8 for the two patients). The controls had a shorter time to arousal during stages 3 and 4 (8.8 ± 0.3 seconds vs. 20 ± 9.9 and 11.5 ± 0.3). Although there was greater variability during rapid eye movement (REM) sleep, again the controls tended to arouse more rapidly than the asthmatic patients (6.2 ± 3 seconds vs. 9.8 ± 6 and 26.8 ± 20). This suggests that patients with severe nocturnal asthma may have a suppressed arousal response to bronchoconstriction and may not awaken until the process is quite severe. The possibility exists that suppressed arousal contributes to fatal nocturnal asthma.

Ballard et al.[12] determined the effect of sleep deprivation on the arousal response to induced bronchoconstriction during the next night's sleep. Airway resistance (measured on a breath-by-breath basis) was measured during sleep on each of two nights. One night was a "normal" sleep night and a second followed 24 hours of sleep deprivation. Nebulized methacholine (not to be confused with a methacholine challenge test) via a mask was used to induce bronchoconstriction during sleep. Sleep following sleep deprivation produced a doubling in airway resistance before the subjects awoke complaining of chest tightness and/or wheezing compared to the "normal" sleep night. Thus, arousal responses are depressed by prior sleep deprivation. In patients whose asthma is progressing over time, worsening sleep patterns probably develop that can possibly lead to sleep deprivation and blunted arousal mechanisms.

For respiratory arrests due to asthma, similar patterns as discussed above for mortality can be seen. In a 2½-year survey of 1,169 consecutive hospital admissions for asthma, a low percentage of respiratory arrests occurred as a result of asthma.[3] In nine patients, there were 10 successful resuscitations. Again, the crisis period was during the sleep hours between midnight and 6:00 AM with 8 of the 10 respiratory arrests occurring during that 6-hour period. A feature common to all of these patients was the presence of large circadian swings in peak expiratory flow rates, with an early morning fall. This early morning decrease was not limited to the patient with an impending respiratory arrest because it was seen in approximately 30% of all admissions. However, it correlated best with the risk of ventilatory arrest and also with sudden death out of all the variables analyzed. Interestingly, unlike other signs on the examination, the circadian swings in peak flow did not show a relationship to the designation of acute or subacute asthma.

Circadian Rhythms and Pathophysiology

The understanding of the pathophysiology of nocturnal asthma brings forth a very important topic called *chronobiology*. This is the understanding of biological processes that have time related rhythms. These rhythms may occur yearly, monthly, or on a 24-hour cycle. For asthma, the 24-hour cycle (*circadian*) is extremely important. Although circadian rhythms have been studied mostly in the asthmatic population, in the coming years many disease processes will need to be understood on the basis of rhythm changes that occur over a 24-hour cycle.

Even in the normal population, there are changes in pulmonary function values between day and night (Fig. 2). The best lung function levels occur at approximately 4:00 PM and the nadir at 4:00 AM. The peak-to-trough difference is approximately 8%. This difference is statistically, but not clinically, significant. For the asthmatic patient, the same circadian pattern in pulmonary function is seen.

Circadian Rhythm - Lung Function

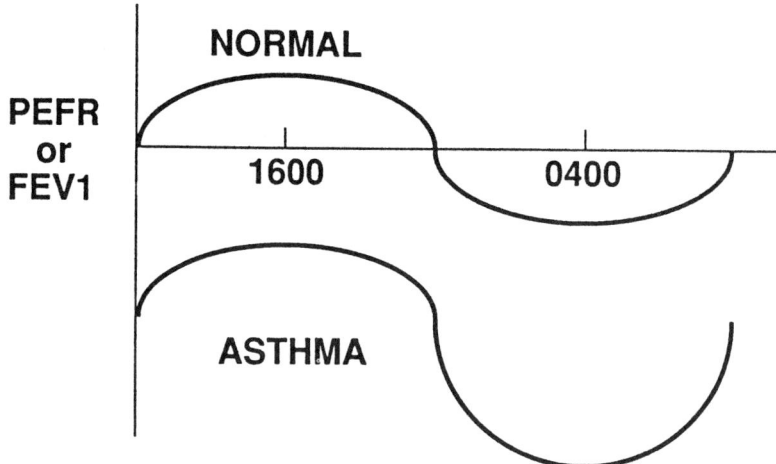

Figure 2. Both normal subjects and asthmatic patients have circadian alterations in lung function with nadirs occurring at approximately 4 AM.

The best function still occurs around 4:00 PM and the worst at 4:00 AM, but the peak-to-trough swing can be as much as 50%.[13]

The problem of evaluating lung function during the night in asthmatic patients (or anyone) was the inability to do so while the patient slept. That is, the investigator was forced to wake the patient or wait for spontaneous awakenings. This obviously interfered with the understanding of events on a continuous basis during sleep. Although difficult to use, techniques have been developed to measure airway resistance continuously throughout sleep. It is of interest that lower airway resistance has a circadian rhythm independent of sleep and it progressively increases while awake and supine from midnight to 6:00 AM (Fig. 3). But over the same time interval the resistance markedly increases further while asleep.[14] This means that a circadian rhythm to lung function exists while awake, but sleep has a dramatic impact on this rhythm. This study additionally showed that the individual sleep stages have little effect on the resistance measurements. This is different from what was previously guessed at by waiting for the patient to spontaneously awaken before measuring lung function.

All biological functions, not just respiration, have circadian rhythms associated with them. However, in asthmatics the respiratory rhythm peak-to-trough variation can be great. When this variation approaches the 50% level, it is an important indicator for potential respiratory arrest and mortality.

Recumbency, Solar Time, Sleep Alterations

To evaluate the effect of recumbency on circadian variation in lung function, five asthmatic patients remained in the sleeping position for over 24 hours.[15] With peak expiratory flow as the measure of change, it was determined that recumbency was not the factor producing the sleep related fall found in this group on control days. This observation did not relate to the presence or absence of atopy. To further evaluate this phenomenon, specifically in day versus night workers, an additional five asthmatic patients who worked night shifts were studied. The results showed that the circadian variation in peak flow was related to sleep rather than to the clock hour. In addition, it was found that when shifting from day to night work and back again, the change in circadian variation occurred rapidly and was complete by the end of the first new sleep

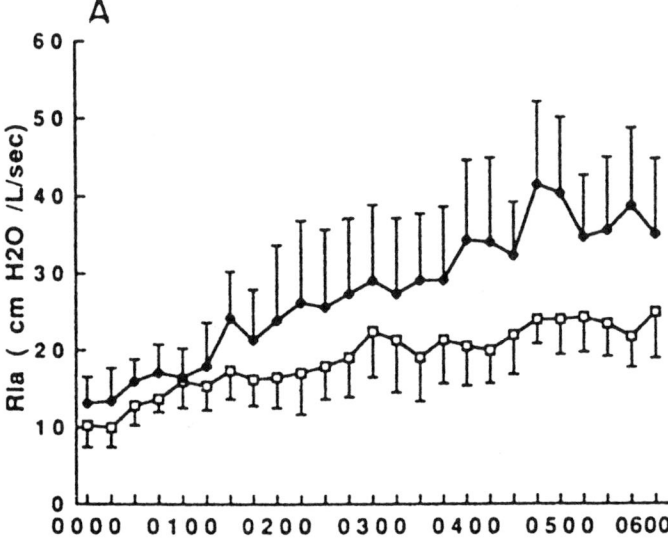

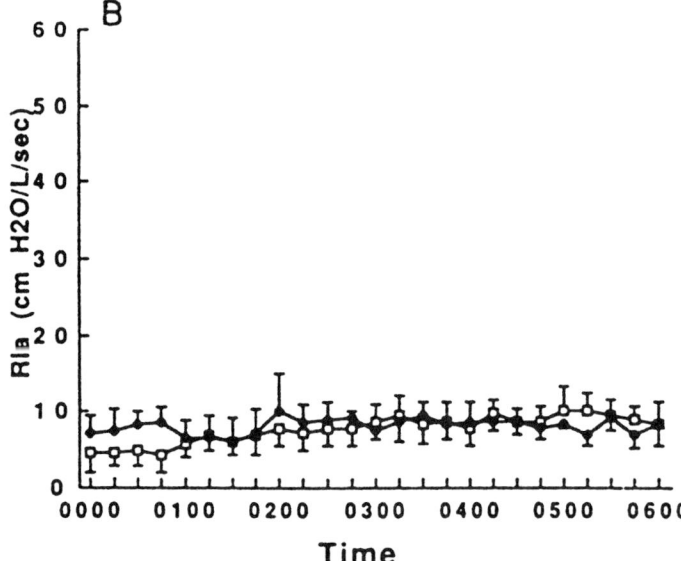

Figure 3. A. In asthma from midnight to 6 AM lower airway resistance (R_{IA}) progressively increases if awake (open squares) or asleep (closed diamonds), but sleep has a much more profound effect on R_{IA}. B. Normal subjects demonstrate minimal increase in R_{IA} during the night. From reference 14 with permission.

cycle following the shift change. This rapid change in peak flow variation upon shift change is similar to the observed changes seen in the basic circadian rhythm, for example, body temperature.[16] Contrary to the peak flow changes, circadian variation in forced expiratory volume in one second in normal shift workers held a constant relationship to solar time.[17] However, it appears that none of these workers were asthmatic.

To further determine the effect of sleep on asthma, the effect of sleep interruption and sleep deprivation was investigated.[18] Twenty-one patients who had at least a 25% decrement in peak expiratory flow rates at 6:00 AM (in comparison to the highest recorded daytime value) were placed in the protocol. The study was divided into three parts to assess the effects of waking the patients once during the night or every 2 hours, and the effect of total sleep deprivation. Seven of the 21 patients slept a control night and were then awakened at 2:00 AM for 15 minutes of activity before resuming sleep on the study night. In six of the seven patients, there was essentially no change in the 6:00 AM decrement in peak flow measurements, although, interestingly, in one patient the peak flow decrement was blocked by this nocturnal awakening. Overall, there was no significant change in early morning peak flow resulting from the interruption of sleep at 2:00 AM. Another important point that this aspect of the study revealed was that the fall in peak expiratory flow was often well established by 2:00 AM with progressive deterioration to 6:00 AM.

Five different patients had measurement of peak flow rates every 2 hours during the night. The fall in this variable occurred between midnight and 2:00 AM in one patient, between 2:00 AM and 4:00 AM in three patients, and between 4:00 AM and 6:00 AM in another patient. Depending on the time the peak flow fell, these individuals were awakened 1 hour before this fall on subsequent nights for 15 minutes of exercise to determine if the 6:00 AM peak flow value would be improved. As before, no improvement in the peak flow decrement was seen.

Eleven patients took part in the last phase of this protocol. Initially, the peak expiratory flow was evaluated during the night to determine at what time the latest fall occurred; then on a subsequent night the patients were kept awake in the supine position until that time before being allowed to sleep. Two distinct groups emerged from this study. Six patients sustained virtually all of their nocturnal fall in peak flow rates while still awake. Five patients showed little or no fall while kept awake, but their usual fall ensued while asleep for the short interval before being awakened at 6:00 AM. This sleep interval lasted for less than 3 hours. Two patients from this group had a repeat deprivation night several days later, and at that time most of the fall in peak flow rates occurred during wakefulness.

This study reveals two important points. The first is that arousal and exercise for 15 minutes does not alter the continuous drop in peak expiratory flow rates experienced throughout the night. The second is the question of whether sleep itself is the causative factor in the decremental changes of expiratory flow. Certainly one group of patients experienced the decreased flow rates while awake (electroencephalographic tracings were not used and the possibility, although slight, exists that the patients may have been in a drowsy state or in and out of sleep during the deprivation period), but this does not necessarily exclude any causal relationship between sleep and nocturnal asthma. The relationship, at best, is complex, and it would be difficult to construct experiments that could alter the time of sleep without affecting the phase of the peak expiratory flow rhythm (asthma) as well. As discussed previously, those same investigators showed that shift workers altered their peak flow responses to sleep time and not to solar time.

One investigation studied the relationship between sleep stages and "asthmatic episodes" in 12 adult patients with nocturnal symptoms.[19] The asthmatic episodes were conveyed by subjective means without the aid of any form of pulmonary function testing. There was a total of 35 subject nights in which 93 asthmatic incidents occurred. Except for the first hour or so of sleep, the incidences of these attacks were distributed fairly evenly throughout the night. This pattern did not depend upon the time of night the patients fell asleep or if corticosteroids were administered during the day. When relating these asthmatic incidents to sleep stages, it was shown that the percentage of episodes arising from each sleep stage approximated the percentage of sleep time spent in that period. Each patient was studied for at least two nights, and on both nights considerable awake time was recorded because of the awakenings from the asthmatic incidents. When comparing the sleep patterns of six of these patients to an age-matched group, the mean sleep time was significantly less as a result of the frequent awakenings.

The results of this study do not support the notion that asthmatics' attacks are related either to sleep stages or to the time of the night. In addition, since dreaming is related to rapid eye movement (REM) sleep, the theory that disturbing dreams are primarily responsible for precipitating asthmatics' symptoms is not supported. However, what is not evident from the study is the severity and duration of each episode. Thus, these factors may be related to sleep stages to some degree. The finding of marked awake time throughout the night supports the complaints of many patients that the inability to sleep well is a major problem.

The problem with the above studies is that the investigators needed to awaken the patients or have them spontaneously awaken. As referred to above, using the measurement of breath-by-breath airway resistance (see Chapter 4), asthmatic patients were allowed uninterrupted sleep while pulmonary function was being evaluated.[14] The asthmatic subjects were studied both during "normal" sleep and during the night, but kept awake in the recumbent position. In the asthmatic group there was a progressive and significant increase in airway resistance while awake during the night. However, on the sleep night there was a more marked increase in resistance (Fig. 3). This was independent of sleep stage. Thus, in asthma, there is a circadian rhythm to airway resistance (increased at night) that sleep potentiates. This potentiation of resistance is related to the duration one sleeps and not to sleep stages.

Bronchial Reactivity

As has been discussed extensively above, lung function worsens during the sleep related hours. In the asthmatic patient, this would imply that bronchial reactivity must be changing during sleep. Martin et al.[20] demonstrated that the greater the overnight fall in peak expiratory flow rates, the larger the circadian change in bronchial reactivity. The bronchial reactivity was tested at 4:00 PM and 4:00 AM with marked increase in reactivity occurring at 4:00 AM. Those patients with the greater change in peak flow rates demonstrated such increased bronchial reactivity that their forced expiratory volume in 1 second (FEV_1) decreased by greater than 20% to inhaled normal saline alone. This effect was not seen at 4:00 PM. Since the starting FEV_1 may or may not (see below) have an influence on bronchial reactivity, separating the overall asthma group into those with and without nocturnal asthma (Fig. 4) still shows a significant increase in reactivity at night in those subjects without a change in the FEV_1.

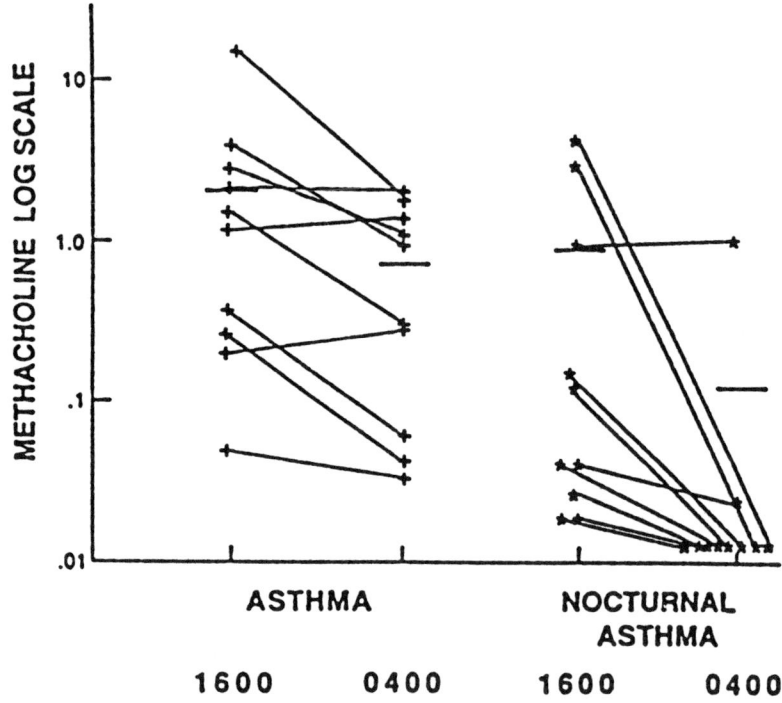

Figure 4. Circadian variation in bronchial reactivity between 4 PM and 4 AM occurs in asthmatics with minimal alterations in overnight FEV_1 (left, twofold increase) or in those with marked decrements in FEV_1 (>20%) (right, tenfold).

The above data are in agreement with those reported by Ryan and coworkers,[21] who found that increased circadian variation in peak expiratory flow rates (PEFR) was related to the bronchial reactivity to histamine (tested between 12 noon and 5:00 PM) and the response to an inhaled β_2 agonist.

It is recognized that there are some inherent problems with the interpretation of changes in bronchial reactivity to methacholine or histamine, and that differences in reactivity may reflect differences only in baseline FEV_1,[22,23], although others have shown that reactivity does not appear to be related to the FEV_1.[24,25] In Martin and colleagues' subjects there was no relationship between the bronchial reactivity and the baseline FEV_1 at 4:00 PM but there was at 4:00 AM. In four subjects whose baseline FEV_1 was similar at 4:00 PM and 4:00 AM (-7%, $+6\%$, -8%, and 0% difference, respectively) there was a greater than threefold increase in bronchial reactivity. Thus, the data suggest that there was a true increase in airway reactivity during the night.

The marked response to normal saline that Martin et al. found at 4:00 AM is particularly interesting and also suggests a marked increase in bronchial reactivity at night. This does not imply that saline and methacholine act by the same mechanism, but rather that the airways are exquisitely sensitive. The eight patients who demonstrated marked increases in airway responsiveness to inhaled saline at 4:00 AM had a greater overnight fall in PEFR (P<0.001) as well as greater nocturnal or awakening symptoms (P<0.001) than the 12 patients who did not respond adversely to saline. Although those who responded adversely had a lower percentage predicted FEV_1 at 4:00 PM (P<0.05), there was no significant difference in the methacholine PC_{20} at 4:00 PM from those without an adverse response (P>0.05).

The change in bronchial reactivity could affect or be affected by the action of other factors that are discussed below. The decrease in serum cortisol, circulating epinephrine, and cyclic AMP, as well as an increase in histamine, during the night[26] could have a direct effect on the airways, as well as an interactive effect with bronchial reactivity. Similarly, cholinergic tone, which increases at night,[27] could affect bronchial responsiveness and produce nocturnal effects. Gastroesophageal reflux[28–32] and airway cooling,[33–37] both of which can increase vagal tone, may induce nocturnal worsening of lung function in asthma, particularly when airway reactivity is increased. Finally, the demonstration that neutrophils and eosinophils in bronchoalveolar lavage fluid are increased at 4:00 AM in patients with nocturnal worsening of asthma, but not in those without nocturnal worsening,[38] suggests that airway inflammation may contribute to the increase in bronchial reactivity.

Sleep Oxygen Desaturation, Respiratory Pattern Abnormalities, Sleep Architecture (see Chapter 4)

Nocturnal awakenings in asthmatics occur more frequently during REM sleep, and oxygen desaturation with irregular nonapneic breathing is worse during this stage[39,40] (Table 2). This rapid eye movement finding is not surprising because normals, sleep apnea patients, and chronic obstructive pulmonary disease patients

Table 2

Sleep Stages and Respiratory Patterns in Nocturnal Asthma

I. Sleep Stages
  A. Increases in airway resistance during the night are related to duration of sleep and not sleep stages
II. Respiratory patterns
  A. Hypopneas and/or prolongation of expiration (reflecting increased airway resistance) are most frequent abnormalities
  B. Apneas—not frequent occurrences
    1. Central
    2. Upper airway obstruction
    3. Mixed

Adapted from: Martin RJ. Cardiorespiratory Disorders During Sleep. Futura Publishing Company, Mount Kisco, New York, 1984.

all have more cardiopulmonary abnormalities that are related to REM sleep.

Does the nonapneic respiratory pattern with oxygen desaturation cause a worsening of expiratory flow rates in asthmatics during sleep? This question has not been directly answered, but mild isocapnic hypoxia (oxygen saturation about 87%) during daytime studies significantly increases bronchial responsiveness to aerosolized methacholine in mild asthmatic patients.[41] This was assessed by the heightened FEV_1 and airway resistance responses to methacholine, as well as the steeper slope of the dose-response curve. The mechanisms involved in the enhancement of nonspecific bronchial responsiveness by mild hypoxia have not been clarified.

What would be the effect on the overnight lung function if the mild oxygen desaturation was corrected only by supplemental oxygen? This has not been fully investigated. Martin and Pak,[42] in a nasal continuous positive airways pressure (CPAP) study in nonapneic nocturnal asthma patients, did place two individuals on supplemental oxygen at 2 L/min during sleep. These two individuals were the only ones who improved their overnight lung function to any clinically significant degree with nasal CPAP. Both the nasal CPAP and supplemental oxygen nights improved the overnight change in FEV_1 to the same extent within each patient. The improvement in oxygen saturation was also similar on both intervention nights. Whereas on the nasal CPAP night the sleep architecture was markedly worsened by CPAP, supplemental oxygen alone maintained the baseline night sleep characteristics. Thus, it appears that improvement in the overnight lung function occurred in these two subjects who had the most time spent, compared to the other subjects, below an oxygen saturation of 90%, but still overall with oxygen desaturation in the mild range. Caution must be stressed here as to the interpretation of these findings in this very small population of two subjects. Further investigation is needed in regard to the potential of hypoxia as a contributor to the nocturnal worsening of lung function in asthma.

Sleep Apnea

Since sleep apnea occurs in approximately 2% to 5% of the general population, one would expect to find a subset of patients who have both obstructive apnea and asthma. Chan et al. evaluated the use of nasal CPAP in nine sleep apnea patients with concomitant asthma.[43] These patients all suffered from frequent nocturnal asthma attacks, which in three patients had resulted in respiratory arrests. Despite maximal bronchodilator therapy, including oral corticosteroids, the frequency and severity of the nocturnal asthma attacks remained unchanged. These patients additionally had symptoms of sleep apnea with heavy snoring during sleep. It was noted that this sonorous snoring started prior to the onset of the unstable asthma in most patients. After the diagnosis of obstructive sleep apnea was made (apnea/hypopnea index between 5 and 67 per hour), these patients were enrolled into a 6-week study. This included a 2-week baseline evaluation of morning and evening PEFR, a nasal CPAP period with PEFR measurements, and then another 2-week period with no nasal CPAP. Of great interest was their findings that nasal CPAP significantly improved both the morning and evening PEFR pre- and postinhaled bronchodilator (Fig. 5). Additionally, although the postnasal CPAP 2-week evaluation showed a decline in PEFRs, there continued to be a beneficial effect on the asthma compared to the initial baseline period.

This study documented that patients with obstructive sleep apnea and coexisting asthma can be safely treated with nasal CPAP therapy. Perhaps of greater interest and importance is that the relief of sleep apnea in some manner improves the asthma not only during the night, but also has a carryover daytime effect. The mechanism(s)

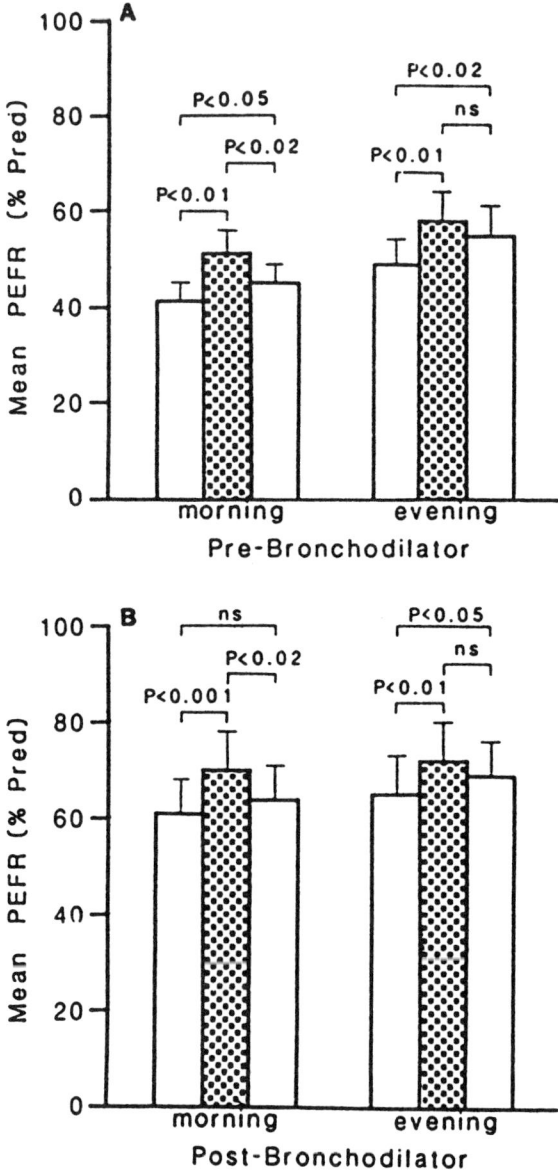

Figure 5.A-B The open bars are 2-week periods off nasal CPAP, and the checkered bars are 2-week periods on nasal CPAP. Both preinhaled bronchodilator (A) and postinhaled bronchodilator (B) PEFR values improve in the morning and evening, while these asthmatic patients with associated sleep apnea used nasal CPAP. With permission from reference 43.

by which this occurs is not known. Chan et al.[43] postulated that the recurrent episodes of upper airway obstruction and snoring act as chronic irritants which, when eliminated by nasal CPAP therapy, improved the asthma. Neural receptors at the glottic inlet and in the laryngeal region have been shown to have potent bronchoconstrictive reflex activity.[44] With the repeated stimulation from heavy snoring and apnea of the oropharynx and glottic inlet or larynx during the night, a neural reflex arc could be initiated producing bronchoconstriction. These investigators[43] suggested that the use of nasal CPAP may stabilize the upper airway and remove the chronic nightly irritation to the oropharyngeal area with subsequent elimination of the reflex bronchoconstriction.

Guilleminault et al. felt that nasal CPAP reduced vagal tone in asthmatic patients with sleep apnea or loud snoring, and this was the cause for the improvement in their nocturnal asthma.[45] These investigators make an assumption that the measured increases in inspiratory esophageal pressure during the night represented a Müeller maneuver from the narrowed or obstructed upper airway and not the result of worsening of asthma during sleep.[14] Assuming the former is correct, then a consequence of inspiratory effort against a partial or complete upper airway obstruction, i.e., a Müeller maneuver, is increased intrathoracic pressure and reflex bradycardia via vagal innervation.[46]

The Müeller maneuver with resulting vagal stimulation apparently is involved in the sleep related hemodynamic changes noted with obstructive apnea or hypopnea.[47–49] Anticholinergic medication such as atropine, or autonomic nervous system lesions eliminate the cardiovascular alterations seen with these obstructive processes during sleep.[48,49] Thus, the vagal cholinergic system is important in the sleep apnea syndrome, and there is a possibility that vagal tone may play a key role in the nocturnal worsening of asthma associated with upper airway obstruction during sleep.

If elimination of partial or complete upper airway obstruction by nasal CPAP improves associated nocturnal asthma, can nasal CPAP also improve nocturnal asthma in those individuals without sleep apnea? Martin and Pak evaluated this question in nonsnoring, nonapneic asthmatic patients with reproducible nocturnal asthma.[42] There were several interesting observations made in this study. First, each individual had markedly worse sleep when using nasal CPAP as compared to baseline. How well the group slept (sleep efficiency) was 83.1% ± 4.9% on the baseline night and only 66.4% ± 4.3% on the nasal CPAP night (P = 0.007). That is, on the baseline night there was significantly less "awake" recorded. Also shown in regard to sleep staging was greater REM sleep on the baseline night (14.1% ± 1.5%) versus the nasal CPAP night (3.4% ± 1.3%; P = 0.003). To determine if better adaptation to nasal CPAP would improve the sleep architecture, two patients used nasal CPAP for 1 week (chronometer verification) and were restudied. Again, their sleep was poor.

For the study group, the overnight change in lung function was variable among patients. However, the overnight decrement in the FEV_1 for the group was similar between the baseline and nasal CPAP nights. Using heart rate as an indicator of changes in vagal tone, it appeared in this particular patient population that nasal CPAP did not decrease vagal input, as the mean heart rates were essentially the same on both nights.

The above studies focus our attention on an important subset of asthmatics with worsening of lung function overnight, i.e., those with corresponding sleep apnea. The exact percentage of patients with the two processes occurring simultaneously is not known, but is probably 1 to 3% of the asthmatic population. Consideration of this possibility should be made with each asthmatic patient so as not to miss a relatively easy treatable cause of nocturnal asthma. Thus, if the asthmatic patient has a history of loud snoring, observation by the bed partner of pauses in respiration during sleep, daytime somnolence, restless sleep, morning headaches, or several of the other numerous signs and symptoms of sleep apnea, then a full polysomnographic evaluation should be undertaken in a sleep laboratory. Included in this study should be bedtime and any awakening spirometry so as to track lung function off and on nasal CPAP.

Allergic Factors

Exposure to bedding or room allergens, particularly house dust, is a simple and attractive explanation for sleep related asthma. This may be true in some cases,[50] but, unfortunately, it is not an appropriate explanation for the vast majority. The frequency of nocturnal asthma in the hospital environment, which is essentially free of common environmental antigens, does not make an allergic etiology plausible. Alternatively, exposure to certain antigens during the day can result in delayed bronchospasm many hours later.[51-54] This might correspond to the nocturnal hours in some patients. Another type of external exposure to consider is the cyclic type of asthma that can occur after such variables as bronchial challenge tests.[55] In these situations, asthma occurs not only during the initial night after exposure but recurs over several nights without further exposure. This pattern may be brought about because the allergen remains in the bronchi or the airways become unstable and, then, in some manner, sleep plays a role in triggering the bronchospasm. This can occur by nocturnal reflex mechanisms, or chemical or humoral substances.

Results by Mohiuddin and Martin[56] document that the time of day a patient receives an inhalation of an allergen will determine if a late (delayed) asthmatic response will occur (Fig. 6). That is, most

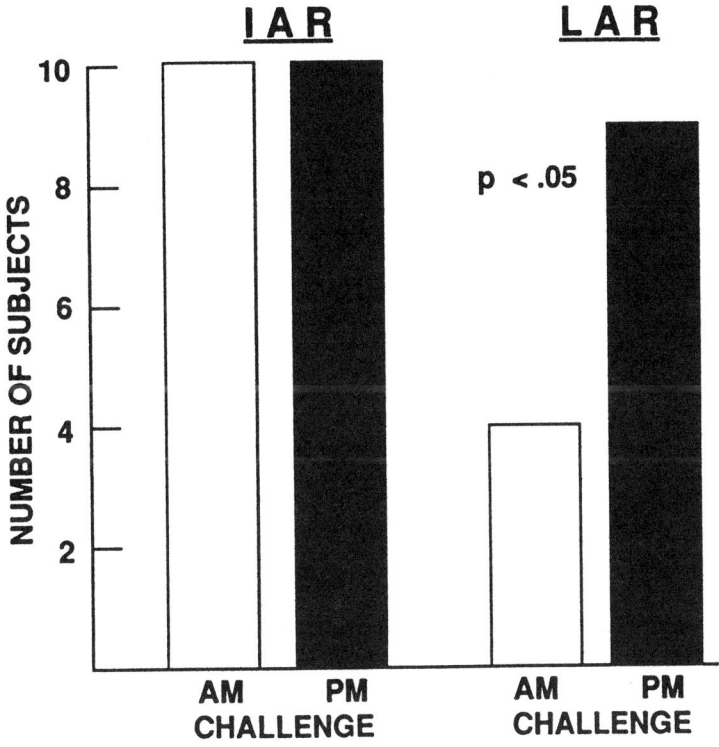

Figure 6. The late asthmatic response (LAR) to an allergen occurs almost 100% of the time if the individual is exposed in the evening compared to the morning. IAR = immediate asthmatic response. With permission from reference 56.

asthmatics develop an immediate response to an inhaled antigen to which they are sensitive. This is seen by a decrement in lung function, which is usually measured by the FEV_1. This response occurs within 15 to 20 minutes and dissipates within an hour. A second or late asthmatic response has been stated to occur in approximately 40% of these patients at 3 to 8 hours after the antigen challenge. As is customarily the case, the testing times have been only during the day. It now appears that the time of day that the patient receives the antigen is very important in determining if a late asthmatic response will occur. This then can lead to severe nocturnal bronchoconstriction.

Airway Secretions

Although asthmatics do not commonly complain of a productive cough, as do patients with chronic bronchitis, secretions can still play a role in some patients with nocturnal asthma. At one extreme, autopsies have shown extensive mucus plugging of the airways in asthmatics.[3] In addition, animal studies have shown that the cough reflex mechanisms during sleep are suppressed, particularly during REM sleep.[57] With a suppressed clearance mechanism and potentially irritable airways, secretions may in some manner trigger or be involved in nocturnal asthma.

Not uncommonly, we see patients who develop upper respiratory infections and cough who tolerate the process better during the waking hours than during sleep. Upper respiratory infections can produce reflex bronchoconstriction, but this does not explain the observed awake-to-sleep difference. The exact role of secretions in this situation is not certain but is probably significant.

Asthma is a disease of the airways and this includes all airways, both intrathoracic and extrathoracic. Chronic sinusitis and/or postnasal drip are frequent problems in asthmatics. Not only will daytime symptoms improve as the sinus inflammation is cleared, but nocturnal symptoms can also dramatically improve.

The exact reason for worsening of symptoms in asthmatics because of sinus problems or extrathoracic secretions is not known. One possibility is a nasal or laryngeal irritation reflex producing bronchoconstriction. Another possibility is the difference between nasal breathing and mouth breathing in asthmatics. Exercise-induced bronchoconstriction is much greater with mouth breathing than nasal breathing.[58] Thus, if nasal congestion is present and the patient is mouth breathing, worsening of nocturnal symptoms may develop. A third possibility is that these secretions are aspirated, which can set off direct or reflex mechanisms worsening the asthma. This last possibility has support in an animal model of Brugman et al.[59] Inducing inflammatory changes in the sinuses only produced increase in airway resistance when the lower airway was left unprotected from the potential for aspiration. Thus, in this model, sinusitis with aspiration of the inflammatory components was responsible for the increase in airway resistance.

Gastroesophageal Reflux

The relationship between gastroesophageal reflux and nocturnal asthma is an intriguing topic. Obviously, in the recumbent position and with many of the antiasthma medications producing a decrease in the lower gastroesophageal sphincter tone, reflux would be potentiated. An important question exists in regard to reflux alone producing reflex bronchoconstriction or if aspiration is needed. If reflux is an important factor in nocturnal asthma, then the worsening would actually be due to a reflex mechanism. That is, the gastric contents irritate the esophageal mucosa, and a reflex bronchoconstriction occurs via the vagal system.

Aspiration of gastric contents is another possible mechanism. Several surgical reports relate that gastroesophageal reflux with possible tracheobronchial aspiration is a trigger factor in asthma.[28–31] The predominant mechanism initiating the condition identified in these reported cases was an incompetent lower esophageal sphincter with or without an associated hiatal hernia. The medications that asthmatics are treated with, e.g., bronchodilators and steroids, tend to cause or potentiate the decreased tone in the gastroesophageal sphincter. The cause and effect of reflux asthma was stated to occur in several of these reports[28–30] when asthmatic symptoms were abolished following the surgical restoration of effective lower esophageal sphincter function.

The relationship between asthma, reflux, and nocturnal symptoms is further supported by the study of Goodall et al.[60] Eighteen of 20 patients with nocturnal asthma completed a double-blind–crossover study using cimetidine. Significant improvement was seen in reflux and nighttime symptoms with cimetidine.

During sleep, Martin et al.[61] infused saline (control) and 0.1 N hydrochloric acid into the distal third of the esophagus. The asthmatic patients were divided into two groups based on the presence or absence of esophagitis as determined by the Bernstein test. The respiratory pattern was monitored continuously using an inductance plethysmograph vest. Saline infusion had no effect on the respiratory pattern. In patients with esophagitis, the acid infusion did produce changes in the respiratory pattern. It was felt that acid in the esophagus alone triggered this altered respiratory pattern and indirectly was indicative of bronchoconstriction. This apparently occurs only in those patients who have preexisting esophagitis. However, other studies have shown only minimal increase in total respiratory resistance[62] or no association between low esophageal pH and worsening of asthma.[63,64]

Tan et al. studied nocturnal asthma patients during sleep with clinical esophagitis using simultaneous and continuous measurements of lower airway resistance and esophageal pH.[65] The goals were to establish if acid in the esophagus triggered bronchoconstriction, to see if the presence of esophagitis was necessary for such an effect, and to determine if there was a difference between the airway responses to spontaneous reflux and intraesophageal acid infusion.

This study suggests that gastroesophageal reflux does not have a significant role as a direct trigger of asthma, either in general or in the nocturnal exacerbation of the disease. This study appears to be the first in which airway patency was measured directly with simultaneous monitoring of the esophageal pH during episodes of spontaneous and simulated reflux. Prior weaknesses of previous investigations have thus been taken into account. One aspect was not studied by Tan et al. This would be aspiration of gastric contents which, although uncommon, could play a role in a small subset of asthmatics. The entire area of gastroesophageal reflux with and without aspiration needs further investigation to determine the exact relationship between reflux and asthma.

Airway Cooling

Daytime studies have shown that a fall in body temperature of 0.7°C secondary to a short duration cold exposure produces acute asthma attacks in the majority of asthmatic patients.[33] These investigations postulated that body cooling leads to vasoconstriction and cooling of the respiratory mucosa. This decrease in temperature of the mucosa consequently initiates bronchoconstriction in predisposed patients. This sequence of events was blocked when airway mucosa temperature was maintained by breathing warm, humidified air during body cooling.[35] Since body temperature can normally decrease by approximately 1°C during sleep,[36] this could be a potential mechanism of nocturnal asthma. Chen and Chai blocked the temperature drop in asthmatic subjects by having them breathe warm humidified air.[37] This improved, but did not eliminate, the overnight decrement in lung function. Although breathing warm humidified air may produce improvement in the overnight lung function of many asthmatic patients, caution is needed as some asthmatics may have exacerbation of symptoms by either heat or humidification. Thus, this modality needs to be tested initially under supervision before the patient uses it at home. Additionally, in over 30 patients tested at our center, none would use this form of therapy at home, even though some improvement was seen in lung function.

Vagal Tone

Vagal tone is increased at night as defined by the regulation of heart rate during sleep.[66] As changes in cardiac vagal tone reflect changes in bronchomotor tone,[67] this has led investigators to hypothesize that heightened vagal activity at night may be of importance in patients with nocturnal asthma. High-dose inhaled anticholinergic agents taken at night have been shown to reduce nocturnal asthma to varying degrees.[68–70] Since inhaled anticholinergics predominantly have an effect on the central airways, Morrison et al.[71] studied the effect of vagal blockade by intravenous atropine given to asthmatics at 4:00 PM and 4:00 AM. Vagal blockade caused significant bronchodilatation at both 4:00 PM (peak expiratory flow 400 to 440 L/min) and 4:00 AM (260 to 390 L/min) (Fig. 7). The marked bronchodilator improvement in peak flows at 4:00 AM was a more prominent change and brought the PEFR toward the 4:00 PM level.

The effects of the vagus nerve may be enhanced during the night in asthma patients by upregulation of the muscarinic receptors due to the inflammation and mediator release (see below) seen at night. This may also exaggerate the effects of the normal circadian variation in vagal tone.

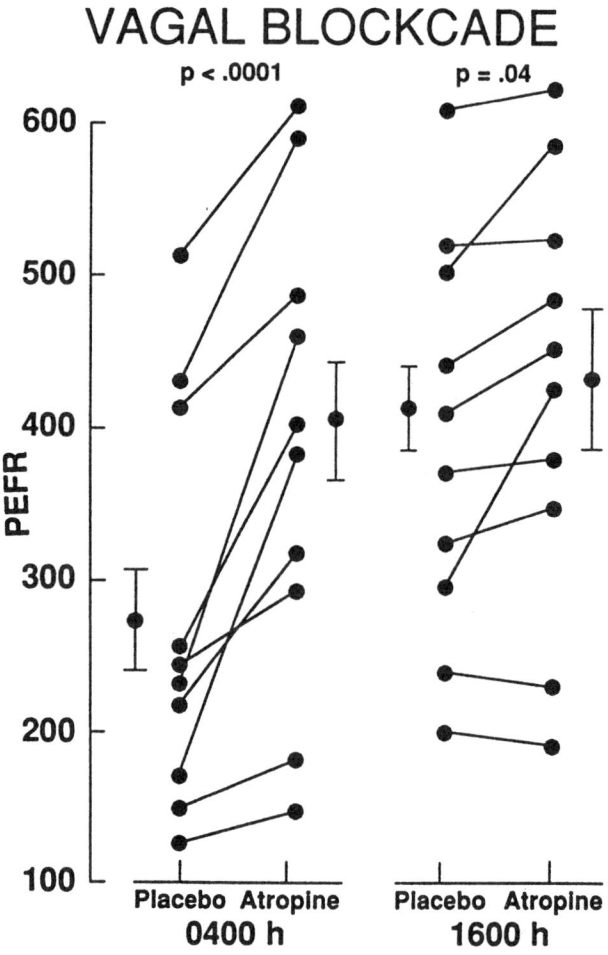

Figure 7. Vagalytic therapy has a much greater effect in improving lung function at 4 AM compared to 4 PM in asthmatic subjects. With permission from reference 71.

Circulating Mediators (*see* Chapters 5 and 6)

Much has been written on mediators and asthma for studies conducted mainly during the daytime hours. Nocturnal asthma is actually an important model of naturally occurring, nonchallenged inflammation (see below) and edema. Thus, the wide range of diurnally studied mediators will eventually be analyzed at night. Two such mediators that have been analyzed are histamine and catecholamine.

There appears to be a temporal relationship between the nadir of circadian catecholamine and pulmonary function levels.[26,72] The low point for either urinary or plasma catecholamine levels is between 3:00 AM and 4:00 AM. This may be of importance, as a fall in circulating epinephrine could directly reduce the stimulation of beta-adrenergic receptors in bronchial smooth muscle, inducing bronchospasm. In addition, an indirect effect of decreased circulating epinephrine is the release of histamine from mast cells because of diminished stimulation of beta-adrenergic receptors. This is important as histamine levels in venous plasma can correlate with the degree of bronchospasm.[73]

Barnes et al.[26] showed, in five asthmatic patients, that circadian changes in peak expiratory flow, circulating epinephrine, and cyclic AMP correlate positively with each other, and inversely with plasma histamine. The levels of these products during the night favor bronchoconstriction. In addition, these investigators infused epinephrine in low doses and demonstrated decreased circulating histamine levels and improved peak expiratory flow rates. Interestingly, in normal subjects the circadian changes in plasma epinephrine occurred as in the asthmatics, but no significant rise in plasma histamine was observed.

A study by Szefler et al.[74] showed somewhat different results. These investigators evaluated three subject groups. A normal control group, and asthmatic groups with and without the nocturnal worsening of their asthma. Between 4:00 PM and 4:00 AM there was a twofold higher plasma histamine concentration than at 4:00 AM in all three groups. There were no changes seen between these time intervals for plasma epinephrine or cyclic AMP concentrations.

Conclusions that circulating histamine is the cause of nocturnal airway obstruction cannot be drawn. Certainly inhaled histamine is a potent bronchoconstrictor, whereas intravenously infused histamine is not.[75] Thus, other mediators may increase as histamine increases and may be the active substance(s). This is an area of interest that needs further investigation.

Inflammation (see Chapters 5 and 6)

Martin et al.[38] have demonstrated that there is a significant increase in bronchoalveolar lavage fluid inflammatory cells in patients with asthma whose lung function was worse during sleep. Neutrophils and eosinophils increase (Fig. 8) while lymphocytes and epithelial sloughing appears to be more prominent in these subjects at 4:00 AM. These changes are not related to differences in sleep pattern, as sleep efficiency and sleep staging were similar to that seen in patients who did not demonstrate such severe overnight changes in lung function. These circadian alterations in bron-

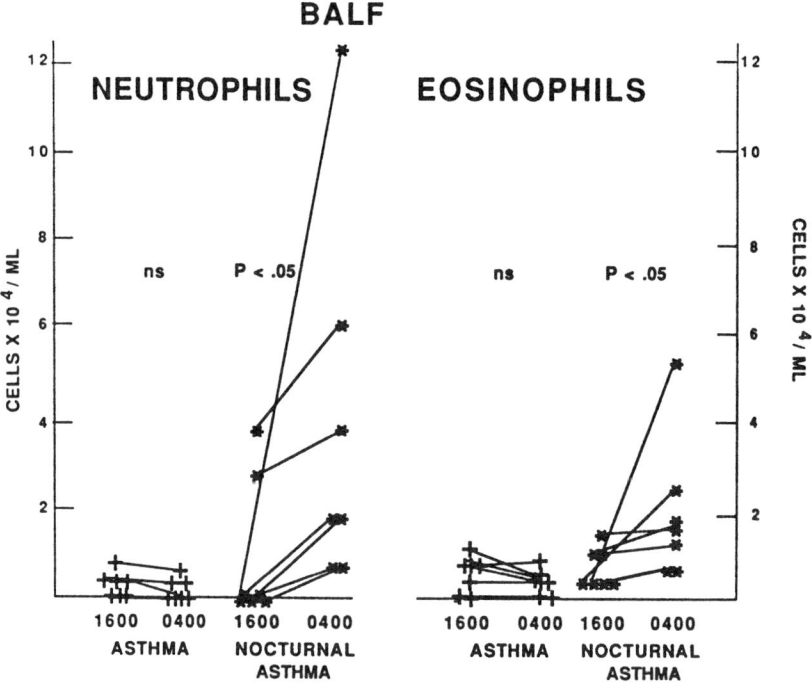

Figure 8. In subjects with nocturnal asthma compared to asthmatics without nocturnal worsening, the number of neutrophils and eosinophils increase from 4 PM to 4 AM in the bronchoalveolar lavage fluid (BALF). With permission from reference 38.

choalveolar lavage fluid total cells, neutrophils, and eosinophils are not apparent in the peripheral leukocyte counts.

Inflammatory cells in bronchoalveolar lavage fluid have been studied in asthmatics during the daytime hours with and without bronchial allergen challenge. It is important to note that techniques and cell counts vary from study to study. It appears that compared to normal subjects prior to any antigen challenge, asthmatic subjects have a higher percent of eosinophils[76-79] but not neutrophils, lymphocytes, or macrophages. With antigen challenge, the eosinophilia increases further.[77,80] One study observed that both eosinophils and neutrophils were elevated 48 hours after local antigen challenge, but at 96 hours only the eosinophils remained elevated.[81] Thus, it appears that at least the eosinophil and perhaps the neutrophil are involved in the asthmatic process following antigen challenge.

Although, in the study of Martin et al., an antigen challenge occurring in the nocturnal asthma population on the night of the lavage cannot absolutely be ruled out, daytime skin testing results were similar to an asthma control group without nocturnal asthma or increased lavage inflammatory cells,[38] thus making the possibility of an antigen challenge remote. Therefore, the issue of nocturnal worsening of asthma in the absence of an antigen challenge and the possibility of circadian variations of airway inflammatory cells in this population appears to be important and needs further investigation.

One possible explanation for the marked increase in cells at 4:00 AM in the nocturnal asthma group[38] is the degree of bronchoconstriction; that is, bronchoconstriction precipitated the influx of cells. Thus, these investigators induced bronchoconstriction in three patients with nocturnal asthma at 4:00 PM to the same degree as they were at 4:00 AM. In each individual the cellular components were similar between the 4:00 PM lavages and markedly different from their 4:00 AM lavages. This suggests that the cells are playing a role in the worsening of lung function during sleep and not coming into the airways as a result of bronchoconstriction.

Another possible explanation for the increased cells in the nocturnal asthma group at 4:00 AM is that this group is a more severe asthmatic population. The investigators specifically divided patients into two groups by a marked separation in their overnight lung function.[38] Although the 4:00 PM FEV_1 measurements were not different between the groups, the nocturnal asthma group tended to have a lower percent predicted FEV_1 and higher 4:00 PM cellular lavage components. It is unlikely, however, that the cellular changes at 4:00 AM in the nocturnal asthma group are due only to more severe disease because the nonnocturnal asthmatics with the lowest FEV_1 (≤71% predicted) did not show increases in cellularity and because inducing bronchoconstriction for as long as 3 hours (simulating severe bronchospasm) does not alter the lavage cell counts. Thus, although the severity of the disease process may still play a role, the increased number of inflammatory cells seen at 4:00 AM in the nocturnal asthma group appears to be of importance in understanding the pathogenesis of nocturnal asthma.

The circadian variation seen in the bronchoalveolar lavage fluid cells of the nocturnal asthma subjects is not observed in the peripheral blood if measured only at 4:00 PM and 4:00 AM.[38] In one study of an asthmatic population, the nadir of the circulating eosinophil count was seen at 10:00 AM and the maximum concentration occurred at midnight.[82] In that study, the 4:00 PM and 4:00 AM counts were similar. This raises the possibility that the peak blood eosinophil and perhaps neutrophil levels precede the increase in airway cells. Alternatively, the maximal influx of these cells into the lung may occur at an earlier time point than when lavage was performed by Martin et al.

In summary, patients with nocturnal asthma have an associated cellular inflammatory response seen in the bronchoalveolar lavage fluid at 4:00 PM and to a greater degree at 4:00 AM. This circadian change is not observed in patients who do not have as marked nocturnal worsening of lung function. In addition to increases in neutrophils and eosinophils, both of which play prominent roles in asthma, there is an increased number of epithelial cells in the bronchoalveolar lavage fluid. It is possible that the inflammatory changes and epithelial damage result in increased bronchial reactivity and play an important role in the production of nocturnal asthma.

Corticosteroid Levels

Multiple investigators have attempted to establish a relationship between circadian variations in airway caliber and plasma cortisol. Reinberg et al. demonstrated synchrony in the timing of nocturnal bronchoconstriction and the lowest urinary excretion of 17-hydroxycorticosteroid over a 24-hour period.[83] Additionally, Barnes et al. found that the nadir of plasma cortisol occurred at midnight, while the PEFR trough was at 4:00 AM in a group of five asthmatics.[26] In an attempt to eliminate the circadian variation in plasma cortisol, Souter et al. infused variable physiological doses of hydrocortisone in a group of six nocturnal asthmatics.[84] In five of the six subjects, these physiological doses of corticosteroid did not totally block the nocturnal decrements in PEFR. Thus, at a physiological dose range, airways inflammation and possible resultant mucosal edema may still play an important role in the development of nocturnal asthma. Nocturnal asthmatics have increased inflammatory cells in bronchoalveolar lavage fluid at 4:00 AM compared to 4:00 PM reflecting elevations in both neutrophil and eosinophil numbers (see above). In an attempt to override the nocturnal inflammatory effect, Beam and colleagues[85] used large intravenous doses (100 mg per unit). In nine of the 11 nocturnal asthmatics studied, this supraphysiological cortico-steroid infusion during sleep resulted in greater than 40% improve-ment in the overnight decrement in FEV_1. As a group, the mean overnight decrement in FEV_1 significantly improved from a baseline value of 46% to 12% concurrent with hydrocortisone infusion. Addi-tionally, fewer subjects awoke from sleep requiring beta$_2$-agonist therapy while receiving the steroid infusion. These results emphasize the contribution of corticosteroid sensitive factors to nocturnal wors-ening of asthma in a setting of clinically stable daytime symptoms, and in spite of daytime therapy with corticosteroids. Nine of 11 subjects were maintained on daytime prednisone therapy with a mean daily dose of 27.2 ± 5.3 mg, yet still demonstrated overnight decrements in FEV_1 between 23% and 66%. Daytime prednisone dose did not correlate with the overnight decrement in FEV_1 on baseline nights or the response to overnight hydrocortisone infusion. Seven of the nine subjects receiving daytime prednisone therapy improved with hydrocortisone infusion. This observation suggests that in addition to dose, timing of corticosteroid therapy may be important in achieving optimal effectiveness in managing nocturnal asthma. Reinberg and colleagues have suggested that the time of corticosteroid administra-tion during the day may be relevant in attenuating nocturnal worsening of asthma.[86,87] Additionally, further data (see below under Therapy) suggest that timing of prednisone dose may be relevant in altering the airways inflammation of asthmatics prone to nocturnal worsening.[88]

A spectrum of response to corticosteroids at pharmacological dose was evident with two of 11 subjects demonstrating less than 10% change in their nocturnal drop in FEV_1, two of 11 showing complete abolishment of nocturnal bronchoconstriction, and the remaining seven subjects demonstrating between 45% and 87%

improvement compared to baseline.[85] This heterogeneity among asthmatics prone to nocturnal bronchoconstriction supports the concept of asthma as a syndrome rather than a distinct disease.[89] Souter el al. demonstrated similar heterogeneity in their population of nocturnal asthmatics treated with physiological doses of corticosteroid.[84] Although Beam et al. demonstrated a greater ability to attenuate the nocturnal fall in FEV_1 with a mean improvement of 67% compared to 20% by Souter et al., the dose of hydrocortisone in the former study exceeded the latter by a factor between 10 and 100. Differences in ability to attenuate nocturnal asthma in regard to steroids may represent greater bronchoalveolar concentrations of corticosteroid, directly related to recirculating blood levels.[90]

The ability of high dose corticosteroids to significantly attenuate nocturnal asthma in a majority of subjects directs attention again to the role of inflammation in the pathogenesis of this phenomenon. Corticosteroids moderate inflammation by directly inhibiting cellular activation and via inhibition of mediator formation.[91-95] Consequently, local recruitment, proliferation, and activation of inflammatory cells may be influenced by high-dose corticosteroids. Bronchoalveolar lavage analysis in asthmatics following daytime antigen challenge reveals increased eosinophilia[77] and, more recently, bronchoalveolar lavage fluid from unchallenged nocturnal asthmatics demonstrated increased numbers of total leukocytes, eosinophils, and neutrophils.[38] It is likely that this inflammatory milieu is the backdrop for the bronchial epithelial damage and resultant airways hyperreactivity that has been demonstrated in asthmatics.[76,96]

Beta$_2$-adrenergic Receptors

If beta$_2$ receptors are adversely altered on a circadian basis, then the asthmatic patient would be at further risk for bronchoconstriction at that particular time. In general, asthmatic patients have been shown to have lower leukocyte β-adrenergic receptor density in the morning compared with daytime measurements.[97] Szefler et al. studied normals, asthmatic patients without nocturnal worsening of lung function, and a third population with nocturnal asthma.[74] These investigators showed that only the nocturnal asthma group demonstrated a difference in the peripheral leukocyte beta$_2$ receptor density and response to isoproterenol at 4:00 AM. That is, at 4:00 AM there was a significantly lower receptor density by 33% and impaired cyclic AMP response to isoproterenol (17% ± 7.3% versus 80.2% ± 21.3% for normals and 69.4% ± 13.7% for the asthma control group).

The reason for the altered $beta_2$ receptor data has not been elucidated. Various physiological conditions can regulate transmembrane signaling elements. The interaction between beta-adrenergic receptor and adenylate cyclase is an example. The beta-adrenergic receptor can be upregulated with glucocorticoids (or conversely downregulated by decreased levels) and downregulated by beta agonists through regulation of receptor mRNA.

Summary

The exact cause of nocturnal exacerbation of asthma is far from understood. More than likely, a combination of multiple effects controls the airways of these susceptible patients (Table 3). These

Table 3

Nocturnal Asthma: Potential Mechanisms

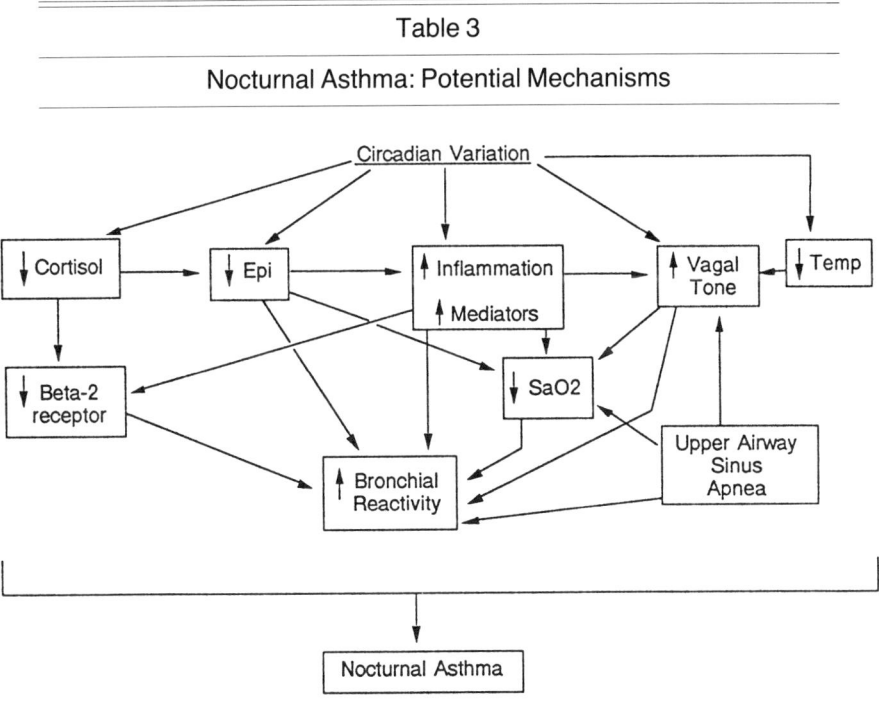

processes would include the net effect of catecholamines, beta-adrenergic receptor responsiveness, mucociliary clearance, reflux mechanisms, vagal tone, inflammation/mediators, corticosteroids, immunologic integrity, and arousal patterns. It appears that the circadian rhythms of all these factors are arranged to strengthen the potential for nocturnal bronchoconstriction, respiratory failure, and death.

Evaluation

To demonstrate the circadian variation in lung function in asthma, a laboratory using complex measurements of lung function during sleep is not needed. A simple and inexpensive peak flow meter is all that is necessary. The patient can record PEFR at bedtime, with any awakening during the night, and in the morning. Additionally, a PEFR measurement can be made late in the afternoon to determine what is usually the best lung function of the day. By this method the physician truly understands the patient's asthma and does not erroneously make assumptions of the clinical status on one measurement during the daytime in the doctor's office. The effect of any therapeutic intervention (see below) can also be determined by the recordings the patient makes at home. Finally, once the patient is stable, he/she has an objective parameter to follow as to the status of the disease. That is, the vast majority of asthmatic exacerbations occur over days. The gradual decrease in PEFRs or increase in circadian swings can alert the patient to impending problems that are much easier to deal with prior to the usual time when the emergency room is finally visited. If an additional diagnosis of sleep apnea is suspected in this population, then a polysomnographic evaluation is needed.

Therapy

The treatment of nocturnal asthma and asthma in general is based on the understanding of the circadian rhythms and the appropriate knowledge of how each medication truly works. Thus, chronopharmacology is the appropriate approach. That is, one must direct more intense therapy when the disease is worse. The "standard" asthma therapeutic interventions assume that the patient is actually a homeostatic system. This is incorrect; as discussed above, many dynamic alterations take place over a 24-hour period.

Potentially Reversible Factors

One should always approach the problem of treating nocturnal asthma by attempting to rule out any reversible causative factors. Investigation into the possibility of an extrinsic causative agent, particularly in the work environment, should be considered. Even if the patient is asymptomatic during the day, nocturnal exacerbation of asthma can be seen as a delayed effect. Be certain that the patient is not sleeping with the family dog or cat, an unfortunate but common environmental "agent." Another factor is the origin of secretions whether intra- or extrathoracic. If the asthmatic patient also has a degree of bronchitis or bronchiectasis, decreased clearance that is associated with sleep can worsen the problem. Bronchial hygiene, including postural drainage and percussion during the day and before sleep, has been beneficial at our institution in those occasional patients fitting this picture. As described above (see Airway Secretions), asthmatics commonly have sinusitis, nasal congestion, and accumulation of secretions in the posterior pharynx. This causes poor sleep and can contribute to nocturnal asthma. Correcting this with oral decongestants, nasal steroid preparations, and saline nasal washes can greatly improve the nocturnal asthma in this subset of patients. If antibiotics are needed, then 3 weeks of therapy is indicated due to the poor circulation in the sinus area. Rarely does a surgical procedure need to be done to correct a particularly difficult-to-control sinus condition.

Reflux

Gastroesophageal reflux may or may not play an important role in nocturnal exacerbations. Certainly if the patient complains of "heartburn," retrosternal spasms, or a sour taste in the mouth upon awakening in the morning, the clinical symptom complex is established. Occasionally, a patient is not symptomatic but still has the problem. A trial of 4- to 6-inch wood or brick blocks under the head of the bed (raising the head with pillows, etc. will not work) and bedtime antacids may give rapid relief. Another beneficial therapeutic intervention is the use of H-2 blockers. In our experience, as well as that of others,[31] these acid secretion blockers improve the symptoms of reflux and perhaps the related asthmatic problems. The dosage is variable, but the bedtime dosing should be the highest and, at times, it is all that is necessary. Surgery to solidify the gastroesophageal junction is rarely indicated.

If recurrent aspiration is documented, a more involved program needs to be undertaken. In addition to blocks under the head of the bed, no food or drink (except for medications) should be taken for several hours before bedtime. Motility medications that increase the rate of gastric emptying, e.g., metoclopramide, appear to be beneficial in this situation on a short-term basis. If further treatment is needed, a surgical fundoplication should prevent the reflux from occurring.

Pharmacological Interventions

The majority of patients will not benefit from the above therapeutic interventions and will need some form of bronchodilator therapy. This may include simply maximizing a subtherapeutic program to ensure optimal bronchodilatory effects throughout the entire 24-hour day, or additional maneuvers. These additional processes would be changes in the program of oral or inhaled agents. Inhaled agents are usually short-lived and do not give therapeutic effects throughout the entire night. However, an occasional patient may have difficulty only at the 2- to 3-hour mark after sleep onset. In this situation, optimal inhaled treatments at bedtime can improve the sleep period.

Beta$_2$-adrenergic Agonists

As longer acting oral and inhaled beta agonists are becoming available, the use of these agents will be of benefit in treating nocturnal asthma. Presently, only studies in patients with relatively mild decrements in lung function have been carried out using these agents. These studies have shown an improvement in the mild overnight decrements in lung function. Szefler et al. have demonstrated that beta-adrenergic receptor function decreases at night in patients with the nocturnal worsening of their asthma,[74] and this may indicate that higher doses of beta$_2$ agents may be needed during the night.

Fitzpatrick et al. used a newer long-acting inhaled beta$_2$ agonist (10 to 12 hour duration), solumeterol, in a group of asthmatic subjects with and without nocturnal asthma.[98] In doses of 50 mcg and 100 mcg inhaled twice daily, there was an overall improvement in morning PEFR compared to a placebo. There was no difference between the 50 or 100 mcg of solumeterol. Only the 50 mcg dose objectively improved sleep quality, spending less time awake or in light sleep, and more time in stage 4 slow-wave sleep compared to placebo. It appears that the higher dose of solumeterol may have had a stimulatory effect on the central nervous system. However, on the 50 mcg dose there was no significant change for the nocturnal requirement of rescue use of salbutamol.

Theophyllines

Zwillich et al.[99] showed the superiority of a sustained-release theophylline preparation (Theo-Dur) in treating asthmatics with mild to moderate nocturnal asthma compared to a long-acting (8-hour duration) inhaled beta$_2$ agonist (bitolterol, Tornalate). In this study it is important to note that sleep quality and architecture were unchanged between the different preparations, demonstrating that the presence of theophylline did not alter sleep compared to the beta$_2$ agonist. Additionally, not only was the FEV$_1$ improved in the morning, but there was less nocturnal oxygen desaturation while on theophylline.

In subjects with more significant nocturnal asthma, higher serum theophylline concentrations (STC) during the night and lower STC during the daytime, where it is easier to control the bronchoconstriction, i.e., chronopharmacology, has been demonstrated to be successful. Comparing two theophylline preparations with different pharmacokinetics, Martin et al. showed that higher STC (about 16 μg/mL) from 3:00 to 5:00 AM produced a marked improvement in the overnight worsening of lung function versus lower therapeutic levels (about 11.5 μg/mL).[100] During the daytime, the agent that gave a higher nocturnal STC (Uniphyl, given once daily at 7:00 PM) had a progressive fall, reaching a nadir prior to the next dose. The other agent (Theo-Dur, given b.i.d.) had higher daytime STC compared to nighttime. The FEV$_1$ measured every 2 hours during the daytime was not significantly different between the agents even though the STC was different. This reinforces the importance of delivering higher concentrations of medication when most needed, i.e., at night. D'Alonzo et al. also showed a significant relationship between increasing STC and improvement in FEV_1 from 2:00 to 6:00 AM, but not from 2:00 to 6:00 PM.[101] This study documents that chronotherapeutic principles of higher therapeutic levels are needed when the disease is worse, i.e., during the night. Also, it helps to explain why previous studies on theophylline carried out during the daytime have not shown that increasing the serum theophylline concentration above approximately 10 μg/mL improves lung function any further. Furthermore, sleep quality and architecture are not altered between higher and lower therapeutic STC, but there is less oxygen desaturation with higher STC.[100]

Corticosteroids

Another example of chronopharmacology is the use of corticosteroids. If an asthma patient needs steroids, then every effort should be made to use inhaled steroids. If oral steroids are needed for long periods of time, every other day would be preferable to a daily morning dosing. However, there are many steroid-dependent asthmatics with nocturnal asthma. An important fact to consider is that increasing a steroid dose tenfold will only increase the length of action twofold. As a result, increasing the morning dosing in these individuals usually leads to more steroid complications without improvement in nighttime asthma control. Consideration of the use of an afternoon dosing should be undertaken. Since the nocturnal worsening of asthma is associated with an increase in inflammation and mediator release (see above), it would be reasonable to assume that therapy directed at this aspect would be of benefit. Reinberg and colleagues have suggested that the time of corticosteroid administration during the day may be relevant in attenuating the nocturnal worsening of asthma.[86,87] Beam et al. have begun to clarify the contribution of timing of corticosteroids to their ability to block the circadian recruitment of inflammatory cells into the lung.[88] The results of their data highlight the relevance of prednisone dose timing in attenuating the nocturnal worsening of asthmatic lung function and decrement in airways inflammation. A 3:00 PM dose produced significant improvement in overnight spirometry (placebo control $-28\% \pm 7\%$, steroid $-10\% \pm 4\%$) as well as a sustained reduction of blood eosinophils during both the early evening and the sleep related hours. Additionally, the 3:00 PM dose of prednisone produced a pancellular reduction in the 4:00 AM bronchoalveolar lavage cytology. These effects were not demonstrated with either an 8:00 AM or 8:00 PM dose phase. Neither alternative time produced an improvement in overnight spirometry or reduction in any bronchoalveolar lavage cellular profile. Alterations in blood eosinophil number by the 8:00 AM and 8:00 PM phases suggest that eosinophil recruitment or activation occurs prior to spirometric decline.

Overall, the results suggest that the 3:00 PM dose of prednisone interrupts the inflammatory cascade at one or more critical steps in its genesis. Corticosteroids are known to influence both the function and kinetics of all the inflammatory cells represented in the bronchoalveolar lavage.[102] It is noteworthy that Martin and colleagues demonstrated elevations in total white cell number, neutrophil, eosinophil, and lymphocyte counts in the bronchoalveolar lavage fluid of the nocturnal asthma cohort when compared to asthmatics without nocturnal worsening.[38] These observations support a collaborative cellular mechanism of inflammation that is corticosteroid sensitive, yet dependent on timing in addition to dosage.

The use of inhaled steroids is of interest in patients with nocturnal asthma. One would think that these agents would be ideal for this problem. However, studies have given mixed results. Horn et al. showed that in 14 asthmatic patients with nocturnal symptoms and morning decreases in PEFR, only eight patients resolved the nocturnal component using inhaled beclomethasone (also on inhaled salbutamol).[103] The dose of beclomethasone was higher than standard, being 400 μg/mL four times per day. Although the other six patients improved daytime lung function, the overnight decrements in function did not improve. The reason for this interesting finding is not immediately apparent, but certainly needs further investigation.

Anticholinergics

As discussed above, vagal tone is increased at night. Unfortunately, there is not a long enough duration anticholinergic presently available if this form of therapy is selected. Higher bedtime dosing compared to the usual daytime dosing is needed to lengthen the duration of effective action of the drug. If the patient wakes during the night, then inhalation of atropine or ipratropium bromide can be of great benefit.

Nonpharmacological Interventions (see Chapter 10)

1. Nasal CPAP. This is a very beneficial form of therapy (see above) for patients who have concomitant sleep apnea and nocturnal asthma. It appears that not only does the nocturnal asthma component improve as the apneas are alleviated, but also the diurnal component improves.

2. Deliberate Nocturnal Awakenings and Treatment. Speelberg et al. studied 10 nocturnal asthmatic patients in regard to deliberate waking followed by use of both an inhaled beta$_2$-adrenergic agonist and inhaled ipratropium bromide.[104] Two actuations of each medication were used. These patients awoke spontaneously at night at least three times per week with symptoms of wheezing, cough, or dyspnea. For 1 full week the number and clock time of spontaneous awakenings, peak expiratory flow rates, and symptom scores were recorded. The deliberate waking during the second week was accomplished once a night with the use of an alarm clock set for 1 hour before the prior week's first expected spontaneous awakening.

   By the end of the second treatment week the patients reported improvement in sleep quality and less disturbance by nocturnal asthma (P = 0.03). Also, morning PEFR improved. However, this was not a double-blind placebo controlled study and the results seen may have been a "placebo" effect.

3. Inspiratory Muscle Training. In an interesting study on specific inspiratory muscle training in asthmatic patients. Weiner et al. hypothesized inspiratory muscle training would result in an increase in strength and endurance, and this would be associated with improvement in symptoms.[105] Thirty patients with moderate to severe asthma were studied. Fifteen received respiratory muscle training and 15 served as a control population with sham training in this double-blind study. The training was performed with a threshold inspiratory muscle device for 30 minutes per day, 5 days per week for 6 months. Both inspiratory muscle strength (expressed by the maximal inspiratory pressure generated at residual volume) and respiratory muscle endurance (the relationship between peak pressure and

maximal pressure from residual volume) significantly improved. Of interest, these objective findings of increased strength and endurance were translated into improvement in nocturnal symptoms, morning tightness, daytime asthma, and cough. Inhaled beta$_2$-adrenergic agonist use, emergency room visits, and sick leave days were all significantly reduced in the treated group versus the sham control group. The mechanism of effect is unknown, but this interesting finding needs further study.

General

With any type of medication or other intervention the clinician must know the individual characteristics of the agent being used. An example is that no two theophylline preparations are the same in regard to makeup, absorption, or when peak effect occurs. An example of this is the theophylline preparation Uniphyl. If given at bedtime, the peak effect will not occur for 10 to 12 hours or when the patient is awake the next day. This agent should be given at 6:00 to 7:00 PM to achieve maximal benefit at 3:00 to 6:00 AM. Similarly, knowledge of steroid kinetics and timing needs to be understood for different agents. Basically, the body is not a homeostatic organism, but has dynamic changes that need to be understood in using medications.

As newer antiinflammatory/mediator inhibitors are developed, this could hold certain keys for treatment. The development of longer-acting anticholinergic agents may also be of help. Studies that continue to unravel the circadian rhythms that favor bronchoconstriction will guide future treatment.

Conclusion

Indeed, the sleeping patient is still a patient, and it is of the utmost importance that clinicians understand the circadian and sleep related events that occur in the patient. To neglect this area of medicine hinders the care of the patient and accelerates the disease process. The future will bring tremendous advances to our understanding and ability to treat the nocturnal aspect of asthma and, thus, the disease itself.

References

1. Robin ED. Some interrelations between sleep and disease. Arch Int Med 1958;102:669–675.
2. Cochrane GM, Clark TJH. A survey of asthma mortality in patients between ages 35 and 65 in the greater London hospitals in 1971. Thorax 1975;30:300–315.
3. Hetzel MR, Clark TJH, Branthwaite MA. Asthma: Analysis of sudden deaths and ventilatory arrests in hospital. Br Med J 1977;1:808–811.
4. Jackson RT, Sears MR, Beaglehole R, Reatt H. International trends in asthma mortality; 1970–1985. Chest 1988;94:914–918.
5. Robertson CF, Rubinfeld AR, Bowes G. Deaths from asthma in Victoria; a 12-month survey. Med J Aust 1990;152:511–517.
6. Floyer J. A Treatise of the Asthma, R. Witkin and W. Inngs, London, 1698, pp 7–8.
7. Salter HH. Asthma: Its Pathology and Treatment, 1st American edition, 1882, p 33.
8. Turner-Warwick M. Epidemiology of nocturnal asthma. Am J Med 1988;85(1B):6–8.
9. Bagg LR, Hughes DTD. Diurnal variation in peak expiratory flow in asthmatics. Eur J Respir Dis 1980;61:298–302.
10. Dethlefsen U, Repgas R. Ein neues therapieprinzip bei nachtilchen asthma. Klin Med 1985;80:44–47.
11. Hudgel DW, Kellum R, Martin RJ, et al. Depressed arousal response to airflow obstruction: A possible factor in near-fatal nocturnal asthma. Am Rev Respir Dis 1982;125(S):202.
12. Ballard RD, Tan WC, Kelly PL, et al. Effects of sleep, before and after sleep deprivation, on ventilatory and arousal responses to induced bronchoconstriction. J Appl Physiol 1990;69:490–497.
13. Hetzel MR, Clark TJH. Comparison of normal and asthmatic circadian rhythms in peak expiratory flow rate. Thorax 1980;35:732–738.
14. Ballard RD, Saathoff MC, Patel DK, Kelly PL, Martin RJ. Effect of sleep on nocturnal bronchoconstriction and ventilatory patterns in asthmatics. J Appl Physiol 1989;67(1):243–249.
15. Clark TJH, Hetzel MR. Diurnal variation of asthma. Br J Dis Chest 1977;71:87–92.
16. Bonjer FH. Physiological aspects of shift work. Proc Int Congr Occup Health 1960;13:848–851.
17. Guberan E, Williams MK, Walford J, et al. Circadian variation of FEV_1 in shift workers. Br J Ind Med 1969;26:121–125.
18. Hetzel MR, Clark TJH. Does sleep cause nocturnal asthma? Thorax 1979;34:749–754.
19. Kales A, Beall GN, Bajor GR, et al. Sleep studies in asthmatic adults: Relationship of attacks to sleep stage and time of night. J Allergy 1968;41:164–173.
20. Martin RJ, Cicutto LC, Ballard RD. Factors related to the nocturnal worsening of asthma. Am Rev Respir Dis 1990;141:33–38.
21. Ryan G, Latimer KM, Dolovich J, Hargreave FE. Bronchial responsive-

ness to histamine: relationship to diurnal variation of peak flow rate, improvement after bronchodilator, and airway calibre. Thorax 1982;37:423–429.

22. Benson MK, Bronchial hyperreactivity. Br J Dis Chest 1975;69:227–239.
23. Greenspon LW, Morrissey WL. Factors that contribute to inhibition of methacholine-induced bronchoconstriction. Am Rev Respir Dis 1986;133:735–739.
24. Yan K, Salome CM, Woolcock AJ. Prevalence and nature of bronchial hyperresponsiveness in subjects with chronic obstructive pulmonary disease. Am Rev Respir Dis 1985;132:25–29.
25. Bhagut RG, Grunstein MM. Comparison of responsiveness to methacholine, histamine, and exercise in subgroups of asthmatic children. Am Rev Respir Dis 1984;129:221–224.
26. Barnes P, Fitzgerald G, Brown M, et al. Nocturnal asthma and changes in circulatory epinephrine, histamine and cortisol. N Engl J Med 1980;303:263–267.
27. Gaultier C, Reinberg J, Gerbeaux J, et al. Circadian changes in lung resistance and dynamic compliance in healthy and asthmatic children: Effects of two bronchodilators. Respir Physiol 1977;31:169–182.
28. Overholt RH, Voorhees RJ. Esophageal reflux as a trigger in asthma. Dis Chest 1966;49:464–466.
29. Urschel HC Jr, Paulson DL. Gastroesophageal reflux and hiatal hernia. Complications and therapy. J Thorac Cardiovasc Surg 1967;53:32–37.
30. Davis MV. Evolving concepts regarding hiatus hernia and gastroesophageal reflux. Ann Thorac Surg 1969;7:120–133.
31. Babb RR, Notarangelo J, Smith VM. Wheezing: A clue to gastroesophageal reflux. Am J Gastroenterol 1970;53:230–233.
32. Mays EE. Intrinsic asthma in adults: Association with gastroesophageal reflux. J Am Med Assoc 1976;236:2626–2628.
33. Chen WY, Horton DJ, Weiser PC. Airway obstruction induced by body cooling in asthmatics. Physiologist 1977;20:16.
34. Chen WY, Horton DJ. Airways obstruction in asthmatics induced by body cooling. Scand J Respir Dis 1978;59:13–20.
35. Horton DJ, Chen WY. Effects of breathing warm humidified air on bronchoconstriction induced by body cooling and by inhalation of methacholine. Chest 1979;75:24–28.
36. Petersdorf RA. Disturbance of heat regulation. In: Wintrobe MM, Thorn GW, Adams RD, et al. (eds). Harrison's Principles on Internal Medicine, 7th ed. McGraw-Hill, New York, 1974.
37. Chen WY, Chai H. Airway cooling and nocturnal asthma. Chest 1982;81:675–680.
38. Martin RJ, Cicutto LC, Smith HR, et al. Airway inflammation in nocturnal asthma. Am Rev Respir Dis 1991;143:351–357.
39. Catterall JR, Douglas NJ, Calverley PMA. Irregular breathing and hypoxaemia during sleep in chronic stable asthma. Lancet 1982;1:301–314.
40. Montplaisir J, Walsh J, Malo JL. Nocturnal asthma: Features of attacks, sleep and breathing patterns. Am Rev Respir Dis 1982;125:18–22.

41. Denjean A, Roux C, Herve P, Bunniot JP, Comoy E, Duroux P, Gaultier C. Mild isocapnic hypoxia enhances the bronchial response to methacholine in asthmatic subjects. Am Rev Respir Dis 1988;138:789–793.
42. Martin RJ, Pak J. Nasal CPAP in non-apneic nocturnal asthma. Chest 1991;100:1024–1027.
43. Chan CS, Woolcock AJ, Sullivan CE. Nocturnal asthma: Role of snoring and obstructive sleep apnea. Am Rev Respir Dis 1988;137:1502–1504.
44. Nadel JA, Widdicombe JG. Reflex effects of upper airway irritation on total lung resistance and blood pressure. J Appl Physiol 1962;17:861–865.
45. Guilleminault C, Quera-Salva MA, Powell N, et al. Nocturnal asthma: Snoring, small pharynx and nasal CPAP. Eur Respir J 1988;1:902–907.
46. Hanly PJ, George CF, Millar TW, Kryger MH. Heart rate response to breath-hold, Valsalva and Müeller maneuvers in obstructive sleep apnea. Chest 1989;95:753–739.
47. Cuccagna G, Mantovani M, Brignani F, Purchi C, Lugaresi E. Continuous recording of the pulmonary and systemic arterial pressure during sleep in syndrome of hypersomnia with periodic breathing. Bull Eur Physiop Thol Respir 1972;8:1159–1172.
48. Guilleminault C, Tiklian A, Lehrman K, Forno L, Dement WC. Sleep apnea syndrome: States of sleep and autonomic dysfunction. J Neurol Neurosurg Psych 1977;40:718–725.
49. Guilleminault C, Winkle R, Melvin K, Tilkian A. Cyclical variation of the heart rate in sleep apnea syndrome, mechanisms and usefulness of 24-hour electrocardiography as a screening technique. Lancet 1984;1:126–136.
50. Gervais P, Reinberg A, Gervais C, et al. Twenty-four-hour rhythm in the bronchial hyperreactivity to house dust in asthmatics. J Allergy Clin Immunol 1977;59:207–213.
51. Davies RJ, Hendrick DJ, Pepys J. Asthma due to inhaled chemical agents: Ampicillin, benzyl penicillin, 6-amino penicillanic acid and related substances. Clin Allergy 1974;4:227–247.
52. Siracusa A, Curradi F, Abbritti G. Recurrent nocturnal asthma due to tolylene diisocyanate: A case report. Clin Allergy 1978;8:195–201.
53. Davies RJ, Green M, Schofield N MC. Recurrent nocturnal asthma after exposure to grain dust. Am Rev Respir Dis 1976;114:1011–1019.
54. Gandevia B, Milne J. Occupational asthma and rhinitis due to western red cedar (Thuja plicata), with special reference to bronchial reactivity. Br J Ind Med 1970;27:235–244.
55. Newman Taylor AJ, Davies RJ, Hendrick DJ, et al. Recurrent nocturnal asthma reactions to bronchial provocation tests. Clin Allergy 1979;9:213–219.
56. Mohiuddin AA, Martin RJ. Circadian basis of the late asthmatic response. Am Rev Respir Dis 1990;142:1153–1157.
57. Sullivan CE, Murphy E, Kazan LF, et al. Waking and ventilatory responses to laryngeal stimulation in sleeping dogs. J Appl Physiol: Respir Environ Exercise Physiol 1978;45:681–689.
58. Shturman-Ellstein R, Zeballos RJ, Buckley JM, et al. The beneficial

effect of nasal breathing on exercise-induced bronchoconstriction. Am Rev Respir Dis 1978;118:65–73.

59. Brugman SM, Larsen GL, Henson PM, Irvin CG. Mechanism of the increase in lower airways responsiveness associated with sinusitus in a rabbit model. Am Rev Respir Dis 1991, In press.
60. Goodall RJR, Earis JE, Cooper DW, et al. Relationship between asthma and gastro-oesophageal reflux. Thorax 1981;36:116–121.
61. Martin ME, Grunstein MM, Larsen GL. The relationship of gastroesophageal reflux to nocturnal wheezing in children with asthma. Ann Allergy 1982;49:318–322.
62. Spaulding HS Jr, Mansfield LE, Stein MR, Sellner JC, Gremillion DE. Further investigation of the association between gastroesophageal reflux and bronchoconstriction. J Allergy Clin Immunol 1982;69:516–521.
63. Hughes DM, Spier S, Rivlin J, et al. Gastroesophageal reflux during sleep in asthmatic patients. J Pediatr 1983;102:666–672.
64. Berquist WE, Rachelefsky GS, Rowshan N, et al. Quantitative gastroesophageal reflux and pulmonary function in asthmatic children and normal adults receiving placebo, theophylline, and metaproterenol sulfate therapy. J Allergy Clin Immunol 1984;73:253–258.
65. Tan WC, Martin RJ, Pandey R, Ballard RD. Effects of spontaneous and simulate gastroesophageal reflux on sleeping asthmatics. Am Rev Respir Dis 1990;141:1394–1399.
66. Baust W, Bohnert B. The regulation of heart rate during sleep. Exp Brain Res 1969;7:169–180.
67. Morrison JFJ, Pearson SB. The effect of the circadian rhythm in vagal activity on bronchomotor tone in asthma. Clin Sci 1988;74(18):71.
68. Cox ID, Hughes DTD, McConnell K. Ipratropium bromide in patients with nocturnal asthma. Postgrad Med J 1984;60:526–528.
69. Coe CI, Barnes PJ. Reduction of nocturnal asthma by an inhaled anticholinergic drug. Chest 1986;90:485–488.
70. Rhind GB, Catterall JR, Douglas NJ. Blocking vagal activity does not abolish morning dip in asthmatics. Clin Sci 1985;69(12):166.
71. Morrison JFJ, Pearson SB, Dean HG. Parasympathetic nervous system in nocturnal asthma. Br Med J 1988;296:1427–1429.
72. Todisco T, Grassi V, Sorbini CA, et al. Circadian rhythms of respiratory functions in asthmatics. Respiration 1980;40:128–135.
73. Simon RA, Stevenson DD, Arroyave CM, et al. The relationship of plasma histamine to the activity of bronchial asthma. J Allergy Clin Immunol 1977;60:312–316.
74. Szefler SJ, Ando R, Cicutto LC, et al. Plasma histamine, epinephrine, cortisol, and leukocyte β-adrenergic receptors in nocturnal asthma. Clin Pharm 1991;49:59–68.
75. Brown R, Ingram RH Jr, Wellman JJ, et al. Effects of intravenous histamine on pulmonary mechanics in non-asthmatic and asthmatic subjects. J Appl Physiol 1977;42:221–227.
76. Wardlaw AJ, Dunnette S, Gleich GJ, Collins JV, Kay AB. Eosinophils and mast cells in bronchoalveolar lavage in subjects with mild asthma.

Relationship to bronchial hyperreactivity. Am Rev Respir Dis 1988;137:62–69.

77. Kirby JG, Hargreave FE, Gleich GJ, O'Byrne PM. Bronchoalveolar cell profiles of asthmatic and nonasthmatic subjects. Am Rev Respir Dis 1987;136:379–383.

78. Flint KC, Leung KBP, Hodspith BW, Brostaff J, Pearce FL, Johnson NM. Bronchoalveolar mast cells in extrinsic asthma: a mechanism for the initiation of antigen specific bronchoconstriction. Br Med J 1985;291:923–926.

79. Godard P, Chaintrevil J, Damon M, et al. Functional assessment of alveolar macrophages: comparison of cells from asthmatics and normal subjects. J Allergy Clin Immunol 1982;79:88–93.

80. Seltzer J, Bigby BG, Stulbarg M, et al. O_3-induced change in bronchial reactivity to methacholine and airway inflammation in humans. J Appl Physiol 1986;60:1321–1326.

81. Metzger WJ, Zavala D, Richerson HB, et al. Local allergen challenge and bronchoalveolar lavage of allergic asthmatic lungs. Am Rev Respir Dis 1987;135:433–440.

82. Dahl R. Diurnal variation in the number of circulatory eosinophil leukocytes in normal controls and asthmatics. Acta Allergologics 1977;32:301–303.

83. Reinberg A, Ghata J, Sidi E. Nocturnal asthma attacks: Their relationship to the circadian cycle. J Allergy 1963;34:323–330.

84. Souter CA, Costello J, Ijaduola O, Turner-Warwick M. Nocturnal and morning asthma: Relationship to plasma corticosteroids and response to cortisol infusion. Thorax 1975;30:436–440.

85. Beam WR, Ballard RD, Martin RJ. Spectrum of corticosteroid sensitivity in nocturnal asthma. Am Rev Respir Dis. 1992;145:1082–1086.

86. Reinberg A, Gervas P, Choussade M, Fraboulet G, Duburgue B. Circadian changes in effectiveness of corticosteroids in eight patients with allergic asthma. J Allergy Clin Immunol 1983;71:425–433.

87. Reinberg A, Halberg F, Falliers CJ. Circadian timing of methylprednisolone effects in asthmatic boys. Chronobiologia 1974;1:333–347.

88. Beam WR, Weiner DE, Martin RJ. Timing of prednisone and alterations of airway inflammation in nocturnal asthma. Am Rev Respir Dis 1992;146:1524–1530.

89. Snapper JR. Inflammation and airway function: The asthma syndrome. Am Rev Respir Dis 1990;141:531–533.

90. Braude AC, Rebuck AS. Pulmonary disposition of cortisol. Ann Int Med 1982;97:59–60.

91. Schleimer RP. Effects of glucocorticosteroids on inflammatory cells relevant to their therapeutic applications of asthma. Am Rev Respir Dis 1990;141:S59–69.

92. Guyre PM, Munck A. Glucocorticoid actions on monocytes and macrophages. In: Schleimer RP, Claman HN, Oronsky AR, (eds). Antiinflammatory steroids: Basic and clinical aspects. Academic Press, New York, 1988, pp 199–225.

93. Gillis S, Crabtree GR, Smith KA. Glucocorticoid-induced inhibition of

T cell growth factor production. I. The effect on mitogen induced lymphocyte proliferation. J Immunol 1979;123:1624–1631.

94. Lamas AM, Marcotte GV, Schleimer RP. Human endothelial cells prolong eosinophil survival. Regulation by cytokines and glucocorticoids. J Immunol 1989;142:3978–3984.

95. Petroni KC, Shen L, Guyre PM. Modulation of human polymorphonuclear leukocyte IgG Fc receptors and Fc receptor mediated functions by IFN-gamma and glucocorticoids. J Immunol 1988;140:3467–3472.

96. Beasely R, Roche WR, Roberts JA, Holgate ST. Cellular events in the bronchi in mild asthma and after bronchial provocation. Am Rev Respir Dis 1989;139:806–817.

97. Titinchi S, Al Shamma M, Patel KR, Kerr JW, Clark B. Circadian variation in number and affinity of β_2-adrenoceptors in lymphocytes of asthmatic patients. Clin Sci 1984;66:323–328.

98. Fitzpatrick MF, Mackay T, Driver H, Douglas NJ. Salmeterol in nocturnal asthma: a double blind, placebo controlled trial of a long acting inhaled B2 agonist. Br Med J 1990;301:1365–1368.

99. Zwillich CW, Neagley SR, Cicutto L, White DP, Martin RJ. Nocturnal asthma therapy: Inhaled bitolterol versus sustained release theophylline. Am Rev Respir Dis 1989;139:470–474.

100. Martin RJ, Cicutto LC, Ballard RD. Circadian variations in theophylline concentrations and the treatment of nocturnal asthma. Am Rev Respir Dis 1989;139:475–478.

101. D'Alonzo GE, Smolensky MH, Feldman S, Gianotti LA, Emerson MB, Steudinger H, Steinijans VW. Twenty-four hour lung function in adult patients with asthma. Am Rev Respir Dis 1990;142:84–90.

102. Venge P, Hakansson L, Peterson GB. Eosinophil activation in allergic disease. Int Arch Allergy Appl Immunol 1987;82:333–337.

103. Horn CR, Clark TJH, Cochrane GM. Inhaled therapy reduces morning dips in asthma. Lancet 1984;1:1143–1145.

104. Speelberg B, deMonchy JRG. Is deliberate waking and bronchodilator use a useful therapy in nocturnal asthma? Am Rev Respir Dis 1991;143:A32.

105. Weiner P, Azgad Y, Ganum R. Specific inspiratory muscle training in patients with bronchial asthma. Chest 1992;143:A633.

4

Effects of Sleep on Respiratory Physiology in Nocturnal Asthma

Robert D. Ballard, M.D.

Introduction

Over the last two decades the sleep apnea syndrome has been the focus of attention for those interested in the interaction between sleep and breathing. However, it was long ago recognized that sleep has a variety of effects upon breathing that can occur entirely separately from sleep apnea. Dr. Edward Smith reported in 1860 that ventilation is reduced during sleep in normal man.[1] This observation has been confirmed many times since,[2-4] the reduction in ventilation evidently resulting from a sleep associated reduction in tidal volume (Fig. 1).

Such changes in ventilatory pattern typically result in a reduction in PaO_2 and an increase in $PaCO_2$, which led to hypotheses that the sleep associated reduction in ventilation may be due to reductions in the ventilatory responses to hypoxia and hypercapnia.[5] More recent studies have suggested that sleep is associated with upper airway narrowing that occurs even in normals without sleep apnea,[3,6,7] and may therefore constitute an intrinsic resistive load to

Martin RJ (editor): *Nocturnal Asthma: Mechanisms and Treatment,* © Futura Publishing Co., Inc., Mount Kisco, NY, 1993.

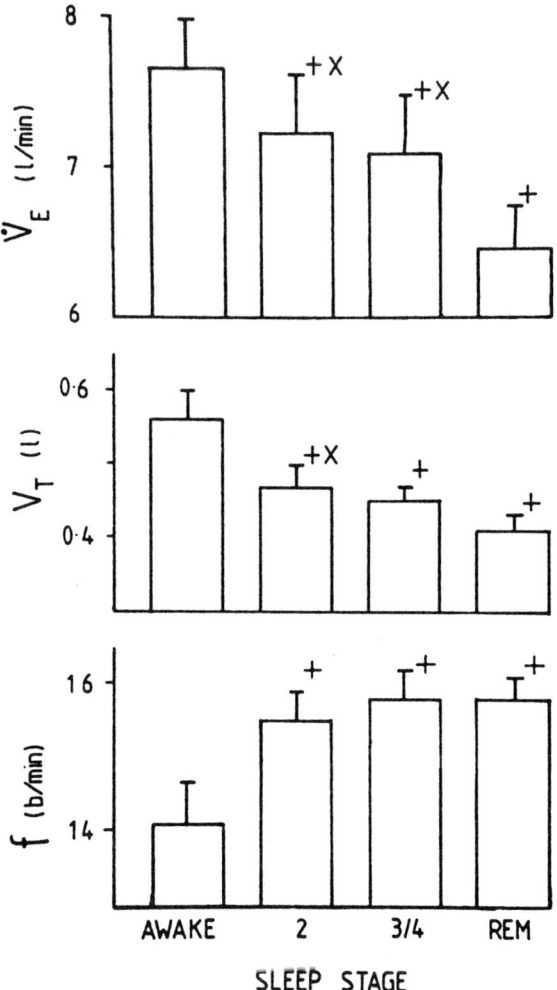

Figure 1. The effects of sleep on ventilatory pattern in normal subjects. $\dot{V}_E$, minute ventilation; V_T, tidal volume; f, respiratory frequency. + $P<0.05$ versus awake, × $P<0.05$ versus REM sleep. From reference 2 with permission.

breathing.[8] As there are well-documented reductions of ventilatory compensatory responses to resistive loading during sleep,[9,10] such upper airway narrowing may be a major contributor to the sleep associated reduction in ventilation.

Fortunately, such effects of sleep upon breathing are of minimal importance in those with normal respiratory physiology while awake. This is apparently not the case for patients with lung disease. The nocturnal worsening of respiratory function is now recognized to occur commonly in patients with a variety of obstructive airways diseases. Several studies have documented nocturnal bronchoconstriction in the majority of patients with asthma.[11,12] This pattern disrupts sleep in affected patients[13] and probably contributes to an excessive nocturnal death rate in the asthmatic population.[14] Patients with COLD (chronic obstructive lung disease) also frequently demonstrate nocturnal worsening manifested by increased airflow limitation,[12] hypoxemia,[15] and disrupted sleep.[16] Similar degrees of hypoxemia have been reported in sleeping patients with cystic fibrosis.[17,18] Although less is known about nocturnal respiratory events in patients with COLD and cystic fibrosis, there is evidence that nocturnal hypoxemia contributes to the morbidity and mortality of these diseases.[19]

Despite the apparent clinical significance of these patterns of nocturnal respiratory dysfunction, we continue to lack a clear understanding of responsible mechanisms. Investigation into multiple potential etiologies of nocturnal asthma has failed to convincingly establish the role of any specific mechanism.[20–26] Studies of patients with COLD[15] and cystic fibrosis[17] have consistently demonstrated sleep associated hypoxemia that is most severe during REM (rapid eye movement) sleep, but the determinants of such changes remain poorly defined. Although we presently lack precise explanations for such nocturnal respiratory changes, it seems virtually certain that sleep significantly alters respiratory function in these patients and is therefore likely to be a major contributor to this pattern of nocturnal worsening. In this chapter we will therefore review current knowledge about the interaction between sleep and breathing in patients with obstructive airways disease and hopefully begin to clarify the role of sleep in nocturnal respiratory dysfunction.

Sleep and Asthma

The nocturnal worsening of asthma has been acknowledged by the medical community for centuries. Aurelianus Caelius described the nocturnal frequency of asthma attacks in the fourth or fifth century A.D.[27] In 1698, Dr. John Floyer observed that his own asthma attacks were exclusively nocturnal.[28] More recently,

Turner-Warwick reported that 64% of 7,729 asthmatic patients surveyed described awakening with symptoms of asthma at least three nights weekly.[11] This pattern of nocturnal worsening almost certainly adversely alters the sleep quality of affected patients. Kales et al.[29] studied 12 young adult asthmatics in the sleep laboratory and observed them to have reduced sleep efficiency (due to frequent awakenings and earlier AM final awakening) and less stage 4 sleep when compared to normal young adult controls. Montplaisir and colleagues subsequently confirmed that nocturnal asthma is associated with a reduction in sleep efficiency.[13] This adverse effect upon the quality of sleep may also explain the recent findings of Fitzpatrick et al.[30] who observed not only a reduction in sleep efficiency in asthmatics with nocturnal worsening, but also impairment of their daytime cognitive performance.

There has also been considerable interest in the effect of sleep upon oxygen saturation in patients with asthma. Smith and Hudgel observed that sleeping asthmatic children had a greater number of oxygen desaturations (decrease in $SaO_2 \geq 4\%$) and greater maximum decreases in SaO_2 than healthy controls.[31] These changes in SaO_2 were correlated with change in FEV_1 over the sleep period. Montplaisir et al.[13] also monitored nocturnal oxygen saturation in asthmatics with nocturnal worsening, noting that while normal controls never desaturated, 80% of asthmatics demonstrated at least transient desaturation (mean duration of 27.3 seconds, mean decrease in SaO_2 of 6.5%) during sleep. Catterall and colleagues studied patients with chronic stable asthma[32] and, although they observed modest desaturation during sleep (mean decrease in SaO_2 to lowest level during sleep was 9.7% in asthmatics vs. 4.3% in controls, $P<0.01$), unlike Smith and Hudgel[31] they were unable to establish a correlation between nocturnal hypoxemia and the degree of nocturnal bronchoconstriction as reflected by overnight change in PEFR (peak expiratory flow rate).

To better assess the relationship between bronchoconstriction and oxygen saturation, we recently developed techniques by which we could monitor lower airway resistance (R_{la}) in sleeping subjects.[33] As demonstrated in Figure 2, decreases in SaO_2 correlated well with increases in R_{la} in six sleeping asthmatics with nocturnal worsening. Although the mean SaO_2 during sleep was significantly lower in the asthmatics than in normal controls (91.3% vs. 94.2%, $P<0.001$), the magnitude of this small difference does not appear to be clinically important. These findings all suggest that although nocturnal

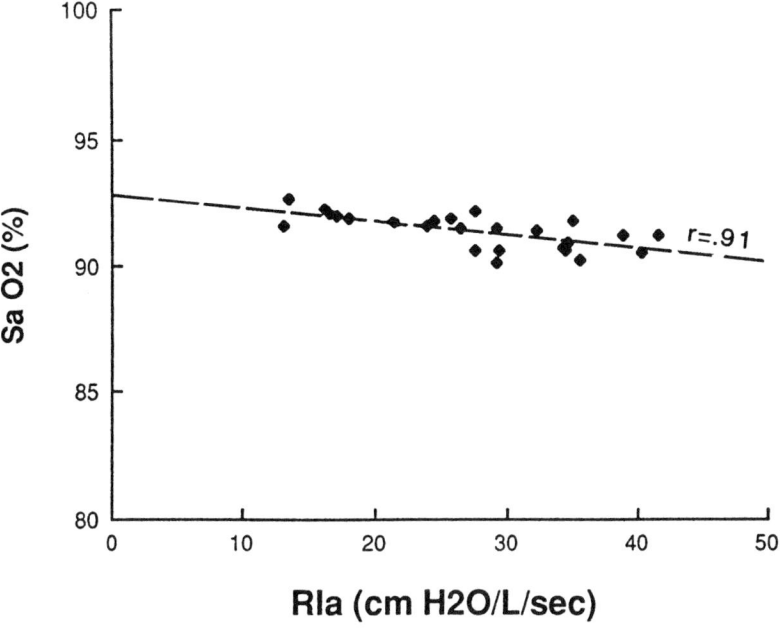

Figure 2. Nocturnal changes in arterial O_2 saturation (SaO_2) as lower airway resistance (R_{la}) increases in six asthmatic patients during sleep. Adapted from reference 33 with permission.

bronchoconstriction contributes to oxygen desaturation in the sleeping asthmatic, clinically significant decrements in SaO_2 are relatively uncommon, most likely occurring in asthmatic patients with very severe airflow obstruction, disorders of ventilatory control, or even sleep apnea.

While studies such as these have established the prevalence of this pattern of nocturnal asthma and documented some adverse effects (disrupted sleep and nocturnal desaturation), they have contributed little to the understanding of the potential mechanism(s) of this process. Several investigators have attempted to clarify the role of sleep itself as a determinant of nocturnal worsening in asthmatics, with somewhat variable results. In studies of asthmatic shift workers,[34] Clark and Hetzel were unable to separate the circadian PEFR rhythm from periods of sleep, as subjects all rapidly changed the phase of their PEFR rhythms on starting a new shift. However, in a subsequent study, the same investigators found that sleep disruption had little effect upon the overnight fall in PEFR, while total sleep prevention eliminated the overnight fall in PEFR in only about 50% of asthmatics studied.[35] These findings led Hetzel and Clark to suggest that circadian changes in airflow obstruction are often in phase with the timing of sleep, but sleep itself does not cause nocturnal asthma.[35] In still another study,[36] Catterall and coworkers evaluated the effect of overnight sleep deprivation on airflow obstruction in twelve adult asthmatics with recurrent nocturnal worsening. As shown in Figure 3, PEFR decreased significantly overnight whether the patients were allowed to sleep or kept awake all night. However, both absolute and percentage overnight falls in PEFR were greater while the AM PEFR was lower when the patients slept. The authors subsequently suggested that sleep is in fact an important determinant of nocturnal asthma.

The findings of such studies seem to be somewhat contradictory, but it is likely that this may result from technical limitations in their assessments of airflow obstruction. Sleeping subjects must of course be awakened for the measurement of FEV_1 and PEFR, and the effects of such awakenings upon pulmonary function were relatively unknown. In addition, previous studies failed to control well for posture, with patients typically assuming the supine posture during sleep but remaining upright during wakefulness. It was therefore apparent that to adequately address the role of sleep in nocturnal asthma, one must first be able to assess airflow obstruction *during* sleep, while at the same time controlling for variables such as posture.

Airflow Resistance in Sleeping Asthmatics

To assess the effect of sleep on airflow resistance, we initially monitored five asthmatics overnight utilizing esophageal balloons and face masks with an attached pneumotachograph.[37] The subsequent esophageal pressure and flow data allowed us to calculate total pulmonary resistance (R_L) during sleep on a breath-by-breath basis. As illustrated by Figure 4, R_L increased progressively in all five patients during sleep, eventually leading to a $51.8\% \pm 10.7\%$ mean increment ($P<0.01$) in R_L from bedtime to morning awakening. Although such changes in R_L appeared to be consistent with sleep associated bronchoconstriction, we were concerned that our measurements of R_L could reflect sleep associated changes in upper airway resistance, which may be substantial even in subjects without sleep apnea.[3]

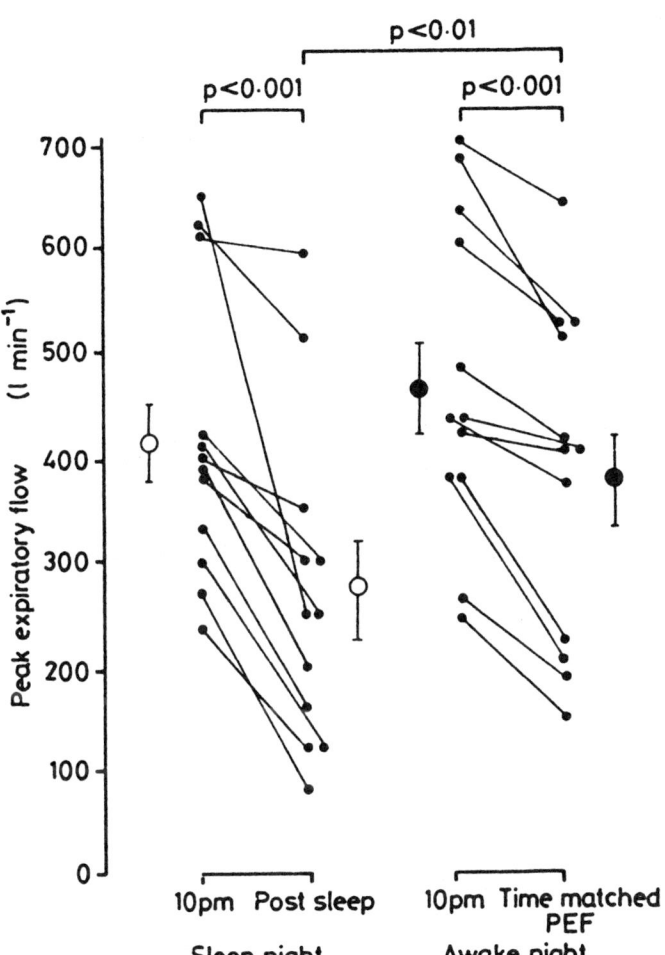

Figure 3. Overnight change in peak expiratory flow rate (PEFR) in 12 adult asthmatics after sleeping normally and after a night of sleep deprivation. From reference 36 with permission.

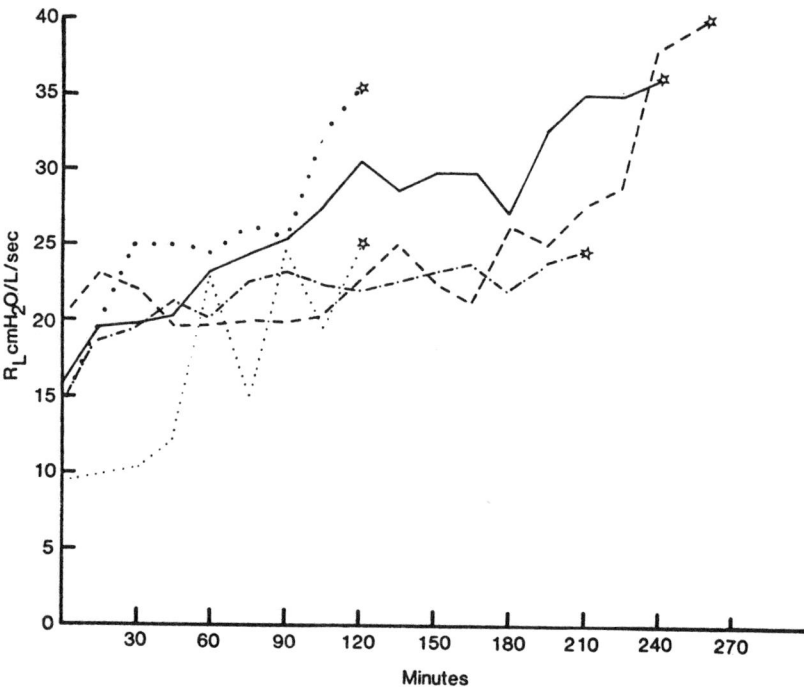

Figure 4. Changes in pulmonary resistance (R_L) measured at 15-minute intervals in five asthmatic patients. Time 0 indicates sleep onset time. Stars indicate final R_L prior to awakening. From reference 37 with permission.

We therefore modified our investigative techniques by adding a supraglottic pressure catheter for the determination of supraglottic resistance (R_{sg}). We could then subtract R_{sg} from R_L to attain a measure of lower airway resistance, R_{la}. We subsequently performed two overnight studies each in six asthmatics with recurrent nocturnal worsening.[33] One study was carried out during routine sleep, while the other was performed during sustained wakefulness, both studies taking place overnight while only in the supine posture. Figure 5A demonstrates that R_{la} increased in the asthmatic patients during both nights, but the rate of increase (slope of R_{la} vs. time) was increased twofold ($P<0.0001$) during the sleep night compared with the sleep prevention night. Sleep was therefore associated with a higher mean R_{la} (18.7 ± 5.5 vs. 11.2 ± 2.7 cm $H_2O/L/s$, $P<0.001$) and a higher peak R_{la} (35.8 ± 9.4 vs. 23.7 ± 5.5 cm $H_2O/L/s$, $P<0.001$) than that observed during sustained wakefulness. Figure 5B demon-

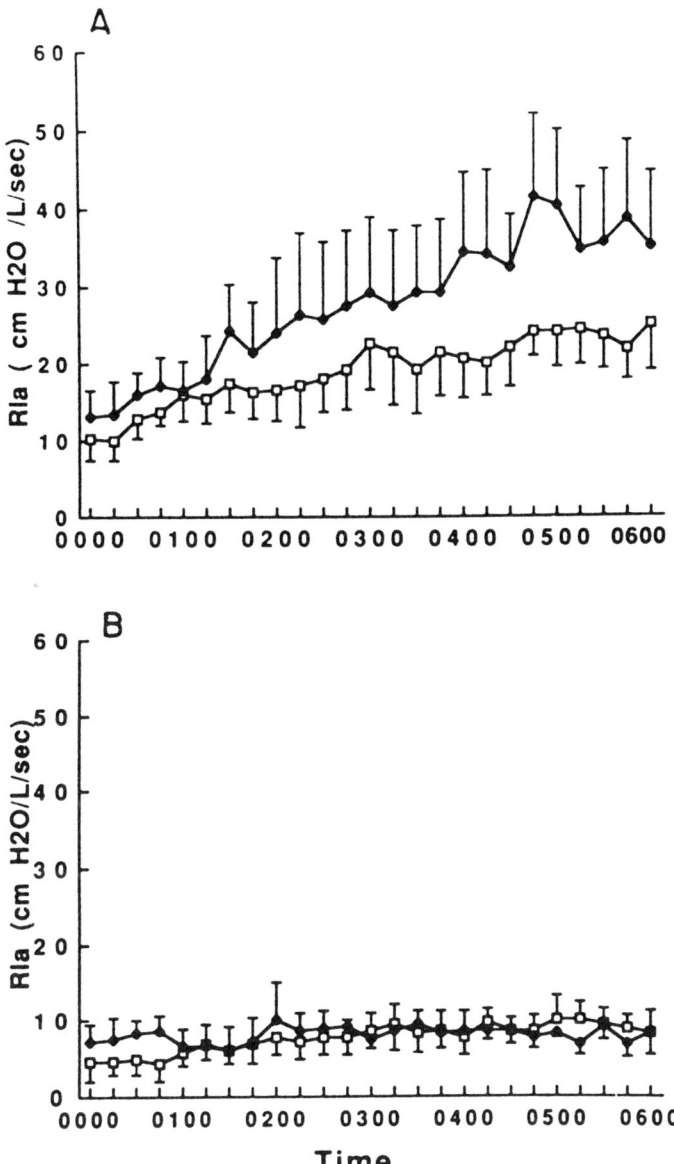

Figure 5. Nocturnal changes in lower airway resistance (R_{la}) during sleep (closed diamonds) and with prevention of sleep (open squares) in six asthmatic patients (A) and four normal subjects (B). From reference 33 with permission.

strates that in four normal controls R_{la} did not significantly change during either the sleep night or sleep prevention night.

These results confirm that in asthmatic patients with nocturnal worsening, airflow obstruction increases overnight when supine regardless of sleep state, but sleep apparently further augments such changes. Our findings were therefore similar to those of Catterall et al.[36] who had earlier reported sleep to be an important determinant of overnight decrements in PEFR. In a study of eight sleeping asthmatics that employed similar techniques,[38] Bellia and colleagues also observed overnight increments in R_{la} (Fig. 6) similar

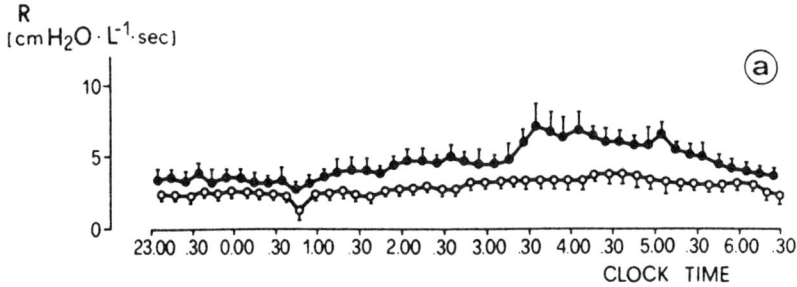

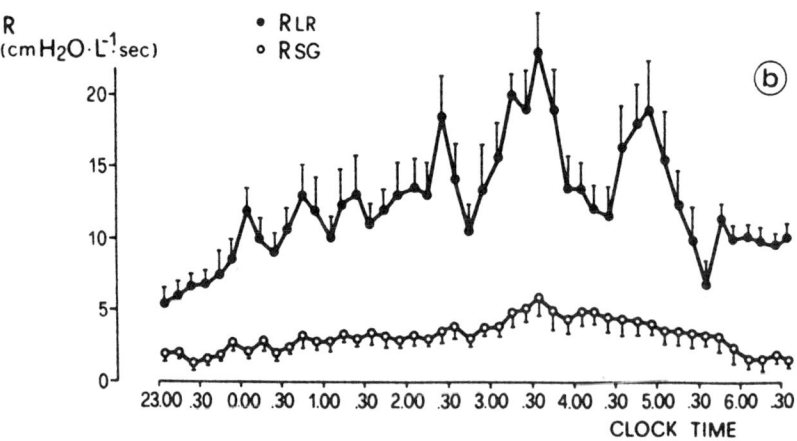

Figure 6. Overnight changes in supraglottic resistance (RSG) and lower respiratory resistance (RLR) measured from four normal subjects (a) and eight asthmatic patients (b) during sleep. From reference 38 with permission.

Table 1

Sleep and Resistance Characteristics during Sleep for Six Asthmatics
and Four Normals (from Ref. 33, with permission)

	Asthmatics	Normals
Sleep latency, min	8.4 ± 1.9	7.4 ± 5.8
Sleep efficiency, %	70.0 ± 5.2	66.1 ± 7.7
Sleep stage, % sleep time		
1	28.0 ± 4.2	22.3 ± 2.7
2	44.0 ± 6.6	48.9 ± 8.5
3–4	16.7 ± 5.6	15.1 ± 4.2
REM	11.3 ± 2.0	13.7 ± 2.3
Mean R_{la} per sleep stage, cm $H_2O/L/s$		
1	18.2 ± 5.4*	2.4 ± 0.9
2	17.3 ± 6.4*	2.7 ± 1.2
3–4	16.6 ± 4.9*	2.3 ± 0.7
REM	16.8 ± 7.3*	3.5 ± 1.2

Values are means ± SE for six asthmatic patients and four normal subjects. REM = rapid eye movement; R_{la} = lower airway resistance. *$P < 0.001$, asthmatic versus normal.

to those observed in our laboratory. While Bellia and coworkers reported that the highest values of R_{la} were observed during stages 3–4 sleep, in our own study[33] we were unable to detect a significant difference in mean R_{la} between sleep stages (Table 1).

The cause(s) of such sleep associated increments in R_{la} have not yet been firmly established. One potential contributor to such a sleep effect is the sleep associated alteration in autonomic activity. The increase in vagal tone and decrease in sympathetic activity previously reported during sleep[39] could contribute to sleep associated bronchoconstriction, although this remains to be established. An alternative possibility is that reductions in functional residual capacity (FRC) similar to those previously demonstrated in sleeping normals could occur in sleeping asthmatic patients.[40] As measurements of resistance are inversely related to lung volume,[41] one would expect sleep associated decrements in FRC to increase measurements of R_{la}.

Lung Volume in Sleeping Asthmatics

To assess the effect of sleep on lung volume, we constructed a horizontal volume-displacement body plethysmograph (Fig. 7) for

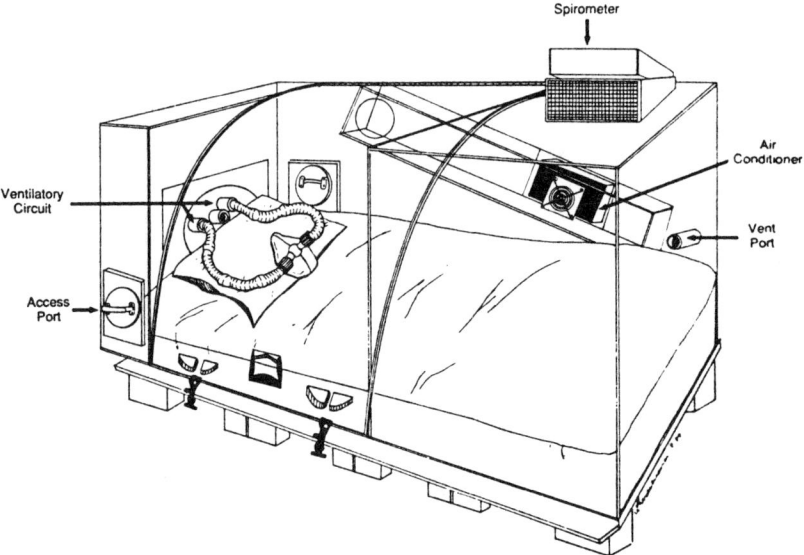

Figure 7. Horizontal volume-displacement body plethysmograph. From reference 42 with permission.

the study of supine subjects.[42] To measure FRC we used the previously described single inspiratory occlusion technique.[43] This allowed the measurement of FRC plethysmographically from a single inspiratory effort against an airway that was occluded at the end of the previous expiration and was, therefore, applicable to sleeping subjects. We subsequently studied the effect of sleep on FRC in five normal subjects and five asthmatic patients with nocturnal worsening.[42] As demonstrated in Figure 8, when supine and awake, asthmatic patients were hyperinflated relative to normal controls (FRC = 3.46 ± 0.18 and 2.95 ± 0.13 liters, respectively, P<0.05). During sleep FRC decreased in both groups, but the decrease was significantly greater in asthmatic patients such that during REM sleep FRC was equivalent between the asthmatic and normal groups (FRC = 2.46 ± 0.23 and 2.45 ± 0.09 liters, respectively). These data therefore suggest that the hyperinflation typically observed in the awake asthmatic is attenuated or even eliminated during sleep.

Having access to measurements of both lung volume and pulmonary resistance (R_L) also allowed us to calculate specific conductance ($sG_L = 1/R_1/FRC$) throughout the sleep studies. As

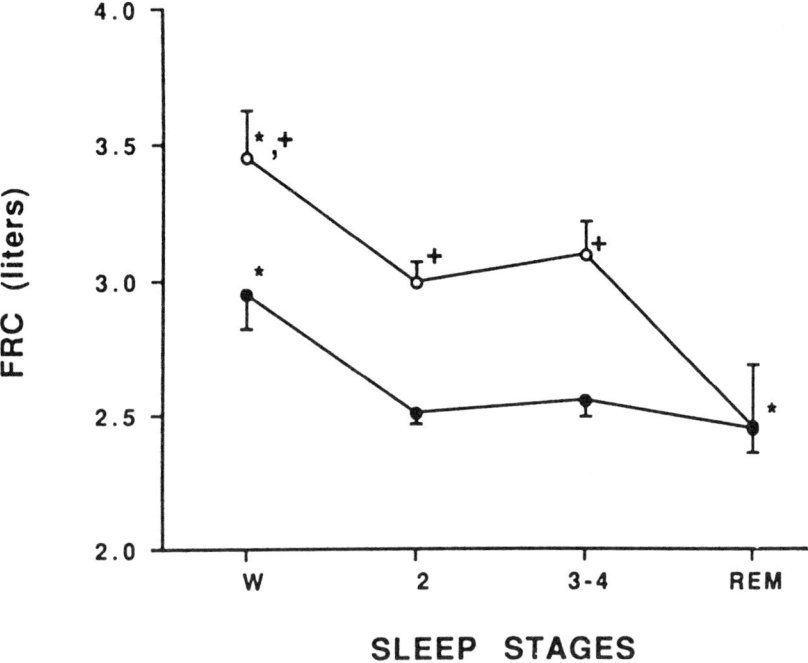

Figure 8. Effect of different sleep stages on FRC in normal subjects (closed circles) and asthmatic patients (open circles). Error bars, SE. *$P<0.05$, wakefulness (W) and REM versus all other stages in asthmatic patients, and W versus all other stages in normal subjects; $+P<0.05$, asthmatic patients versus normal subjects. From reference 42 with permission.

illustrated in Figure 9, sG_L decreased progressively and significantly in the asthmatic patients during the night. This suggests that although sleep associated reductions in FRC may allow lung volume-dependent reductions in airway caliber,[41] they do not apparently account for all of the nocturnal increase in airflow resistance observed in asthmatic patients with nocturnal worsening.

Another factor that is usually associated with sleep and has received considerable attention is the supine posture. Bouhuys observed that the supine posture enhanced the bronchoconstrictor response to inhaled histamine, an effect that was postulated to be at least partly explained by a postural decrease in lung volume.[44] More recently, Mossberg and Jonsson proposed that the supine posture may induce progressive airflow obstruction in asthmatic patients.[45]

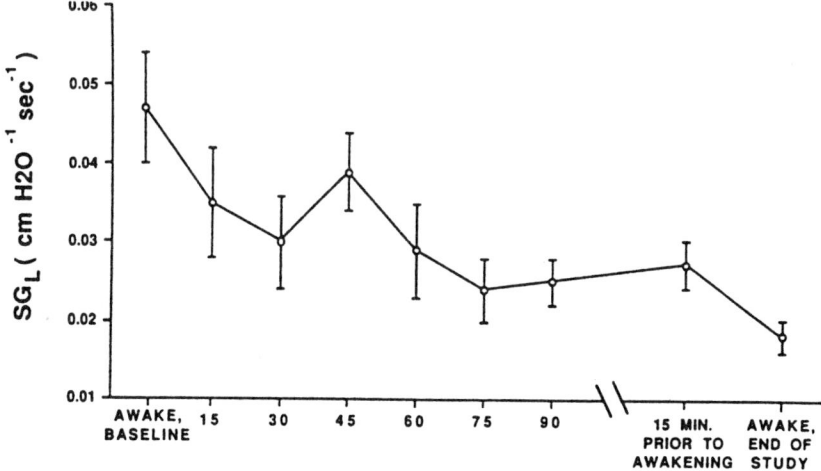

TIME (min) FROM SLEEP ONSET

Figure 9. Changes in sG_L measured from five asthmatic patients during overnight sleep studies. Abscissa is interrupted because of varying study durations. Error bars, SE. From reference 42 with permission.

Although such an effect remains to be clearly established, our previous findings[42] suggest that the reduction in lung volume associated with the supine posture[44,46] is further augmented when the subject sleeps. We therefore questioned whether the prolonged combination of supine posture and substantial decrements in lung volume, as occurs during sleep,[42] might lead to progressive and sustained worsening of airflow obstruction.

We subsequently monitored eight asthmatic patients with nocturnal worsening overnight in the horizontal body plethysmograph to determine mean FRC during sleep.[47] We then compared, during wakefulness, the effect on FEV_1 from chest wall and abdomen strapping (to maintain FRC at mean sleep levels) for six hours in the supine and upright postures. As illustrated in Figure 10A, FEV_1 was significantly decreased after strapping in the supine posture (2.54 ± 0.36 vs. 3.38 ± 0.29 liters on control day, P = 0.0001). However, Figure 10B demonstrates that FEV_1 was not significantly altered after strapping in the upright posture (3.07 ± 0.30 vs. 3.34 ± 0.31 liters on control day).

These observations suggest that the sleep associated reduction of lung volume in conjunction with supine posture may contribute to the nocturnal worsening of asthma, although a precise mechanism by which this could occur remains uncertain. One potentially important factor could be alterations in intrapulmonary blood volume. It has been demonstrated that the intrapulmonary pooling of blood can increase airflow obstruction in both animal models[48] and humans.[49] This effect has been suggested to occur via a cholinergic reflex triggered from the activation of intrapulmonary C-fiber nerve endings[48] and/or from bronchial wall edema.[49] It has also been documented that moving from the upright to the supine posture may increase intrapulmonary blood volume,[50,51] an effect which presumably accounts for at least part of the supine posture-dependent reduction in thoracic gas volume at FRC.[44,46] Findings from studies of the effects of general anesthesia[52] and submaximal paralysis[53] suggest that additional intrapulmonary pooling of blood could occur during sleep, although this has yet to be confirmed.

Additional circumstantial evidence to support such an effect of sleep on intrapulmonary blood pooling can be derived from recent studies in which we evaluated the relationships between sleep associated changes in inspiratory muscle tonic activity and lung volume. As it was previously demonstrated that hyperinflation in awake asthmatic patients results at least partly from increased tonic activity of inspiratory (diaphragm, inspiratory intercostal, and accessory) muscles,[54,55] we questioned if the onset of sleep might cause a reduction in such inspiratory muscle activity and subsequently lead to a decrement in lung volume. In our overnight studies of twelve asthmatic patients with nocturnal worsening,[56] we observed that onset of the nonREM sleep was associated with a rapid reduction ($P<0.05$) in the EMG (electromyographic) tonic activity of all three inspiratory muscles (Fig. 11). These lower levels of activity remained unchanged during subsequent sleep, but activity returned and was actually augmented ($P<0.001$) with awakening at the conclusion of the study.

Figure 12 demonstrates that the effect of sleep upon FRC differed somewhat. Although 15 minutes into sleep we observed a reduction in FRC, over the next 45 minutes of sleep there occurred additional, progressive decrements in FRC, such that FRC had decreased from 3.63 ± 0.27 liters presleep to 3.15 ± 0.27 liters after 60 minutes of nonREM sleep ($P<0.005$). FRC did not change further during subsequent sleep, but increased to 4.69 ± 0.48 liters ($P<0.01$) with awakening at the end of the study.

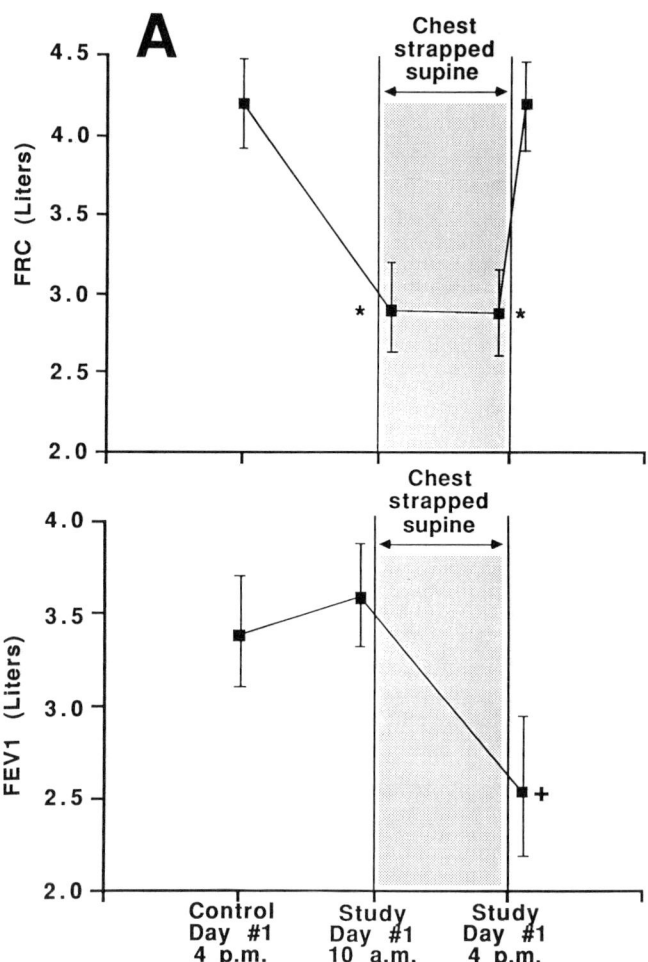

Figure 10. Changes in FEV_1 and FRC in response to a 6-hour period of chest wall and abdomen strapping in the supine posture (A) and upright posture (B). * P = 0.0001, strapped FRC versus unstrapped FRC; +P = 0.0001, poststrapping FEV_1 versus control Day 1 FEV_1. From reference 47 with permission.

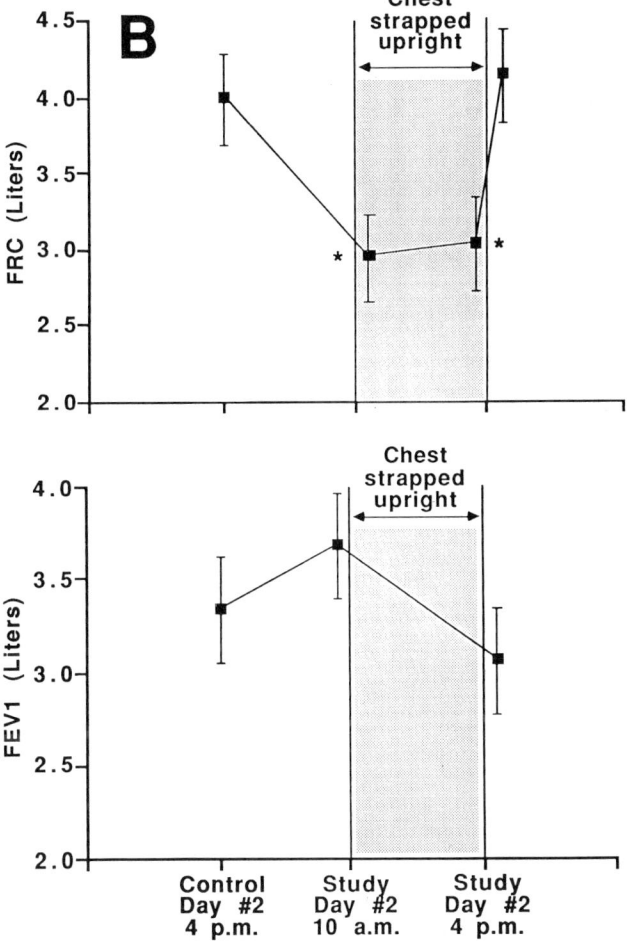

Figure 10B.

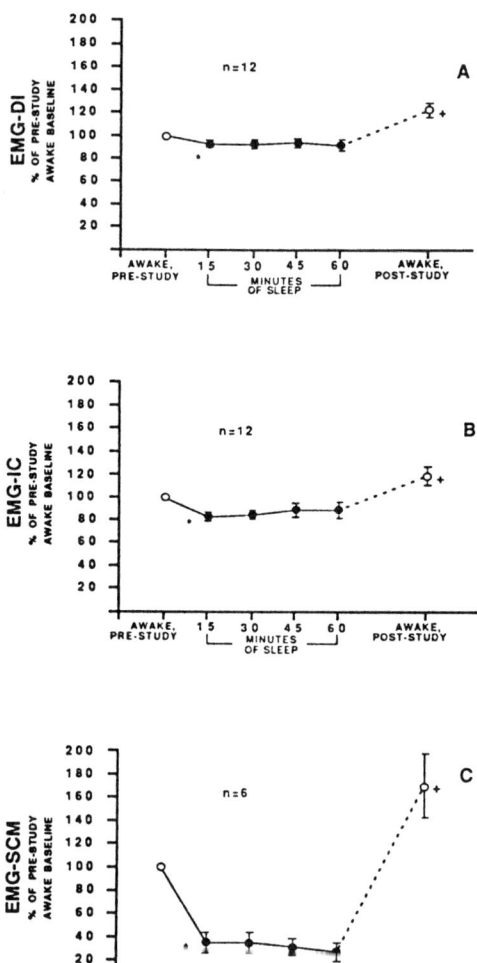

Figure 11. Effect of 60 minutes of nonREM sleep on tonic EMG activities of diaphragm, DI (A); intercostal, IC (B); sternocleidomastoid, SCM (C), measured from 12 asthmatic patients with nocturnal worsening. Open circles measured during wakefulness. Closed circles measured during continuous nonREM sleep. Error bars, SE. * $P<0.05$, 15 minutes of sleep versus awake prestudy. $+P<0.0001$, awake poststudy versus all previous levels of activity. From reference 56 with permission.

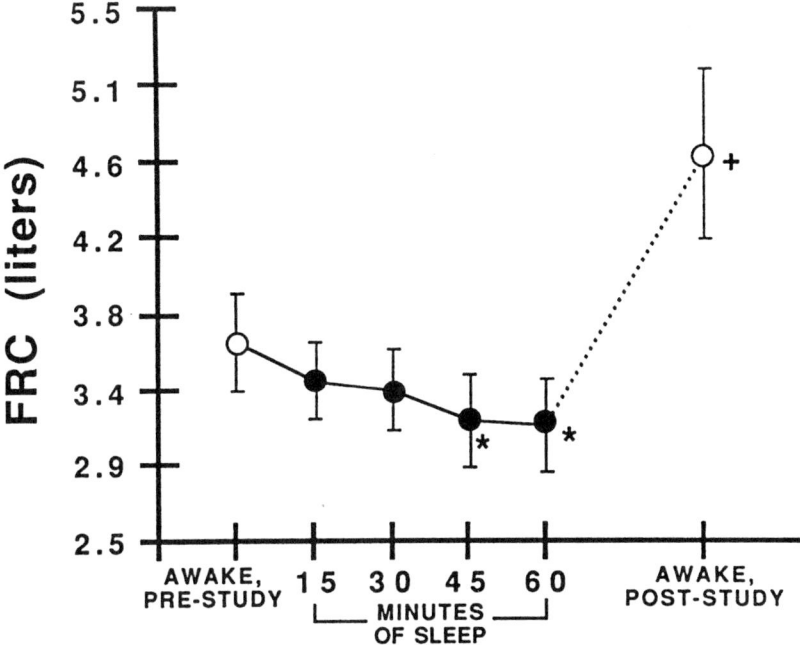

Figure 12. Effect of 60 minutes of nonREM sleep on FRC in 12 asthmatic patients. Error bars, SE. * P<0.005, 45 and 60 minutes of sleep versus awake prestudy. + P<0.0001, awake poststudy versus all previous levels of activity. From reference 56 with permission.

These results suggest that while changes in inspiratory muscle tonic activity may contribute importantly to the relatively abrupt changes in FRC observed with sleep onset and with subsequent awakening, other processes must contribute to the progressive decrease in FRC that occurs over 60 minutes of sleep. It seems likely that the sleep associated progressive decrease in FRC represents, at least partly, the intrapulmonary pooling of blood. The abrupt decrement in FRC associated with sleep onset allows rapid airway narrowing,[41] which necessitates increasingly negative inspiratory pleural pressure swings in order to maintain airflow and ventilation.[57] Such changes in intrathoracic pressure would promote the intrapulmonary pooling of blood,[58] displacing thoracic gas volume and resulting in a further decrement in measured FRC, as well as contributing to additional airway narrowing.[48,49] Although such a sequence of events seems plausible and could potentially account for the nocturnal (sleep associated) worsening of asthma, we await additional studies of the effect of sleep upon intrapulmonary blood volume to establish its validity.

Nocturnal Changes in Bronchial Responsiveness

The current widely accepted definition of asthma includes the following characteristics: 1) airway obstruction, 2) airway inflammation and 3) airway hyperresponsiveness to a variety of stimuli.[59] We have already discussed at length the patterns of nocturnal and sleep associated changes in airflow obstruction, while other chapters in this book are dealing thoroughly with the topic of nocturnal changes in airway inflammation. The potential contribution of bronchial hyperresponsiveness to this pattern of nocturnal worsening also warrants consideration. Indeed, Ryan and associates previously suggested that airway hyperresponsiveness may play a major role in the nocturnal worsening of asthma, as they observed that asthmatics with greater bronchial responsiveness to inhaled histamine during the daytime also had more pronounced nocturnal reductions in PEFR.[60]

It now seems likely that nocturnal changes in bronchial responsiveness could be an important contributor to the nocturnal worsening of asthma. De Vries and colleagues reported circadian variability in bronchial responsiveness in an early study of eleven young asthmatic patients and five older patients with COLD.[61] These investigators observed that the lowest aerosolized histamine concentration causing a 10% reduction in FEV_1 (PC_{10}) decreased markedly in both groups of patients during the night. Reinberg et al.[62] subsequently reported a similar circadian rhythm in bronchial responsiveness to acetylcholine in both normal subjects and asthmatic patients.

In a more recent study,[63] we evaluated bronchial responsiveness to aerosolized methacholine and its circadian variation in a population of 20 asthmatics. Similar to Ryan et al.,[60] we demonstrated a significant correlation between overnight fall in PEFR and the methacholine concentration causing a 20% reduction in FEV_1 (PC_{20}) at both 4 AM (r = 0.57, P<0.01) and 4 PM (r = 0.48, P<0.05). Furthermore, we observed that the patients with the greatest overnight decrements in PEFR became so hyperresponsive at 4 AM that inhalation of the aerosolized saline control alone triggered more than a 20% reduction in FEV_1. These findings therefore suggest that overnight deterioration of respiratory function in asthmatics may be related to both daytime measures of nonspecific bronchial responsiveness as well as the circadian variation in bronchial responsiveness.

Most recently, Bonnet and colleagues evaluated nine mild asthmatic patients for circadian rhythms in bronchial responsiveness to both aerosolized methacholine and histamine.[64] In this study, the PC_{20} for either methacholine or histamine was determined every 4 hours for 13 consecutive measurements. These investigators found significant and reproducible circadian variations in all cases to both methacholine and histamine. Maximum responsiveness to methacholine was detected at 3:06 AM while maximum responsiveness to histamine was found at 4:54 AM.

Gervais and associates found that there are similar circadian variations in bronchial responsiveness to common antigens such as house dust.[65] In a study of four allergic asthmatics who received bronchial challenges with house dust at varying times (8 AM, 3 PM, 7 PM, and 11 PM, these investigators observed that the largest decrement in FEV_1 occurred after the 11 PM challenge. Gervais et al. also determined that when such challenges were administered at 11 PM they resulted in greater persistence of airflow obstruction than the same challenges administered at other times. These findings take on added significance if one considers that exposure to common antigens such as house dust may be highest at night when the asthmatic patient is in bed.

In a more recent study,[66] Mohiuddin and Martin also studied the sequelae of challenging ten asthmatics with inhaled antigens at different times of day. They observed that an antigen administered in the evening was much more likely to trigger a late asthmatic response than a similar challenge administered during the morning (Fig. 13). These investigators also determined that bronchial responsiveness to aerosolized methacholine was greater 24 hours after the evening antigen challenge than 24 hours after the morning challenge. The findings of these last two studies suggest that exposure to an antigen during the nocturnal hours is much more likely to result in both immediate and sustained obstruction, as well as a prolonged increase in bronchial responsiveness, than exposure to the same antigen during the day. Such nocturnal increases in bronchial responsiveness could therefore contribute significantly to nocturnal worsening in many asthmatic patients.

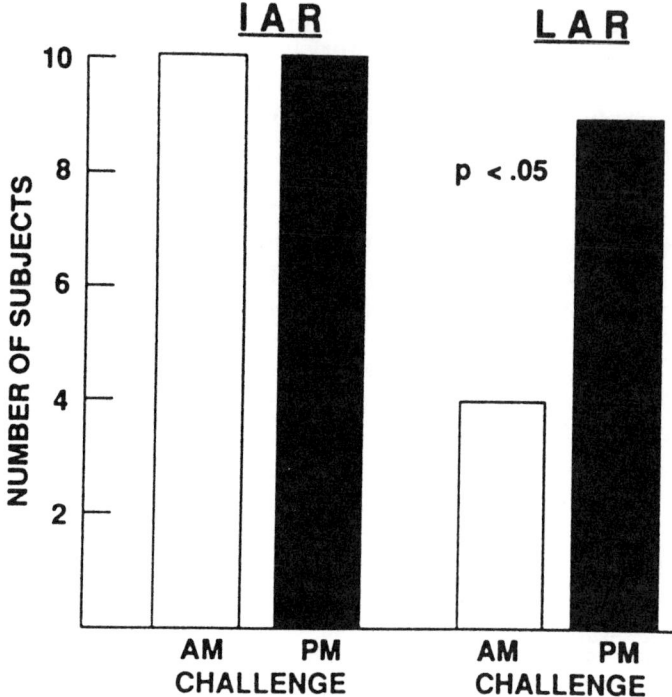

Figure 13. The occurrence of the immediate asthmatic response (IAR) and late asthmatic response (LAR) in 10 asthmatic patients after antigen challenges administered in the morning (AM) and evening (PM). From reference 66 with permission.

Bronchial hyperresponsiveness in asthma is presently an area of intense interest and investigational effort. As summarized in recent reports,[59,67] there is an increasing consensus that airway inflammation is a major determinant of bronchial hyperresponsiveness, although precise mechanisms remain uncertain. It is therefore possible that nocturnal increases in airway inflammation occur in many asthmatics (as described elsewhere in this book), leading to transient increments in bronchial responsiveness. Other nocturnal changes that could increase bronchial responsiveness include alterations in serum cortisol,[22] circulating epinephrine,[21] and cholinergic tone.[26]

It has also been suggested that the apparent nocturnal increase in bronchial responsiveness may result at least partly from nocturnal decreases in baseline airway caliber,[68] as the asthmatics who demonstrate nocturnal increments in responsiveness also typically demonstrate nocturnal reductions in prechallenge FEV_1.[62-66] However, support for such an "artifactual" change in responsiveness is not overwhelming. De Vries et al.[61] addressed this question in their study of circadian alterations in bronchial responsiveness, but were unable to find a consistent relationship between changes in measures of airflow obstruction (FEV_1, FVC) and changes in histamine PC_{10}. In our recent study of twenty asthmatics,[63] we identified four subjects with little change in baseline FEV_1 between 4 PM and 4 AM who demonstrated at least a threefold increase in responsiveness to methacholine over the same period. In their study of circadian variability in responsiveness to both histamine and methacholine,[64] Bonnet et al. specifically compared circadian changes in FEV_1 and PC_{20}. These researchers were able to find significant correlations between variations in FEV_1 and PC_{20} in only a few of the nine subjects (two during the histamine study and three during the methacholine study). Finally, Mohiuddin and Martin[66] were unable to demonstrate a difference in prechallenge FEV_1 between morning and evening antigen challenges in their study of ten asthmatics. These investigators were also unable to find significant correlations between change in FEV_1 and change in methacholine PC_{20} following antigen challenge. All of these data suggest that the observed nocturnal increases in bronchial responsiveness do not necessarily result from concurrent changes in airway caliber.

Another sleep associated change that could potentially contribute to the nocturnal increase in bronchial responsiveness is the intrapulmonary pooling of blood. As we have already discussed, there is strong evidence that the supine posture promotes intrapulmonary blood pooling,[50,51] and there is substantial circumstantial evidence that sleep further augments this effect.[52,53,56] Bouhuys demonstrated that the supine posture enhances the bronchoconstrictor response to inhaled histamine,[44] a possible result of posture-dependent alterations in intrapulmonary blood volume. Recent studies have also demonstrated bronchial hyperresponsiveness in patients with impaired left ventricular function, a condition that presumably contributes to intrapulmonary blood pooling.[69] Finally, Regnard et al.[49] recently observed that using antishock trousers to shift blood to the lungs increased bronchial responsiveness to aerosolized methacholine in healthy subjects. These investigators proposed that the increase in intrapulmonary blood volume triggered an increase in vagal bronchomotor tone,[48] as discussed earlier in this chapter, and that this accounted for the subsequent increase in bronchial responsiveness. There are therefore multiple reasons to suggest that a sleep associated increase in intrapulmonary blood volume may contribute to the frequently observed nocturnal increase in bronchial responsiveness. As before, we await additional studies of the effect of sleep upon intrapulmonary blood volume to better establish the plausibility of this concept.

Ventilation in Sleeping Asthmatics

There is substantial evidence that during wakefulness asthmatic patients typically increase inspiratory effort[70,71] as well as respiratory frequency[70,72,73] to maintain or even increase ventilation during acute bronchoconstriction. In the awake asthmatic, hypoventilation therefore usually occurs only in the presence of very severe airflow obstruction and/or respiratory muscle fatigue. This was not clearly the case during sleep, a condition which has been shown to reduce ventilatory compensatory responses to extrinsic resistive loading.[9,10] If sleep had a similar adverse effect upon ventilatory responses to bronchoconstriction, the asthmatic patient could possibly be at risk for developing life-threatening sleep associated hypoventilation.

Fortunately, recent evidence suggests that sleep does not significantly impair ventilatory responses to bronchoconstriction. Earlier studies[13,31,32] demonstrated greater sleep associated oxygen desaturation in asthmatics than normal controls, changes that presumably result from sleep associated increases in airflow obstruction. Such desaturations are relatively mild, which suggests that ventilation is fairly well-maintained during sleep. Neagley et al.[74] subsequently monitored in a qualitative manner the breathing of twelve stable asthmatic patients during sleep. These investigators found that although the patients spent nearly 10% of sleep time below an oxygen saturation of 90%, there was little evidence of disordered ventilation. Morgan and associates used similar methods to study ventilatory patterns during sleep in fifteen asthmatic patients with a history of recurrent nocturnal worsening.[75] These researchers were also unable to identify changes in ventilatory pattern that might result from bronchoconstriction during sleep, although their subjects did have nocturnal worsening as demonstrated by a 31% overnight reduction in FEV_1.

In a study of ten sleeping asthmatic patients,[76] Issa and Sullivan employed similar study methods in conjunction with the monitoring of electromyographic (EMG) activity from respiratory muscles. These investigators identified six acute exacerbations of asthma during the ten sleep studies and observed that respiratory muscle activity was augmented during these acute exacerbations. Issa and Sullivan interpreted these findings as an indication that ventilatory compensatory responses enable sleeping asthmatic patients to maintain stable ventilation despite significant bronchoconstriction. In fact, all of these studies suggest that ventilation is well-maintained despite bronchoconstriction during sleep, although no previous investigators actually measured ventilation in sleeping asthmatic patients.

To more accurately assess the impact of sleep on breathing in asthmatics, we utilized esophageal balloons, supraglottic pressure catheters, and face masks with an attached pneumotachograph, as previously described in this chapter. These methods allowed us not only to quantitate lower airway resistance (R_{la}) during sleep, but also to determine the impact of changes in R_{la} upon ventilatory pattern.[33] As demonstrated in Figure 14, although tidal volume (V_t) remained stable in the presence of increasing R_{la}, respiratory frequency (f) increased slightly as R_{la} increased (slope of f vs. $R_{la} = 0.0397$, $P < 0.01$). This resulted in an actual increase in minute ventilation (V_I) as R_{la} increased during sleep (slope of V_I vs. $R_{la} = 0.0222$, $P < 0.01$). It was therefore readily apparent that ventilation was well-maintained during sleep despite spontaneously developing bronchoconstriction.

In an effort to better define the mechanisms that so well maintain ventilation in sleeping asthmatic patients, we subsequently employed the previously described methods in another study of eight mild asthmatics.[57] In this study we used aerosolized methacholine to induce bronchoconstriction during wakefulness and normal sleep while we monitored ventilatory pattern and occlusion pressure ($P_{0.1}$) as an indicator of inspiratory drive.[77] As already discussed, asthma often leads to the disruption of sleep,[29,30] and prior sleep deprivation has been previously observed to depress ventilatory responses to hypoxia[78] and hypercapnia,[79] plus arousal responses to various respiratory stimuli.[80] We therefore also studied our patients responses to induced bronchoconstriction during sleep after a 36-hour period of sleep deprivation. Supplemental oxygen was administered during all studies to prevent the development of hypoxemia.

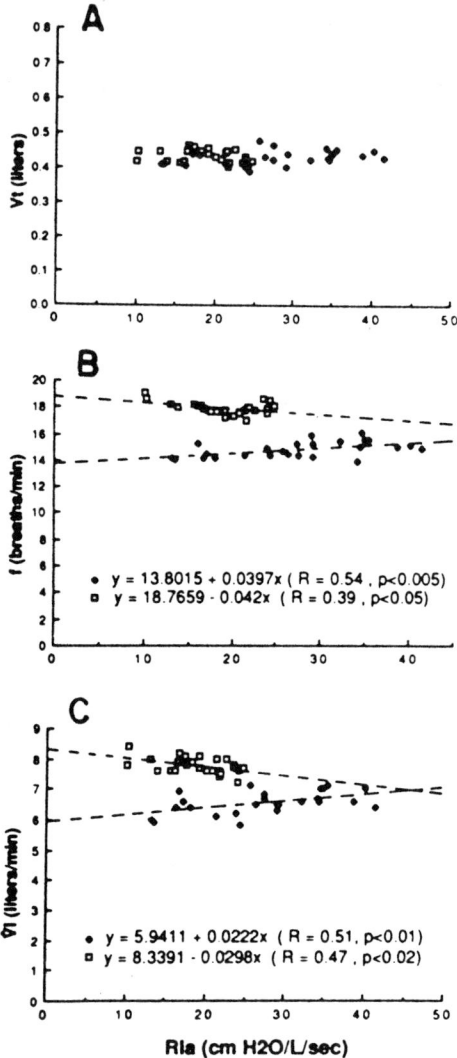

Figure 14. Comparison of nocturnal ventilatory changes in association with lower airway resistance (R_{la}) changes in six asthmatic patients during sleep (closed diamonds) and with prevention of sleep (open squares). Ventilatory parameters include tidal volume (V_t, A), respiratory frequency (f, B), and minute ventilation (V_I, C). Only regression lines associated with a significant R value (P<0.05) are plotted. From reference 33 with permission.

As illustrated in Figure 15, V_t, f, and V_I were all well-maintained as bronchoconstriction (R_{la}) increased in response to methacholine during wakefulness, normal sleep, and sleep after prior sleep deprivation. Figure 16 demonstrates that $P_{0.1}$ increased significantly (slopes of $P_{0.1}$ vs. $R_{la} = 0.249$, 0.112, and 0.154 for awake, normal sleep, and sleep after prior sleep deprivation, respectively, P<0.0006) as R_{la} increased during each of the three studies. The administration of supplemental oxygen during studies prevented state-dependent or intrastudy changes in oxygen saturation (SaO_2), nor were there significant intrastudy changes in expired carbon dioxide ($F_{et}CO_2$) (Table 2). These findings suggest that the preservation of ventilation in all states was primarily attributable to increases in ventilatory drive ($P_{0.1}$), to which chemostimuli do not apparently contribute. Our observations were therefore consistent with previous reports of respiratory responses to methacholine-induced bronchoconstriction in awake asthmatic patients.[72,73]

One additional observation from this study that may have significance is that prior sleep deprivation apparently raises the arousal threshold to bronchoconstriction. During wakefulness the asthmatic patients tolerated a mean maximal R_{la} of 22.0 ± 1.7 cm H_2O/L/sec before asking for the study to be terminated because of discomfort (Fig. 17). During normal sleep the same patients tolerated a mean maximal R_{la} of only 8.3 ± 2.1 cm H_2O/L/sec before awakening spontaneously and asking for a bronchodilator. During sleep after a 36-hour period of sleep deprivation, the asthmatic patients tolerated a mean maximal R_{la} of 23.7 ± 2.1 cm H_2O/L/sec before four of the eight awakened spontaneously and asked for a bronchodilator, and four were awakened by the investigators because their R_{la} was in excess of our predesignated safety limit of 25 cm H_2O/L/sec. The mean maximal R_{la} tolerated during normal sleep was therefore significantly lower than that tolerated either during wakefulness or during sleep after prior sleep deprivation (P<0.01). The findings were similar when we assessed the changes in FEV_1 tolerated during each of the three studies (Table 3). We therefore found that while prior sleep deprivation had no effect upon ventilatory and $P_{0.1}$ responses to bronchoconstriction during subsequent sleep, it did allow the asthmatic patients to tolerate greater bronchoconstriction before awakening from subsequent sleep. These observations were similar to those of Bowes et al.,[80] who demonstrated in dogs that although arousal responses to respiratory stimuli were impaired by sleep fragmentation, ventilatory responses

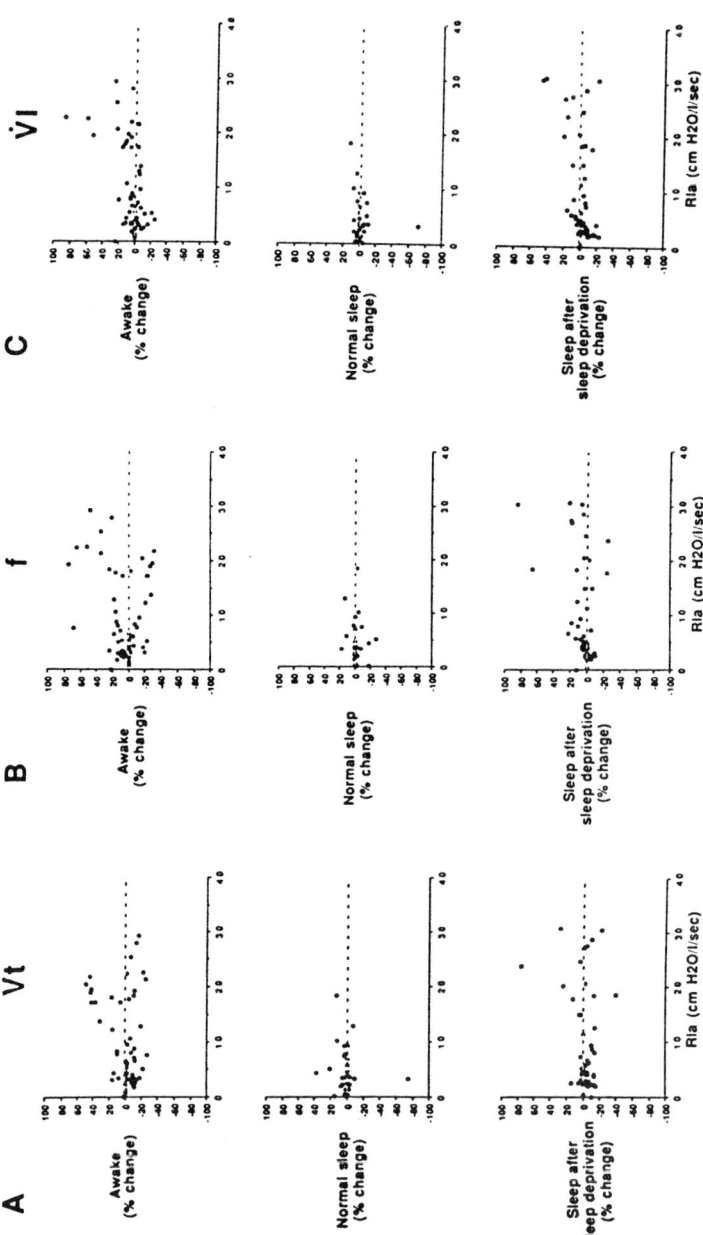

Figure 15. Comparison of ventilatory changes in response to induced increases in R_{la} in eight asthmatic patients while they were awake, during normal sleep, and during sleep after prior sleep deprivation. Ventilatory parameters include tidal volume (V_t, A), respiratory frequency (f, B), and minute ventilation ($\dot{V}_I$). Dashed lines, flat lines of identity included for reference. From reference 57 with permission.

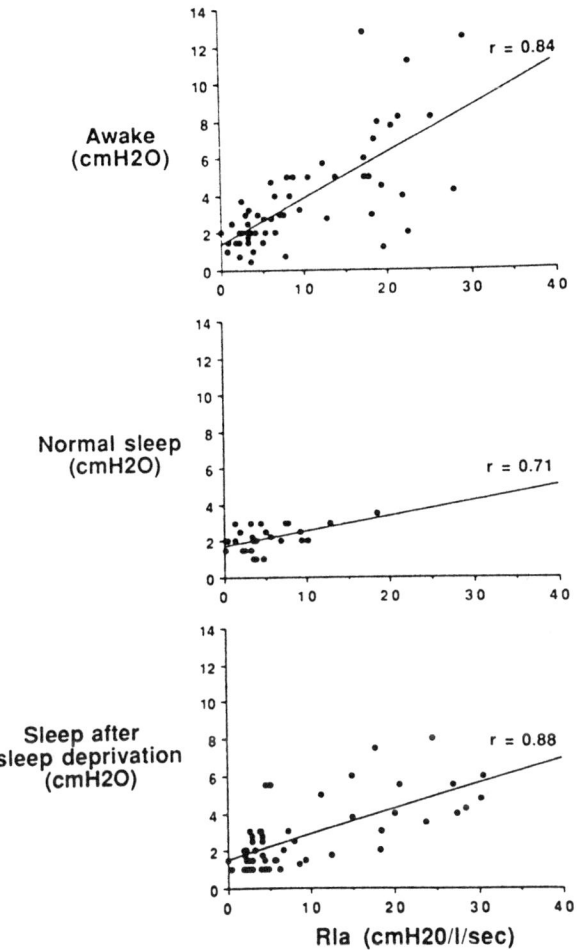

Figure 16. Comparison of $P_{0.1}$ changes in response to induced increases in R_{la} in eight asthmatic patients while they were awake, during normal sleep, and during sleep after prior sleep deprivation. The *r* values are corrected for subject effect. From reference 57 with permission.

Table 2

SaO$_2$ and FetCO$_2$ Levels from Studies of Eight Asthmatic Patients (from Ref. 57, with permission)

	Awake	Normal Sleep	Postdepriv. Sleep
Baseline SaO$_2$, %	93.0 ± 1.7	91.6 ± 1.1	92.8 ± 1.0
Mean SaO$_2$, %	93.0 ± 0.7	91.5 ± 0.7	91.1 ± 0.7
R$_{la}$ Max SaO$_2$, %	91.9 ± 1.5	91.1 ± 0.7	91.5 ± 0.7
Baseline FetCO$_2$, %	5.48 ± 0.39*	5.93 ± 0.24	6.16 ± 0.15
Mean FetCO$_2$, %	5.13 ± 0.27*	5.88 ± 0.17	6.01 ± 0.09
R$_{la}$ Max FetCO$_2$, %	4.95 ± 0.29*	5.66 ± 0.20	5.79 ± 0.23

Values are means ± SE for eight asthmatic subjects. *$P < 0.01$, awake versus normal sleep and sleep after sleep deprivation (postdepriv. sleep).

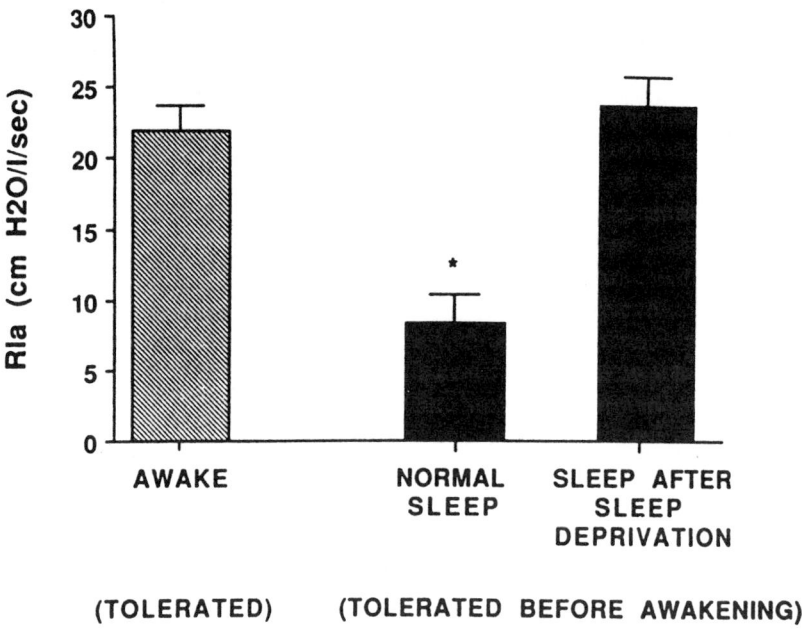

* = p < 0.01

Figure 17. Mean maximal R$_{la}$ attained by eight asthmatic patients before they asked the study to be terminated because of discomfort (awake), awakened spontaneously because of bronchoconstriction (normal sleep), and awakened spontaneously because of bronchoconstriction (4) or were awakened because they exceeded R$_{la}$ of 25 cm H$_2$O/L/sec (4) (sleep after prior sleep deprivation). From reference 57 with permission.

Table 3

Spirometric Response to Methacholine from Eight Asthmatic Patients
(from Ref. 57, with permission)

	Awake	Normal Sleep	Postdepriv. Sleep
Baseline FEV_1, L	2.99 ± 0.38	2.94 ± 0.35	3.03 ± 0.33
Postmethacholine FEV_1, L	1.46 ± 0.31 +	2.04 ± 0.27* +	1.54 ± 0.28 +

Values are means ± SE for eight asthmatic subjects. *P < 0.01, normal sleep versus awake and sleep after sleep deprivation (postdepriv. sleep). + P < 0.01, postmethacholine versus baseline.

to hypoxia and hypercapnia were not altered during subsequent sleep. As already stated, asthma frequently causes the disruption of sleep and presumably contributes to sleep deprivation.[30–32] A subsequently increased arousal threshold might eventually prevent sleeping asthmatic patients from awakening and receiving treatment when severely bronchoconstricted. Such a sequence of events could possibly contribute to the excessive nocturnal death rate from asthma,[14,81] although that remains speculative.

Sleep and Chronic Obstructive Lung Disease

Although this book is primarily intended to address the topic of nocturnal asthma, it is appropriate for us to discuss briefly what is currently known about the impact of sleep upon respiratory function in other obstructive airway diseases. In particular, COLD (chronic obstructive lung disease) is of considerable importance as if affects 5% of the adult population in the United States.[82] Like asthma, there is also evidence that sleep adversely affects respiration in this disease. McNicholas and Fitzgerald reported in 1984[83] that of 24 COLD patients who died over a four-year period in their hospital (St. Vincent's, Dublin) with respiratory failure (hypoxemia and hypercapnia), 15 (63%) died between 11 PM and 7 AM I also recently surveyed death reports over a five-year period from the Denver Veterans Affairs Medical Center. I found that 63% of the respiratory arrests that subsequently occurred in patients admitted for exacerbation of COLD happened between 10 PM and 6 AM (unpublished data). These findings suggest that COLD, like asthma,[14,81] has an excessive nocturnal death rate.

The reasons for such a nocturnal increase in mortality remain uncertain. There are certainly many reports of severe hypoxemia and hypercapnia in sleeping patients with COLD. In 1962, Trask and Cree reported oximetry data from seven COLD patients that demonstrated oxygen desaturation to as low as 37% during apparent sleep.[84] Pierce et al.[85] used electroencephalographic sleep staging and indwelling arterial catheters in 19 COLD patients to demonstrate a mean maximal fall in PaO_2 of 7.4. mm Hg from wakefulness to sleep. Douglas et al.[15] subsequently studied ten COLD patients and observed that all patients reduced their SaO_2 by at least 10% during sleep, with eight of the ten demonstrating SaO_2s below 50%. In these patients the lowest PaO_2s (measured from arterial catheters) were in the range of 26–44 mm Hg and occurred predominantly during REM (rapid eye movement) sleep.

A multitude of other studies have confirmed the severity of sleep associated hypoxemia in COLD patients. Several studies have demonstrated both episodic and relatively fixed pulmonary hypertension associated with nocturnal desaturation in these patients, suggesting also that nocturnal oxygen therapy can subsequently improve these abnormalities.[86,87] Other studies have also revealed poor sleep quality in patients with COLD, and suggested that the quality of sleep subsequently improves with nocturnal oxygen therapy.[16] Finally, data from the NOTT (Nocturnal Oxygen Therapy Trial)[19] and MRCWP (Medical Research Council Working Party)[88] Reports demonstrated that long-term survival in hypoxic COLD patients was improved by the administration of supplemental oxygen for 12 to 15 hours (including the sleeping hours) daily. These observations all suggest that sleep associated hypoxemia contributes significantly to the morbidity and mortality of COLD.

The potential mechanism(s) by which sleep causes hypoxemia and hypercapnia remains uncertain due to a previous lack of accurate techniques with which to monitor respiration during sleep. There is some evidence that sleep is associated with increased airflow obstruction in a fashion similar to asthma. Connolly studied the circadian variation of PEFR in a population of 350 patients with asthma, chronic bronchitis, and emphysema.[12] He found that 31% of 118 patients with chronic bronchitis and 54% of 26 patients with emphysema demonstrated a circadian variation of PEFR that was similar to that seen more commonly in asthmatics (highest PEFR in the midday, lowest PEFR in the early morning or at bedtime). There is also some evidence for a circadian variation to airway responsiveness in patients with COLD. De Vries and colleagues included five patients with COLD in their study of circadian variability in airway responsiveness.[61] These investigators observed a nocturnal decrease in the concentration of aerosolized histamine causing a 10% reduction in FEV_1 (PC_{10}) that was similar to that observed in asthmatic patients. The findings from these last two studies suggest that airflow obstruction and airway responsiveness can increase during the night in a manner similar to asthmatics, although there has yet been no linkage established between these mechanisms and sleep associated hypoxemia.

Most attention has focused on sleep associated hypoventilation or ventilation/perfusion mismatch as potential causes of sleep associated hypoxemia in COLD. Littner et al.[89] qualitatively monitored ventilatory pattern (abdominal strain gauge) and inspiratory effort (esophageal balloon) in nine sleeping subjects with COLD. These researchers observed oxygen desaturation of 11% or more during sleep in six of the nine subjects, and that such desaturation was typically associated with transient episodes of reduced ventilation. Hudgel and colleagues subsequently used respiratory inductance plethysmography and respiratory muscle EMG monitoring to study ventilation in seven sleeping COLD patients.[90] These investigators also found evidence for sleep associated hypoventilation that was most severe during REM sleep and was typically associated with oxygen desaturation. Hudgel et al. concluded that hypoventilation is the major determinant of sleep associated hypoxemia in COLD patients, although they also found evidence for a sleep associated reduction in lung volume that they postulated could contribute to oxygen desaturation via ventilation/perfusion mismatch.

Other investigators have emphasized the role of ventilation-perfusion mismatching as the major cause of sleep associated hypoxemia in COLD.[91,92] Although this is a likely contributor to nocturnal hypoxemia, Catterall and coinvestigators demonstrated the inaccuracy of applying steady-state gas exchange equations to the relatively unstable respiratory pattern of REM sleep.[93] We must therefore conclude that currently available techniques are inadequate for the assessment of ventilation/perfusion mismatch during sleep. It should also be remembered that ventilation itself has not previously been measured in a quantitative fashion in sleeping patients with COLD.

In our laboratory we have only recently begun to monitor ventilation and respiratory mechanics in sleeping patients with COLD. Using the techniques previously described for the monitoring of sleeping asthmatics (esophageal balloon, supraglottic pressure catheter, face mask with attached pneumotachograph, horizontal body plethysmograph),[33,42] we have to date completed sleep studies of three patients with laboratory evidence of emphysema.[94] From these studies we observed that patients with COLD were unlike asthmatics in that they did not lose lung volume during sleep (Fig. 18), nor did lower airway resistance (R_{la}) increase over the course of the sleep study (Fig. 19). As illustrated in Table 4, ventilation did decrease during sleep, a result of sleep associated decrements in tidal volume (V_t) that were clearly most marked during REM sleep. These findings support the role of sleep associated hypoventilation as a determinant of sleep associated hypoxemia in COLD patients, but they do not exclude a contribution from ventilation/perfusion mismatch. We await the completion of a larger study of sleep in COLD patients to confirm these preliminary findings.

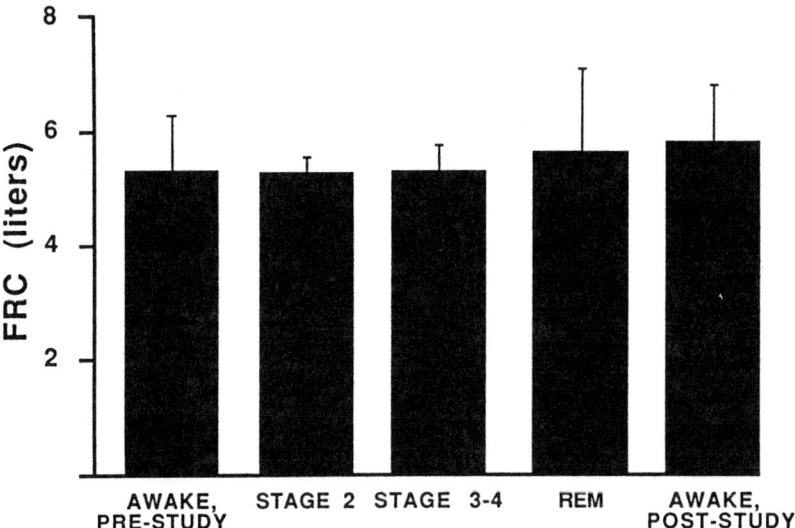

Figure 18. Changes in FRC measured from overnight studies of three sleeping patients with COLD. From reference 94 with permission.

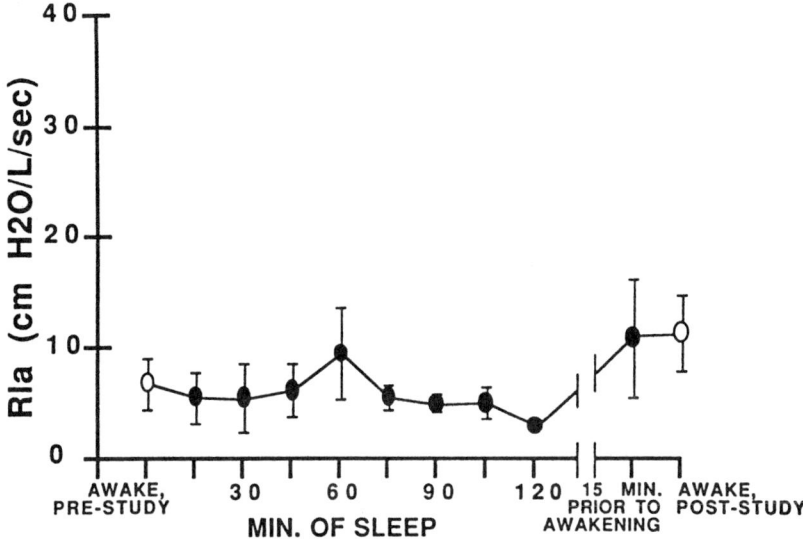

Figure 19. Changes in lower airway resistance (R_{la}) measured from overnight studies of three sleeping patients with COLD. From reference 94 with permission.

Table 4

Ventilatory Changes During Sleep in Three Patients with COLD (from Ref. 94, with permission)

	Awake	*Stage 2*	*Stage 3–4*	*REM*
V_t, L	0.65 ± 0.12	0.44 ± 0.02*	0.44 ± 0.03*	0.32 ± 0.01*
f, bpm	16.3 ± 2.1	17.5 ± 2.1	17.3 ± 1.0	19.0 ± 4.2
$\dot{V}_E$, L	10.17 ± 1.25	7.77 ± 1.43*	7.59 ± 1.11*	6.11 ± 1.63*

Values are means ± SE from three COLD patients. V_t = tidal volume; L = liters; f = respiratory frequency; bpm = breaths per minute; $\dot{V}_E$ = minute ventilation. *P < 0.05, Stages 2, 3–4, and REM versus Awake.

Sleep and Cystic Fibrosis

Cystic fibrosis (CF) is the most common genetic disorder in the Caucasian population, with a prevalence of approximately one case in every 2,500 Caucasian births.[95] When initially described in 1938, CF was associated with a life expectancy of <5 years. However, current newborns with CF can expect to live at least 30 years, primarily due to aggressive nutritional and enzyme supplementation, in addition to intensive respiratory care. Unfortunately, even as the CF population ages, pulmonary dysfunction and eventual respiratory failure continue to be the major contributors to the morbidity and mortality of this disease.

Although little investigated, the effect of sleep upon breathing has been suggested to play a major role in the progression of this disease. In 1980, Francis et al.[17] reported from twenty CF patients that during wakefulness, nonREM, and REM sleep mean SaO_2s fell from 91.7% to 88.2% (P<0.001) to 84.3% (P<0.001), respectively. In a contemporary paper,[18] Stokes and colleagues observed similar changes, noting that oxygen desaturation during sleep was most severe in patients with the most severe airflow obstruction while awake. Muller et al.[96] confirmed these observations and also used chest wall surface magnetometers to relate REM sleep associated desaturations to decrements in lung volume. Tepper and coinvestigators later used similar methods to suggest that the REM sleep associated decrement in SaO_2 was a result of hypoventilation.[97] Whatever the mechanism(s) accounting for nocturnal desaturation in CF, these investigators all speculated that such hypoxemia may contribute to the development of pulmonary hypertension and cor pulmonale, events that have historically been associated with a life expectancy of only months.[98]

We have also recently begun to employ the previously described investigative methods (esophageal balloon, supraglottic pressure catheter, face mask with attached pneumotachograph, horizontal body plethysmograph) to evaluate sleep associated changes in respiratory function in adults with CF. In sleep studies of five adult CF patients,[99] we were unfortunately able to assess only the effects of nonREM sleep, as only two of the subjects achieved REM sleep. During sleep CF patients behaved unlike asthmatics but similar to COLD patients, in that nonREM sleep was not associated with decrements in lung volume (Fig. 20) or increments in R_{la} (Fig. 21). As demonstrated in Table 5, ventilatory patterns during sleep were also

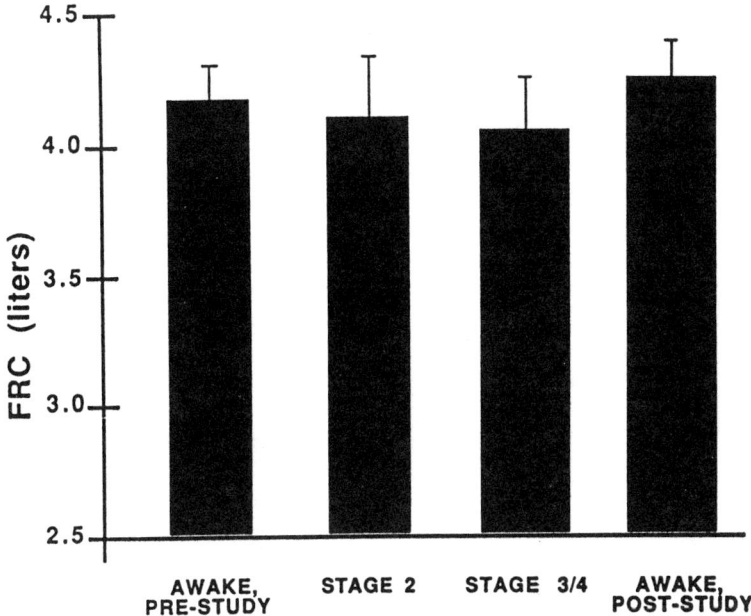

Figure 20. Changes in FRC measured from overnight studies of five sleeping patients with cystic fibrosis. From reference 99 with permission.

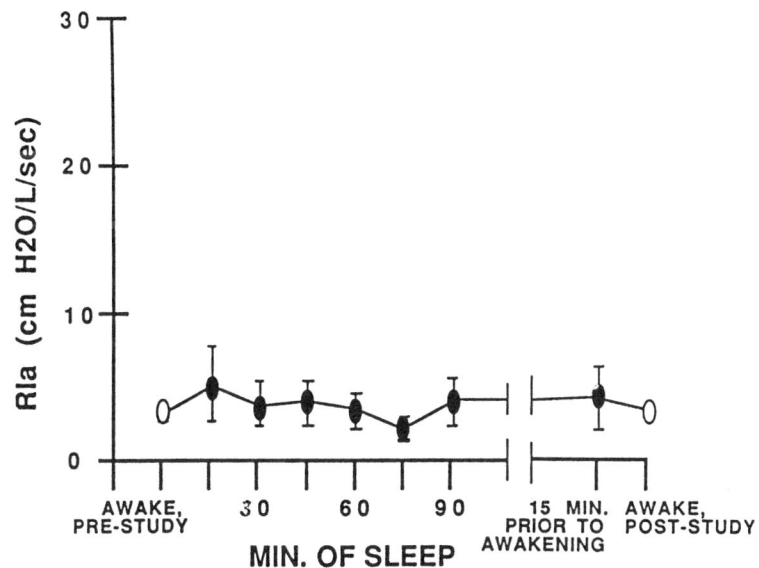

Figure 21. Changes in lower airway resistance (R_{la}) measured from overnight studies of five sleeping patients with cystic fibrosis. From reference 99 with permission.

Table 5

Ventilatory Changes During Sleep in Five Patients with Cystic Fibrosis
(from Ref. 99, with permission)

	Awake	Stage 2	Stage 3–4
V_t, L	0.56 ± 0.04	0.38 ± 0.01*	0.37 ± 0.03*
f, bpm	20.6 ± 1.1	24.9 ± 0.8*	25.8 ± 1.1*
$\dot{V}_E$, L	10.96 ± 0.24	9.27 ± 0.42*	9.24 ± 0.64*

Values are means ± SE from five patients with cystic fibrosis. V_t = tidal volume; L = liters; f = respiratory frequency; bpm = breaths per minute; $\dot{V}_E$ = minute ventilation. *$P < 0.05$, Stages 2 and 3–4 versus Awake.

similar to those seen in COLD patients. NonREM sleep was associated with reductions in ventilation that resulted from decrements in V_t, despite marked increases in respiratory frequency. The implications of these observations remain uncertain, but our findings suggest that sleep is associated with rapid, shallow breathing in patients with CF, a disadvantageous pattern that would likely reduce ventilatory efficiency (the relationship of alveolar ventilation to respiratory work) and impair gas exchange.

Summary

It is readily apparent that although the effects of sleep upon breathing are of minimal importance in those with normal respiratory physiology while awake, this is not the case for patients with obstructive airway disease. Patients with asthma frequently have sleep associated worsening manifested by increases in airflow obstruction. I have presented data that suggest such nocturnal worsening may result from sleep associated decrements in lung volume and the supine posture typically assumed during sleep. However, as discussed elsewhere in this book, there are multiple other potential contributors to nocturnal worsening, including inflammatory changes, neurogenic mechanisms, changes in bronchial responsiveness, and the intrapulmonary pooling of blood. Sleep fortunately does not significantly impair ventilatory compensatory responses to bronchoconstriction, although there is evidence that prior sleep deprivation suppresses arousal responses to bronchoconstriction during subsequent sleep.

This last effect could pose a threat to asthmatics who frequently experience recurrent disruption of their sleep.

There is considerably less known about the effects of sleep upon breathing in patients with COLD and CF. I have presented evidence that sleep is associated with hypoxemia that contributes to morbidity and mortality in both of these populations. We also have data suggesting that, unlike asthmatics, patients with COLD and CF demonstrate neither sleep associated decrements in lung volume nor increases in airflow resistance. However, in both COLD and CF sleep is associated with a reduction in alveolar ventilation which appears to be a major determinant of sleep associated hypoxemia.

All of these findings confirm that sleep has an important and often detrimental effect upon respiratory function in patients with obstructive airway disease. It is also clear that, despite a recent resurgence of interest in this area, we are only beginning to understand the mechanisms of this adverse sleep effect and the subsequent clinical consequences.

References

1. Smith E. Recherches experimentales sur la respiration. J Physiol de l'Homme et des animaux 1860; 3:506–521.
2. Douglas NJ, White DP, Pickett CK, Weil J, Zwillich C. Respiration during sleep in normal man. Thorax 1982; 37:840–844.
3. Hudgel DW, Martin RJ, Johnson B, Hill P. Mechanics of the respiratory system and breathing pattern during sleep in normal humans. J Appl Physiol 1984; 56:133–137.
4. Stradling JR, Chadwick GA, Frew AJ. Changes in ventilation and its components during sleep. Thorax 1985; 40:364–370.
5. Douglas NJ, White DP, Weil JV, Pickett CK, Zwillich CW. Hypercapnic ventilatory response in sleeping adults. Am Rev Resp Dis 1982; 126:758–762.
6. Lopes JM, Tabachnik E, Muller NL, Levison H, Bryan AC. Total airway resistance and respiratory muscle activity during sleep. J Appl Physiol 1983; 54:773–777.
7. Hudgel DW, Hendricks C, Hamilton HB. Characteristics of the upper airway pressure-flow relationship during sleep. J Appl Physiol 1988; 64:1930–1935.
8. Skatrud JB, Dempsey JA. Airway resistance and respiratory muscle function in snorers during NREM sleep. J Appl Physiol 1985; 59:328–335.
9. Iber C, Berssenbrugge A, Skatrud JB, Dempsey JA. Ventilatory adaptations to resistive loading during wakefulness and non-REM sleep. J Appl Physiol 1982; 52:607–614.

10. Wiegand L, Zwillich CW, White DP. Sleep and the ventilatory response to resistive loading in normal men. J Appl Physiol 1988; 64:1186–1195.
11. Turner-Warwick M. Epidemiology of nocturnal asthma. Am J Med 1988; 85:6–8.
12. Connolly CK. Diurnal rhythms in airway obstruction. Br J Dis Chest 1979; 73:357–366.
13. Montplaisir J, Walsh J, Malo JL. Nocturnal asthma: Features of attacks, sleep, and breathing patterns. Am Rev Resp Dis 1982; 125:18–22.
14. Hetzel MR, Clark TJH, Branthwaite MA. Asthma: Analysis of sudden deaths and ventilatory arrests in hospital. Br Med J 1977; 1:808–811.
15. Douglas NJ, Calverley PMA, Leggett RJE, Brash HM, Flenley DC, Brezinova V. Transient hypoxemia during sleep in chronic bronchitis and emphysema. Lancet 1979; 1:1–4.
16. Cormick W, Olsen LG, Hensley MJ, Saunders NA. Nocturnal hypoxemia and quality of sleep in patients with chronic obstructive lung disease. Thorax 1986; 41:846–854.
17. Francis PWJ, Muller NL, Gurwitz D, Milligan DWA, Levison H, Bryan AC. Hemoglobin desaturation: Its occurrence during sleep in patients with cystic fibrosis. Am J Dis Child 1980; 134:734–740.
18. Stokes DC, McBride JT, Wall MA, Erba G, Strieder DJ. Sleep hypoxemia in young adults with cystic fibrosis. Am J Dis Child 1980; 134:741–743.
19. Nocturnal Oxygen Therapy Trial Group. Continuous or nocturnal oxygen therapy in hypoxemic chronic obstructive lung disease; a clinical trial. Ann Intern Med 1980; 93:391–398.
20. Hetzel MR, Clark TJH. Comparison of normal and asthmatic circadian rhythms in peak expiratory flow rate. Thorax 1980; 35:732–738.
21. Barnes P, Fitzgerald G, Brown M, Dollery C. Nocturnal asthma and changes in circulating epinephrine, histamine, and cortisol. N Engl J Med 1980; 303:263–267.
22. Soutar CA, Costello J, Ijaduola O, Turner-Warwick M. Nocturnal and morning asthma: Relationship to plasma corticosteroids and response to cortisol infusion. Thorax 1975; 30:436–440.
23. Reinberg A, Gervais P. Circadian rhythms in respiratory functions, with special reference to human chronophysiology and chronopharmacology. Bull Physiopath Respir 1972; 8:663–675.
24. Bateman JRM, Pavia D, Clarke SW. The retention of lung secretions during the night in normal subjects. Clin Sci 1978; 55:523–527.
25. Tan WC, Martin RJ, Pandey R, Ballard RD. Effects of spontaneous and simulated gastroesophageal reflux on sleeping asthmatics. Am Rev Resp Dis 1990; 141:1394–1399.
26. Catterall JR, Rhind GB, Whyte KF, Shapiro CM, Douglas NJ. Is nocturnal asthma caused by changes in airway cholinergic activity? Thorax 1988; 43:720–724.
27. Aurelianus Caelius. In: De Morbis Acutis et Chronicis. Westeniand, Amsterdam, 1709.
28. Floyer J. A treatise of the Asthma. Wilkin, London, 1698.
29. Kales A, Beall GN, Bajor GF, Jacobson A, Kales JD. Sleep studies in

asthmatic adults: Relationship of attacks to sleep stage and time of night. J Allergy 1968; 41:164–173.

30. Fitzpatrick MJ, Engleman H, Whyte KF, Deary IJ, Shapiro CM, Douglas NJ. Morbidity in nocturnal asthma: sleep quality and daytime cognitive performance. Thorax 1991; 46:569–573.

31. Smith TF, Hudgel DW. Arterial oxygen desaturation during sleep in children with asthma and its relation to airway obstruction and ventilatory drive. Pediatrics 1980; 66:746–751.

32. Catterall JR, Douglas NJ, Calverley PMA, Brash HM, Brezinova V, Shapiro CM, Flenley DC. Irregular breathing and hypoxaemia during sleep in chronic stable asthma. Lancet 1982; 1:301–304.

33. Ballard RD, Saathoff MC, Patel DK, Kelly PL, Martin RJ. Effect of sleep on nocturnal bronchoconstriction and ventilatory patterns in asthmatics. J Appl Physiol 1989; 67:243–249.

34. Clark TJH, Hetzel MR. Diurnal variation of asthma. Br J Dis Chest 1977; 71:87–92.

35. Hetzel MR, Clark TJH. Does sleep cause nocturnal asthma? Thorax 1979; 34:749–754.

36. Catterall JR, Rhind GB, Stewart IC, Whyte KF, Shapiro CM, Douglas NJ. Effect of sleep deprivation on overnight bronchoconstriction in nocturnal asthma. Thorax 1986; 41:676–680.

37. Ballard RD, Kelly PL, Martin RJ. Estimates of ventilation from inductance plethysmography in sleeping asthmatic patients. Chest 1988; 93:128–133.

38. Bellia V, Cuttitta G, Insalaco G, Visconti A, Bonsignore G. Relationship of nocturnal bronchoconstriction to sleep stages. Am Rev Resp Dis 1989; 140:363–367.

39. Postma DS, Keyzer JJ, Koeter GH, Sluiter HJ, De Vries K. Influence of the parasympathetic and sympathetic nervous system on nocturnal bronchial obstruction. Clin Sci 1985; 69:251–258.

40. Hudgel DW, Devadatta P. Decrease in functional residual capacity during sleep in normal humans. J Appl Physiol 1984; 57:1319–1322.

41. Briscoe W, DuBois A. Relationship between airway resistance, airway conductance and lung volume in subjects of different age and body size. J Clin Invest 1958; 37:1279–1285.

42. Ballard RD, Irvin CG, Martin RJ, Pak J, Pandey R, White DP. Influence of sleep on lung volume in asthmatic patients and normal subjects. J Appl Physiol 1990; 68:2034–2041.

43. Desmond KJ, Demizio DL, Allen PD, Beaudry PH, Coates AL. An alternate method for the determination of functional residual capacity in a plethysmograph. Am Rev Resp Dis 1988; 137:273–276.

44. Bouhuys A. Effect of posture in experimental asthma in man. Am J Med 1963; 34:470–476.

45. Mossberg B, Jonsson E. Is asthma at night caused by posture? Chest 1985; 87:2165–2175.

46. Linderholm H. Lung mechanics in sitting and horizontal postures studied by body plethysmographic methods. Am J Physiol 1963; 204:85–91.

47. Ballard RD, Pak J, White DP. Influence of posture and sustained loss of

lung volume on pulmonary function in awake asthmatic subjects. Am Rev Resp Dis 1991; 144:499–503.

48. Chung KF, Keyes SJ, Morgan BM, Jones PW, Snashall PD. Mechanisms of airway narrowing in acute pulmonary oedema in dogs: influence of the vagus and lung volume. Clin Sci 1983; 65:289–296.

49. Regnard J, Baudrillard P, Salah B, Dinh Xuan AT, Cabanes L, Lockhart A. Inflation of antishock trousers increases bronchial response to methacholine in healthy subjects. J Appl Physiol 1990; 68:1528–1533.

50. Hamilton WF, Morgan AB. Mechanism of the postural reduction in vital capacity in relation to orthopnea and storage of blood in the lungs. Am J Physiol 1932; 99:526–533.

51. Sjostrand T. Determination of changes in intrathoracic blood volume in man. Acta Physiol Scand 1951; 22:114–128.

52. Hedenstierna G, Lofstrom B, Lundh R. Thoracic gas volume and chest-abdomen dimensions during anesthesia and muscle paralysis. Anesthesiology 1981; 55:499–506.

53. Kimball WR, Loring SH, Basta SJ, DeTroyer A, Mead J. Effects of paralysis with pancuronium on chest wall statics in awake humans. J Appl Physiol 1985; 58:1638–1645.

54. Muller N, Bryan AC, Zamel N. Tonic inspiratory muscle activity as a cause of hyperinflation in asthma. J Appl Physiol 1981; 50:279–282.

55. Martin J, Powell E, Shore S, Emrich J, Engel LA. The role of respiratory muscles in the hyperinflation of bronchial asthma. Am Rev Resp Dis 1980; 121:441–447.

56. Ballard RD, Clover CW, White DP. Influence of non-REM sleep on inspiratory muscle activity and lung volume in asthmatic patients. Am Rev Resp Dis, in press.

57. Ballard RD, Tan WC, Kelly PL, Pak J, Pandey R, Martin RJ. Effect of sleep and sleep deprivation on ventilatory response to bronchoconstriction. J Appl Physiol 1990; 69:490–497.

58. Steiner SH, Frayser R, Ross JC. Alterations in pulmonary diffusing capacity and pulmonary capillary blood volume with negative pressure breathing. J Clin Invest 1965; 44:1623–1630.

59. Expert Panel on the Management of Asthma. Guidelines for the Diagnosis and Management of Asthma. National Asthma Education Program, National Institutes of Health, Publication No. 91-3042. August, 1991.

60. Ryan G, Latimer KM, Dolovich J, Hargreave FE. Bronchial responsiveness to histamine: relationship to diurnal variation of peak flow rate, improvement after bronchodilator, and airway calibre. Thorax 1982; 37:423–429.

61. De Vries K, Goei JT, Booy-Noord H, Orie NGM. Changes during 24 hours in the lung function and histamine hyperreactivity of the bronchial tree in asthmatic and bronchitic patients. Int Arch Allergy 1962; 20:93–101.

62. Reinberg A, Gervais P, Morin M, Abulker C. Rythme circadien humain

du seuil de la reponse bronchique a l'acetylcholine. C R Acad Sci (Paris) 1971; 272:1879–1881.

63. Martin RJ, Cicutto LC, Ballard RD. Factors related to the nocturnal worsening of asthma. Am Rev Resp Dis 1990; 141:33–38.

64. Bonnet R, Jorres R, Heitmann U, Magnussen H. Circadian rhythm in airway responsiveness and airway tone in patients with mild asthma. J Appl Physiol 1991; 71:1598–1605.

65. Gervais P, Reinberg A, Gervais C, Smolensky M, DeFrance O. Twenty-four-hour rhythm in the bronchial hyperreactivity to house dust in asthmatics. J Allergy Clin Immunol 1977; 59:207–213.

66. Mohiuddin AA, Martin RJ. Circadian basis of the late asthmatic response. Am Rev Resp Dis 1990; 142:1153–1157.

67. International Symposium on Airway Hyperreactivity. Airway hyper-reactivity. Am Rev Resp Dis 1991; 143(3):S1–S82.

68. Moreno RH, Hogg JC, Pare PD. Mechanics of airway narrowing. Am Rev Resp Dis 1986; 133:1171–1180.

69. Cabanes LR, Weber S, Matran R, Regnard J, Richard MO, DeGeorges ME, Lockhart A. Bronchial hyperresponsiveness to methacholine in patients with impaired left ventricular function. N Engl J Med 1989; 320:1317–1322.

70. Mann J, Bradley CA, Anthonisen NR. Occlusion pressure in acute bronchoconstriction induced by methylcholine. Respir Physiol 1978; 33:339–347.

71. Kelsen SG, Fleegler B, Altose MD. The respiratory neuromuscular response to hypoxia, hypercapnia, and obstruction to airflow in asthma. Am Rev Resp Dis 1979; 120:517–527.

72. Kelsen SG, Prestel TF, Cherniack NS, Chester EH, Deal EC. Comparison of the respiratory responses to external resistive loading and bronchoconstriction. J Clin Invest 1981; 67:1761–1768.

73. Millman RP, Silage DA, Peterson DD, Pack AI. Effect of aerosolized histamine on occlusion pressure and ventilation in humans. J Appl Physiol 1982; 53:690–697.

74. Neagley SR, White DP, Zwillich CW. Breathing during sleep in stable asthmatic subjects. Chest 1986; 90:334–337.

75. Morgan AD, Rhind GB, Connaughton JJ, Catterall JR, Shapiro CM, Douglas NJ. Breathing patterns during sleep in patients with nocturnal asthma. Thorax 1987; 42:600–603.

76. Issa FG, Sullivan CE. Respiratory muscle activity and thoracoabdominal motion during acute episodes of asthma during sleep. Am Rev Resp Dis 1985; 132:999–1004.

77. Whitelaw WA, Derenne J, Milic-Emili J. Occlusion pressure as a measure of respiratory center output in conscious man. Respir Physiol 1975; 23:181–199.

78. White DP, Douglas NJ, Pickett CK, Zwillich CW, Weil JV. Sleep deprivation and the control of respiration. Am Rev Resp Dis 1983; 128:984–986.

79. Schiffman PL, Trontell MC, Mazar MF, Edelman NH. Sleep deprivation decreases ventilatory response to CO_2 but not load compensation. Chest 1983; 84:695–698.

80. Bowes G, Woolf GM, Sullivan CE, Phillipson EA. Effect of sleep fragmentation on ventilatory and arousal responses of sleeping dogs to respiratory stimuli. Am Rev Resp Dis 1980; 122:899–908.
81. Cochrane GM, Clark TJH. A survey of asthma mortality in patients between ages 35 and 64 in the Greater London hospitals in 1971. Thorax 1975; 30:300–305.
82. National Center for Health Statistics: Current estimates from the National Health Reviews Survey—United States 1985. Series 10, No. 160.
83. McNicholas WT, Fitzgerald MX. Nocturnal deaths in patients with chronic bronchitis and emphysema. Br Med J 1984; 289:878.
84. Trask CH, Cree EM. Oximeter studies on patients with chronic obstructive emphysema, awake and during sleep. N Engl J Med 1962; 266:639–642.
85. Pierce AK, Jarrett CE, Werkle G, Miller WF. Respiratory function during sleep in patients with chronic obstructive lung disease. J Clin Invest 1966; 45:631–636.
86. Boysen PG, Block AJ, Wynne JW, Hunt LA, Flick MR. Nocturnal pulmonary hypertension in patients with chronic obstructive pulmonary disease. Chest 1979; 76:536–542.
87. Fletcher EC, Levin DC. Cardiopulmonary hemodynamics during sleep in patients with chronic obstructive pulmonary disease: the effect of short- and long-term oxygen. Chest 1984; 85:6–14.
88. Medical Research Council Working Party Report. Long-term domiciliary oxygen therapy in chronic hypoxic cor pulmonale complicating chronic bronchitis and emphysema. Lancet, 1981; 1:681–686.
89. Littner MR, McGinty DJ, Arand DL. Determinants of oxygen desaturation in the course of ventilation during sleep in chronic obstructive pulmonary disease. Am Rev Resp Dis 1980; 122:849–857.
90. Hudgel DW, Martin RJ, Capehart M, Johnson B, Hill P. Contribution of hypoventilation to sleep oxygen desaturation in chronic obstructive pulmonary disease. J Appl Physiol 1983; 55:669–677.
91. Koo KW, Sax DS, Snider GL. Arterial blood gases and pH during sleep in chronic obstructive pulmonary disease. Am J Med 1975; 58:663–670.
92. Fletcher EC, Gray BA, Levin DC. Non-apneic mechanisms of arterial oxygen desaturation during rapid-eye-movement sleep. J Appl Physiol 1983; 54:632–639.
93. Catterall JR, Calverley PMA, Shapiro CM, Douglas NJ, Flenley DC. Mechanism of transient nocturnal hypoxemia in hypoxic chronic bronchitis and emphysema. J Appl Physiol 1985; 59:1698–1703.
94. Ballard RD, Clover CW. Influence of sleep on respiratory function in emphysema. Abstract submitted, Am Rev Resp Dis.
95. Hammond KB, Reardon MC, Accurso FJ, Cotton EK, Sokol RJ, Bonner A. Early detection and follow-up of cystic fibrosis in newborns: the Colorado experience. *In* Genetic Disease: Screening and Management. Alan R. Liss, New York, 1986, pp 81–101.
96. Muller NL, Francis PW, Gurwitz D, Levison H, Bryan AC. Mechanism of hemoglobin desaturation during rapid-eye-movement sleep in nor-

mal subjects and in patients with cystic fibrosis. Am Rev Resp Dis 1980; 121:463–469.

97. Tepper RS, Skatrud JB, Dempsey JA. Ventilation and oxygenation changes during sleep in cystic fibrosis. Chest 1983; 84:388–393.

98. Moss AJ, Dooley RR, Mickey MR. Cystic fibrosis complicated by heart failure. West J Med 1975; 122:471–473.

99. Ballard RD, Clover CW. Influence of non-REM sleep on respiratory function in cystic fibrosis. Abstract submitted, Am Rev Resp Dis.

5

Cellular Mechanisms of Nocturnal Asthma: The Role of the Eosinophil, Neutrophil, and Mast Cell

Nizar N. Jarjour, M.D., Ketan K. Sheth, M.D., and William W. Busse, M.D.

Introduction

Nocturnal worsening of airway obstruction is a common event in asthma. Turner-Warwick found that among nearly 8,000 asthma patients responding to a postal questionnaire, 39% complained of nocturnal awakening due to asthma, and 64% had the problem at least 3 nights per week.[1] In this population of patients, the frequency of nocturnal awakening increased with patient perception of disease severity and the number of asthma medications prescribed by their physicians.[1] Furthermore, nocturnal exacerbations are important facets of asthma because of greater morbidity (obstruction, hypoxia, hospital visits) and increased mortality at night. Despite its prevalence and significance, the exact mechanisms leading to increased nocturnal airway obstruction in asthma are not yet established (Table 1).

In asthma, there is evidence that airway hyperresponsiveness is

Martin RJ (editor): *Nocturnal Asthma: Mechanisms and Treatment,* © Futura Publishing Co., Inc., Mount Kisco, NY, 1993.

Table 1
Proposed Mechanisms of Nocturnal Asthma

Late phase reaction to allergen
Airway cooling
Sleep
Enhanced airway inflammation
Circadian changes in catecholamines and cortisol
Gastroesophageal reflux
Increased airway secretions
Supine posture
Enhanced vagal tone

associated with airway inflammation and may be a critical link to the pathogenesis of this disease. This is derived from various studies that included histopathology of bronchial tissue and bronchoalveolar lavage (BAL). These studies suggest that airway inflammation is a pivotal feature of asthma and possibly the driving force for bronchial hyperresponsiveness in this disease. Moreover, present observations indicate that airway inflammation in asthma is a multicellular process and the cell-cell interactions that lead to inflammation are complex, redundant, and both autocrine and paracrine in nature (Fig. 1). Recognition of such events has stimulated a new direction in identifying mechanisms in the pathogenesis of asthma and its treatment: specifically the study of mechanisms of inflammation and their regulation.[2] Based upon this new information, it is postulated that greater insight into mechanisms of asthma will follow, as will more successful and precise treatment. As will become apparent in our discussion, nocturnal asthma is not an exception to this general trend and, in fact, may be the ideal model for study.

As studies into mechanisms of inflammation in asthma have unfolded, it became more apparent that this phenomenon is dependent on complex cell-cell interaction; however, the lymphocyte, with its ability to modulate immunoglobulin production, and the functions of other inflammatory cells appear to play a leading role. While the contribution of lymphocytes and macrophages is discussed elsewhere (see Chapter 6), the possible role of T-cells in activating other "effector" cells will be briefly reviewed to help the understanding of the interaction between this and other inflammatory cells. T-cells promote antibody synthesis and generate cytokines. A number of recognized cytokines have pivotal roles in the generation of

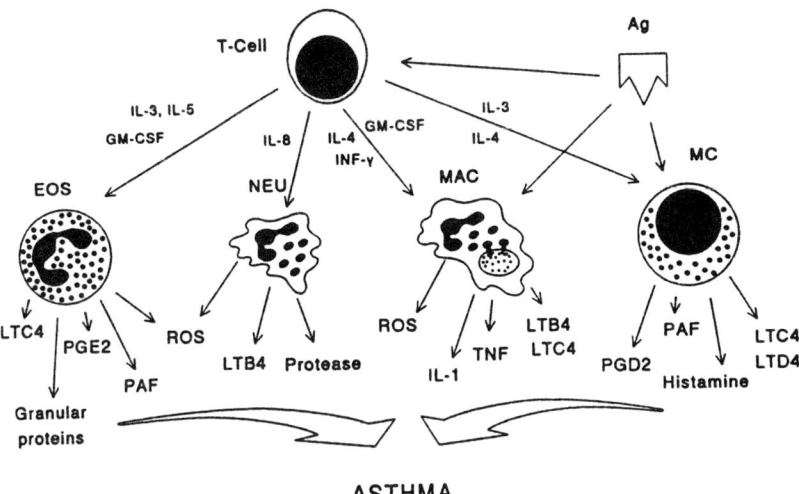

ASTHMA

Figure 1. A schematic representation of possible interactions between T-cells and other effector-cells in bronchial asthma. ROS = reactive oxygen species; TNF = tumor necrosis factor; EOS = eosinophil; Neu = neutrophil; MAC = macrophage; MC = mast cell.

inflammation. For example, interleukin-5 (IL-5) enhances the viability and function of eosinophil; in addition IL-5 may reduce eosinophil density.[3] The hypodense eosinophil appears to be functionally upregulated and, when activated, generates greater quantities of superoxide and LTC_4.[4] Granulocyte-macrophage colony-stimulating factor (GM-CSF) also promotes the development and activation of eosinophil and other leukocytes. Interferon-Gamma (INF-γ) activates macrophages and increases their oxidative metabolism. IgE production by B-cells is promoted by IL-4 and downregulated by INF-γ.[5] From these properties, it is obvious that T-cell cytokines can modulate functions of other inflammatory cells that are likely relevant to the development of airway inflammation.

Several studies suggest T-cell involvement in airway inflammation in asthma. In tissue obtained postmortem, large numbers of lymphocytes are found in airways of patients with asthma.[6] Further, morphological studies of bronchial biopsies in subjects with mild asthma reveal increased numbers of atypical interepithelial lymphocytes.[7] Moreover, when bronchial biopsies in atopic asthma subjects with mild disease are compared to normal and atopic nonasthmatic subjects, asthma subjects are found to have significant increases in T-lymphocyte that express the activation surface marker CD25 (IL-2 receptor).[8]

The eosinophil has also emerged as a critical cell in the pathogenesis of asthma. Evidence for eosinophil involvement in asthma is found on many fronts. First, a clinical association exists between asthma severity and the presence of eosinophils. For example, both peripheral blood eosinophils and eosinophilic infiltration of bronchial biopsies are increased in proportion to airway responsiveness[6]; with treatment of asthma and improvement in airway obstruction, eosinophil numbers decrease. In addition, the biological properties of the eosinophils can be linked to the pathophysiology of asthma including the release of granular proteins, platelet activating factor (PAF), leukotriene C_4 (LTC_4) and oxygen metabolites. Eosinophil associated granule proteins include major basic protein (MBP), eosinophil peroxidase (EPO), and eosinophil cationic protein (ECP), and have many properties that could be central to mechanisms of asthma. For example, they damage human airway epithelium, promote bronchial hyperresponsiveness, cause mast cell and basophil secretion, and directly contract airway smooth muscle.[9-11]. PAF is produced by several inflammatory cells, including eosinophils and neutrophils, and can cause increased bronchial responsiveness, enhanced eosinophil accumulation in the lung, increased microvascular leak, and stimulated mucus secretion. In humans, PAF has been shown to cause bronchoconstriction and enhanced methacholine responsiveness by some investigators.[12] Sulfidopeptide leukotrienes (LTC_4, LTD_4, LTE_4) are produced by eosinophils as well as mast cells and macrophages. They are very potent bronchoconstrictors[13] and stimulants of mucus secretion in human airway. Finally, toxic reactive oxygen metabolites are released by eosinophils and other phagocytic cells, and are capable of causing tissue damage directly or indirectly. Their involvement has been postulated in spontaneous and antigen induced asthma exacerbations. Therefore, present and accumulating data strongly implicate eosinophils and their products as important contributors to the pathophysiology of asthma: airway obstruction, bronchial inflammation, and hyperresponsiveness.

While evidence for a role for the eosinophil in asthma has become quite well-established, mast cell involvement in asthma, and more particularly chronic airway inflammation, is undergoing reevaluation. However, there is also much evidence to suggest a pivotal role for mast cells in asthma. Anatomically, pulmonary mast cells are strategically located in the airway (i.e., in the airway lumen, bronchial epithelium, submucosa, and lung parenchyma) to participate in lung allergic reactions. Furthermore, there is convincing evidence that pulmonary mast cells are activated in asthma following antigen challenge and release their potent airway smooth muscle modifying mediators.[14,15]

Finally, although neutrophils have a well-defined role in many pulmonary diseases, their participation in asthma and airway hyperresponsiveness is less well-established. Evidence exists from animal[16] and human studies[17] that neutrophils are capable of causing airway injury and are likely important in some types of bronchial asthma or as initiators of the inflammatory process.

In nocturnal asthma, the precise contribution of the eosinophil and the mast cell to this process is still emerging, and that of the neutrophil remains controversial. These cells appear to function as effector cells in an inflammatory cascade. In this chapter, we will first review the biology of mast cells, eosinophils, and neutrophils, with attention to how these cells can contribute to asthma. With this information as a background, we will then discuss how these cells may participate in nocturnal asthma.

Mast Cells

Mast Cell Biology

The mast cell contains preformed mediators, such as histamine and tryptase, and has the ability to generate sulfidopeptide leukotriene compounds when activated. Both histamine and mast cell derived leukotrienes can cause immediate airway smooth muscle contraction. Tryptase, like histamine, is exocytosed during mast cell activation and, because it is unique to the mast cell, serves as a more specific marker of activation by allergen.[18,19] In addition to releasing histamine and tryptase, mast cell activation initiates synthesis of the sulfidopeptide leukotrienes: LTC_4, LTD_4, and LTE_4[20]; these arachidonate products are potent airway smooth muscle contractile substances. New information on other mast cell derived products may explain more precisely how they could participate in the late, or inflammatory, allergic response. Murine mast cell activation generates a variety of cytokines, including GM-CSF, IL-3, IL-5 and

IFN-γ.[21–23]. These cytokines, if generated and released by human pulmonary mast cells, could have profound effects on the recruitment, priming, and activation of neutrophils, monocytes, and eosinophils that are recruited to the airway. Thus, the mast cell's role in the allergic reaction should no longer be viewed as limited to the acute phase response, but may also be considered as a contributor to the development and perpetuation of the late inflammatory response and bronchial hyperresponsiveness.

Relationship of Mast Cell to Asthma

Evidence that mast cells are activated in asthma and that their products participate in this process has been derived from many observations. Bronchoprovocation with inhaled antigen, or airway segmental instillation of allergen, causes an acute release of histamine that is detected in the circulation[24] and bronchoalveolar lavage (BAL) fluid.[25] There is also evidence that airway fluid concentrations of histamine correlate to bronchial responsiveness[25] and degree of airflow obstruction[26]; these and other data indicate an early participation of pulmonary mast cell mediators in asthma and a possible role in the pathogenesis of airway disease.

Tryptase is detected in the bronchial lavage fluid of allergic asthma patients at baseline[15] and immediately following airway antigen challenge in allergic subjects.[14] The presence of tryptase substantiates mast cell activation by antigen, and increases the likelihood that mast cell products participate in the development of acute bronchial obstruction. Furthermore, there is evidence that tryptase can enhance bronchial responsiveness, which raises the possibility of an expanded role for mast cell mediators in asthma.

The role of these products in clinical asthma has begun to become more apparent. For example, LTD_4 antagonists prevent exercise induced asthma.[27] Moreover, inhibitors of the 5-lipoxygenase pathway modify the upper airway allergic reaction[28] and diminish bronchoconstriction from isocapnic hyperventilation.[29]. Finally, antagonists of LTD_4 cause immediate bronchodilation and diminish airway obstruction that is not fully modulated by beta-agonists.[30,31] Consequently, it appears that leukotrienes participate in the asthma process and contribute to bronchomotor tone. What quantity of pulmonary tissue leukotrienes is derived from the mast cell versus other cells is, at present, unknown.

The role of pulmonary mast cells in the late or inflammatory asthmatic reaction is less well-established. However, mast cells contain chemotactic substances which attract eosinophils and neutrophils to the airway where they are presumably activated to cause tissue injury and "late" airway obstruction. Evidence for mast cell participation in late phase pulmonary obstruction is presently indirect. Activation of cutaneous mast cells by F(ab)'2 anti-IgE leads eventually to a late phase reaction with the histologic features of inflammation.[32] Although this observation suggests mast cell participation in the cutaneous late phase reaction, there is evidence that macrophages and lymphocytes may also be activated by allergen via Fc_ϵ RII on their surface membranes and contribute to the late reaction.[33,34]

Eosinophils and Asthma

Eosinophil Biology

Eosinophils are bilobed granulocytes that are derived from the bone marrow and distinguished by their cytoplasmic granules, which have an affinity for acid aniline dyes such as eosin.[35] Eosinophils mature in the bone marrow under the influence of a number of factors including interleukin-2 (IL-2), IL-3, IL-5, and GM-CSF.[36,37] The cytoplasmic granules of the eosinophil contain preformed mediators which are released upon stimulation. In addition, the eosinophil is able to synthesize newly generated mediators (Table 2). The biology of these substances as they may relate to asthma are reviewed below.

Table 2

Preformed and Newly Formed Mediators From Eosinophils

Preformed Mediators
    Major Basic Proteins (MBP)
    Eosinophil Cationic Protein (ECP)
    Eosinophil Derived Neurotoxin (EDN)
    Eosinophil Peroxidase (EPO)

Generated Mediators
    Prostaglandin E_2 (PGE_2)
    Prostacyclin (PGI_2)
    Platelet activating factor (PAF)
    Leukotriene C_4 (LTC_4)

Preformed Mediators

Major Basic Protein (MBP)

MBP, a 14,000 dalton protein, is the most abundant eosinophil cationic protein and is localized to the crystalloid core of the granule. MBP is rich in arginine residues and has a large number of sulphydryl groups which increase its affinity to aggregate to surfaces.[38] MBP is toxic to parasites and tumor cells[39,40] and can cause damage and loss of guinea pig[41] and human respiratory epithelium.[42] Moreover, MBP can abolish ciliary activity of respiratory epithelium[41,42] and stimulate airway epithelial chloride and water secretion[43] which could, in turn, contribute to the decreased mucous clearance.

To more definitively establish a role for these products, Gundel et al.[44] measured the effects of direct intratracheal installation of eosinophil granule proteins on airway responsiveness. Various eosinophil granule proteins were instilled into trachea of monkeys; this intervention was followed by measurement of airway resistance and methacholine responsiveness. Only eosinophil MBP caused a significant and dose related increase in airway responsiveness. The other eosinophil proteins (ECP, EDN, EPO) did not affect airway responsiveness. This suggests a direct role for eosinophil MBP in augmenting airway responsiveness.

Eosinophil Cationic Protein (ECP)

ECP is a 21,000 dalton basic protein that is also localized to the matrix of eosinophil granules.[40] The secreted form of ECP has been associated with activated eosinophils in skin, gastrointestinal tract, heart, and spleen, and is detected by staining with a specific monoclonal antibody, EG-2. ECP is toxic to parasites[46] and may also play a role as an inhibitor of lymphocyte proliferation.[47]

Eosinophil Derived Neurotoxin (EDN)

EDN is a protein of 18,000 dalton, and similar to ECP. It is named for its ability to induce cerebrocerebellar dysfunction when injected into the cerebral spinal fluid of rabbits.[48] While toxic to parasites, the role of EDN in pathogenesis of asthma is as yet unestablished.

Eosinophil Peroxidase (EPO)

EPO is a two polypeptide protein with a combined molecular weight of 70,000 daltons. In the presence of H_2O_2, EPO is toxic to parasites, bacteria, tumor cells, and host cells.[36] EPO also inactivates leukotrienes and binds to mast cells and basophils with retention of its enzymatic activity.[49]

Newly Formed Mediators

Eosinophils synthesize and release newly formed mediators in response to cell activation. Activation of arachidonic acid metabolism leads to the generation of prostaglandin E2 (PGE_2) and prostacyclin (PGI_2). Eosinophils also produce LTC_4 via the 5-lipoxygenase (5-LO) pathway after a glutathione conjugation of LTA_4.[50] LTC_4 and its derivatives LTD_4 and LTE_4 are potent stimulators of smooth muscle constriction and mucous secretion. Furthermore, eosinophil secretion of LTC_4 is enhanced by exposure of the eosinophil to LTB_4, PAF, IL-3, IL-5, or GM-CSF.[51] There is also evidence that eosinophils generate PAF; whether this product is released following cell activation has not been established.

Relationship of Eosinophils to Asthma

Antigen challenge of asthma patients provokes airway obstruction shortly after the challenge, i.e., the immediate asthmatic response; this obstructive response peaks in 10 to 30 minutes postchallenge and usually resolves within 1 to 2 hours. A second phase, the late asthmatic response (LAR), appears 3 to 8 hours following antigen challenge in about half of the subjects challenged, when the challenge is done during the daytime. LAR is characterized by bronchoconstriction and an increase in airway responsiveness. Compared to bronchoconstriction in the immediate reaction, airway obstruction in the LAR is often more protracted and less responsive to therapy. Pathologically, these changes correlate with a cellular infiltration into airway mucosa and the lumen, exudation and edema. All of these changes are characteristics of airway inflammation and recognized as hallmarks of chronic asthma. Thus, LAR has served as a model for studying mechanisms of chronic inflammation in asthma. Clearly, there are shortcomings with this approach, and creating clear demarcations between the immediate and late phase may be artificial; nonetheless, the use of allergen challenge with an examination of the late response has been a helpful model. Although LAR manifests physiologically at 3 to 8 hours postchallenge, BAL fluid which is commonly performed 24 to 72 hours following an antigen challenge, continue to show evidence of airway inflammation: increased airway macrophages, eosinophils, total proteins, and mediators (histamine, LTC_4, PGD_2), and cellular activation (e.g., higher superoxide release and chemiluminescence).

The LAR also provides a model for evaluating the contribution of eosinophils to airway inflammation and bronchial hyperresponsiveness. Initially, deMonchy et al.[52] used bronchoalveolar lavage and found striking airway eosinophilia in patients with late reactions to inhaled antigen. Furthermore, Frick et al.[53] demonstrated a rise in peripheral blood low density eosinophils in patients experiencing a LAR following inhaled antigen; presumably, the low density eosinophil phenotype represents an upregulated cell.[54,55] These data suggest that there may be a link between the appearance of the LAR and circulating eosinophils. More specifically, they raise the interesting possibility that this component of the allergic reaction is marked by an increase in an upregulated subpopulation of eosinophils.

To expand upon these observations and focus on events in the lung, Sedgwick et al.[14] used segmental bronchoprovocation (SBP) with antigen to characterize more specifically the allergic response in the airway and eosinophil involvement in this process. Lavage samples obtained immediately following antigen SBP revealed a rise in histamine and tryptase; no change in cellular components (i.e., eosinophils, neutrophils, or macrophages) was found in the immediate reaction. When BAL was repeated 48 hours postantigen challenge, the concentration of tryptase, a mast cell mediator, was close to the baseline concentrations; histamine concentration remained somewhat elevated, although not as dramatically as in the immediate fluid. In contrast, a marked increase in cellular components was found, with the rise in eosinophils being the most striking change. Furthermore, the 48-hour lavage samples had dramatic increases in LTC_4 and eosinophil granular proteins: MBP, ECP, EPO, and EDN.

These observations imply that the late airway response to antigen (48-hour) is characterized by an inflammatory cellular infiltrate, with eosinophils being the predominant cell type. In addition, the presence of eosinophil granular proteins in the lavage fluid is presumptive evidence that not only are cells recruited to the airway, but they are also activated and release their inflammatory products. The finding of these eosinophil products supports the possibility that this cell participates in bronchial inflammation and hyperresponsiveness associated with antigen challenge.

The mechanisms that control selective eosinophil recruitment to the lung have not been fully established, nor has the precise method of eosinophil activation. However, there are several possibilities that could explain cell activation under these settings. First, some eosinophils have low affinity IgE membrane receptors that could be activated by antigen.[56] Second, IgA complexes, which may also be present in the airway, are potent triggers of eosinophil granule protein release and could participate in this process.[57] Third, cell activation can follow attachment of eosinophil to epithelium via specific adhesion proteins.[58] Fourth, the eosinophil chemoattractant, PAF, can prime[59] and also stimulate degranulation.[60] Finally, cytokines, including IL-5, enhance release of eosinophil granule proteins.[56] Most likely it is a combination of various events which actually causes eosinophil activation.

Another important, and perhaps pivotal, step in this process is cell adhesion to airway epithelium. Wegner and Gundel[58] have shown that the preadministration of a monoclonal antibody to intercellular adhesion molecule-1 (ICAM-1) will block both airway eosinophilia and enhanced bronchial responsiveness associated with repetitive antigen challenges in monkeys. From current information it is apparent that the participation of the eosinophil in airway hyperresponsiveness is complex, and involves at least cell migration, upregulation, adhesion, and activation as interrelated events.

From these observations, it is apparent that the eosinophil is likely an important effector cell in allergic airway injury; its participation in this process is influenced, or determined, by a variety of factors, including T-cell products. For example, both GM-CSF and IL-5 activate mature eosinophils, cause increased cytotoxicity, stimulate oxidative metabolism, and prolong survival in culture.[61–63] IL-5 also has major *in vitro* effects on eosinophil differentiation, degranulation,[56] superoxide generation,[62,63] chemotaxis,[62] antibody dependent cytotoxicity,[63] density,[64] and survival in culture.[64] Sedgwick et al.[14] found increased IL-5 in BAL samples that were obtained 48 hours postantigen challenge which correlated with the degree of airway eosinophilia. The site and cell source of IL-5 was not established by Sedgwick.[14] Hamid et al.[65] used *in situ* hybridization to detect message for IL-5 in airway cells. Although T-lymphocytes are the probable source of this cytokine, other cell sources must be considered, including mast cells[21,22,66] and even eosinophils.[67,68] Therefore, the interaction of the airway with allergens and eventual eosinophilia represents a complex interplay with the generation of cytokines as, presumably, a major factor in promoting airway injury (Fig. 1). Moreover, the recognition that even eosinophils produce cytokines to further upregulate function indicates that this cell can establish a self-perpetuating process in asthma.

Neutrophils

Neutrophil Biology

Neutrophils provide effective host defense against bacterial and fungal infections.[69] To accomplish this action, they destroy fungal and bacterial elements through generation and release of oxidative metabolites and lysosomal enzymes. In addition, activated neutrophils can recruit other inflammatory cells to perpetuate the inflammatory cascade. The same oxidative and nonoxidative mechanisms that are beneficial in eliminating invading microorganisms can cause tissue damage, including that of the lung. More specifically, neutrophil induced tissue damage has been implicated in several respiratory diseases including adult respiratory distress syndrome, emphysema, and asthma.[69] To appreciate mechanisms of neutrophil induced oxidative tissue injury, it is helpful to review metabolism of reactive oxygen species in general, keeping in mind that other cells including eosinophil, monocyte, and alveolar macrophage are all rich sources of reactive oxygen species and can contribute to tissue damage.

Reactive Oxygen Species Metabolism

Following activation, neutrophils, eosinophils, and monocytes have marked changes in oxygen metabolism including increased oxygen uptake, production of large amounts of superoxide (O^-_2) and hydrogen peroxide (H_2O_2). Due to the marked increase in oxygen uptake (which can be up to fiftyfold) the process has been called the "respiratory burst."[70] This process requires a series of single electron transfers from nicotinamide-adenine dinucleotide phosphate (NADPH) to O_2 creating superoxide anion (O^-_2); in the presence of superoxide dismutase, most of superoxide dismutates rapidly into hydrogen peroxide (H_2O_2) and oxygen. Both superoxide and hydrogen peroxide have minimal microbicidal activity; however, they are metabolized into active oxidants in one of two pathways. The first involves the oxidation of halogens into hypohalous acids, (e.g., hypochlorite OCl^-) as catalyzed by myeloperoxidase; and the second produces oxidizing radicals via Fe^{++} dependent mechanism (e.g., hydroxyl radical OH·).[65] Of the neutrophil oxidative metabolites, it is believed that the hydroxyl radical is the most damaging toxic oxygen reactant to host tissue.

Oxygen radicals damage all cellular compounds including nucleic acids, proteins, lipids, and carbohydrates, and thus will affect various cell functions.[71] Most tissues contain a variety of compounds that function as antioxidants and limit the deleterious effects of oxygen radicals. Antioxidants include albumin, ceruloplasmin, α_2-macroglobulin, transferrin-lactoferrin, uric acid, glucose, vitamin E and haptoglobin-hemopectin.[71] Tissue injury may occur when the oxidative burden exceeds the available antioxidants.

Relationship of Neutrophils to Asthma

As discussed, airway inflammation is a consistent feature of asthma and is detected even when the disease is mild and the patient asymptomatic. A close association between airway inflammation and hyperresponsiveness in asthma patients has been noted in individuals exposed to ozone,[72] toluene diisocyanate,[73] or allergen.[52] Although mast cells and eosinophils are likely the most relevant cells in this process, neutrophils may also contribute to enhanced airway inflammation and responsiveness. Studies in dogs have shown that the neutrophil infiltration of airway after exposure to ozone is closely associated with, and possibly related to, the development of bronchial hyperresponsiveness.[16] In this model, neutrophil depletion prevented the development of airway hyperresponsiveness to ozone.[74]

Kelly et al. investigated the role of neutrophil in human asthma.[17] BAL was performed in 22 asthma patients and 20 control subjects to investigate the number and function of neutrophils in these individuals. They found that airway neutrophils from asthma patients had a twofold increase in chemiluminescence when compared to control subjects. Furthermore, airway responsiveness (FEV_1 PD_{20}) correlated with percentage of neutrophils in BAL. This study suggested that there is an association between neutrophil numbers and airway responsiveness.

Other studies have found that neutrophils are present in the large airway in significant numbers both in responsive and nonresponsive individuals, and that asthmatics have similar numbers of neutrophils to atopic and nonatopic controls.[75] Thus, despite their potential to cause significant tissue damage and the possible association between neutrophils and bronchial responsiveness, the evidence that neutrophils play an important role in human asthma is as yet not fully substantiated.

The Contribution of Mast Cells, Eosinophils and Neutrophils to Nocturnal Asthma

While the postulated role of the mast cell, eosinophil, and neutrophil in asthma has been evolving over the past few decades, their contribution to nocturnal worsening of asthma has been more difficult to outline because most studies investigating the mechanisms of asthma have been done during the daytime. Direct evaluation of the role of airway cells and mediators awaited the implementation of BAL in investigating asthma. This approach has begun to provide direct access to airway cells and mediators, thus allowing for a more accurate assessment of their contribution to asthma. It is anticipated that the expanded use of BAL and airway biopsy along with cell and molecular biological techniques of analysis will establish better understanding of the mechanism of asthma. The use of these investigative techniques has now begun to improve the understanding of the mechanisms controlling nocturnal asthma (Table 3). The following section will expand upon the basic knowledge of mast cell, eosinophil, and neutrophil biology and extend this information more specifically to nocturnal asthma. From these observations, it is becoming apparent that circadian patterns in cortisol and catecholamines are instrumental in the development of nocturnal asthma in selected individuals (Fig. 2); furthermore, it is likely that these circadian patterns in hormone levels will influence not only airway smooth muscle function but, perhaps more importantly, the function of the mast cell, eosinophil, and neutrophil.

Table 3

Changes in BAL Fluid at 0400 in Patients with Nocturnal Asthma

- Increased total leukocyte counts
- Increased neutrophil counts
- Increased eosinophil counts
- Increased airspace cells generation of superoxide
- Higher BAL fluid histamine levels

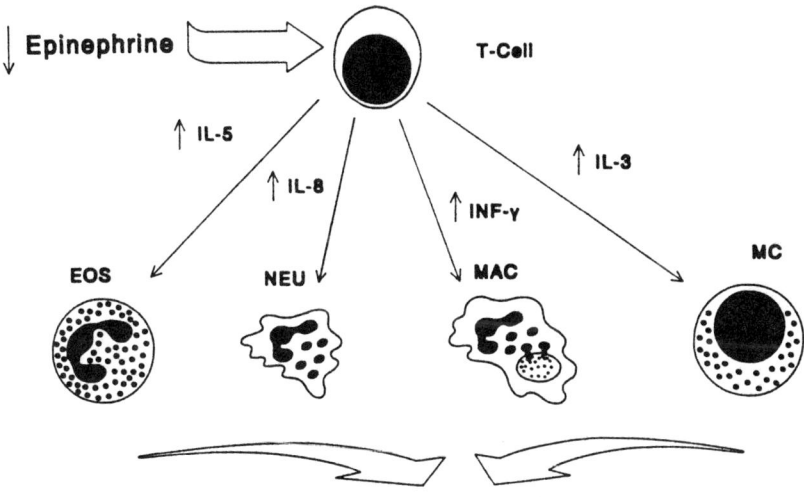

NOCTURNAL ASTHMA

Figure 2. A proposed postulate for inflammatory cell-cell interaction in nocturnal asthma. Lower epinephrine levels might allow for increased cytokine release from pulmonary T-cells with subsequent activation of other inflammatory cells. The products of cell activation lead to smooth muscle contraction, airway edema, and increased secretions, manifesting as nocturnal asthma.

The Role of Mast Cells in Nocturnal Asthma

As discussed above, the mast cell is a pivotal factor in the development of airway flow obstruction to inhaled antigen. Although the principal focus of the mast cell in the allergic reaction has been on the immediate response, its activation is likely important to the initiation of the late, inflammatory response. Furthermore, the airway response to allergen provides a model to assess mechanisms of allergic airway disease and mast cell involvement in this process. In studies of nocturnal asthma, evidence for mast cell involvement has been largely indirect, and often inferred from studies with antigen challenge. With these limitations, there are data to either directly link mast cell activators, or events associated with mast cell activation, to the development of nocturnal asthma.[76–79] When Davies et al.[77] exposed an asthma patient to grain dust during the day, acute airway obstruction developed; importantly, wheezing occurred nightly for the next week even though daytime pulmonary functions were normal. This reaction was felt to represent a recurring late asthmatic response that was only apparent at night.

Likewise, Cockcroft et al.[78] challenged two sensitive patients with Western red cedar extract and caused recurrent nocturnal asthma to develop over the next several days. Data from these studies suggest that a single antigen challenge, and activation of the pulmonary mast cell, can initiate a series of events to cause recurrent nocturnal wheezing in some patients. These observations also indicate that a late response to inhaled antigen is more likely to be manifest at night, but daytime exposure to antigen can serve as the inducer of nocturnal attacks of asthma. Precisely how mast cell function is linked to recurrent nocturnal asthma was not established in those studies.

However, other investigations have expanded upon this observation. Gervais et al.[79] challenged four patients with house dust mite antigen at various times over 24 hours. Airway obstruction was greater when the antigen exposure occurred at night. To further assess the time relationship of antigen provocation to intensity of airway obstruction, and particularly the late asthmatic reaction, Mohiuddin and Martin,[80] compared the effect of morning and evening antigen challenge on the frequency, time of onset, and severity of a LAR. Ten patients with mild asthma had four randomized inhalation challenges, each separated by two weeks: AM placebo

(0800), AM antigen (0800), PM placebo (2000) and PM antigen (2000). None of the patients had immediate or late asthmatic responses following either AM or PM placebo inhalation challenge. When the antigen challenge was performed at 0800, four of 10 developed a LAR; in contrast, when antigen challenge occurred at 2000, nine of 10 developed a LAR ($P<0.05$, Fig. 3). Moreover, the nocturnal (2000) antigen challenge caused a greater fall in FEV_1 during LAR than that associated with 0800 challenge (36% vs. 18%, $P<0.05$). Further, the duration of the LAR, i.e., the time the FEV_1 remained 15% below baseline, was 8.2 hours following 2000 challenge versus 5.3 hours for 0800 challenge. Finally, airway responsiveness postantigen challenge was significantly greater at 24 hours following evening antigen challenge when compared to the morning challenge

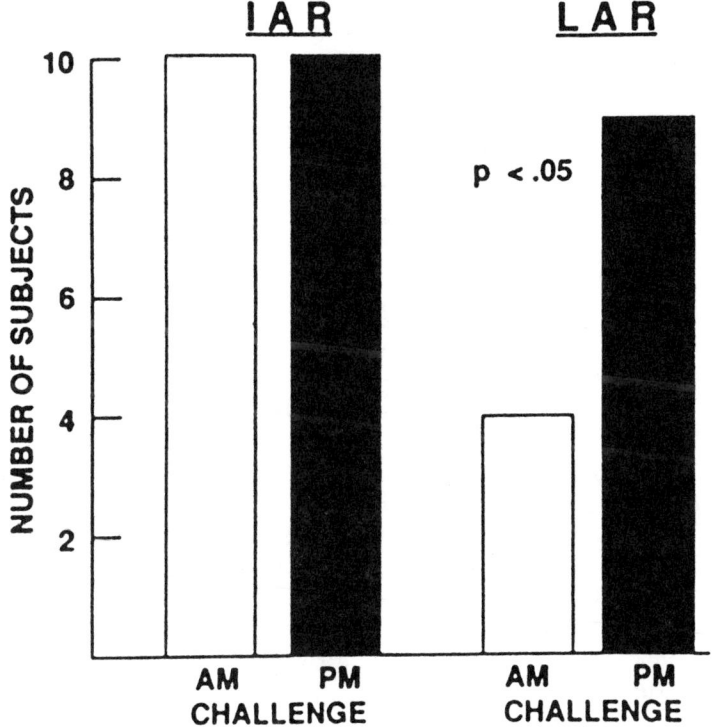

Figure 3. The occurrence of the immediate asthmatic response (IAR) and late asthmatic response (LAR) in the 10 subjects on both the morning (AM) and evening (PM) challenge days. Reprinted with permission.[80]

(P<0.05). Thus, late asthmatic reactions to antigen are seen more frequently, have a more prolonged duration of obstruction, and are more severe when antigen exposure occurs at night. This pivotal observation suggests that factors determined by normal circadian patterns likely promote and enhance the development of the late, inflammatory asthmatic response to antigen.

More direct evidence for the mast cell involvement in nocturnal asthma is suggested by studies that quantitated airway function and serum histamine concentrations over 24 hours. Barnes et al.[81] evaluated five patients with nocturnal asthma by measuring changes in peak expiratory flow (PEF) and concentrations of venous plasma epinephrine, cortisol, and histamine. Plasma epinephrine was lowest at 0400, a time patients also had their lowest PEF values. At 0400, plasma histamine levels were the highest (Fig. 4). Further, plasma histamine concentrations correlated inversely with peak flow (r = −0.94, P<0.001). These observations suggested to Barnes et al.[81] that the rise in plasma histamine (which presumably reflects airway values of this mediator) may be important to nocturnal airway obstruction. Since catecholamines may inhibit mast cell mediator release via the generation of intracellular cyclic AMP, the authors suggest that a fall in epinephrine at night has a permissive effect on sensitized lung mast cells and leads to enhanced release (or leakage) of histamine and other mediators which, in turn, cause nocturnal bronchoconstriction. Furthermore they showed that in unsensitized individuals, with presumably more "stable" mast cells, nocturnal changes in epinephrine were not associated with alterations in plasma histamine concentrations. More recently Szefler et al.[81a] reported finding a twofold higher plasma histamine concentration at 4 AM compared to 4 PM in normal subjects and those with asthma regardless of the presence of nocturnal asthma. The reasons for the different results in these two studies are unclear.

To further evaluate the role of mast cell derived mediators in this process, we performed BAL at 0400 and 1600 in five patients with nocturnal asthma; lavage concentrations of histamine were determined and compared to values detected in 10 asthma subjects *without* nocturnal exacerbations of airway obstruction.[82] In patients with nocturnal asthma, the BAL histamine at 0400 was significantly higher than in subjects without nighttime airway obstruction (Table 4). In contrast, there was no difference in BAL histamine values in these two groups when evaluated at 1600. Plasma histamine also showed a trend to being higher at 0400 in subjects with nocturnal

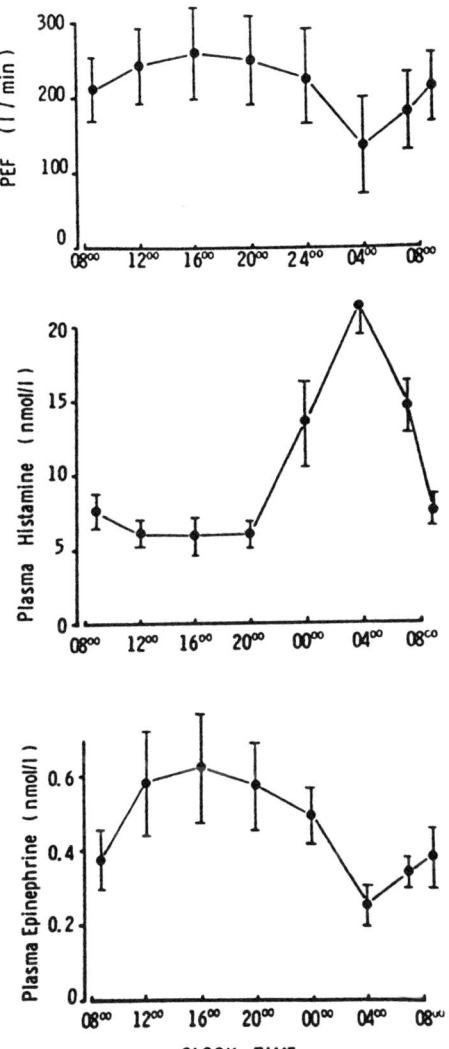

Figure 4. Mean-change (± SEM) in peak expiratory flow (PEF) (upper panel), plasma histamine (middle panel), and plasma venous plasma epinephrine (lower panel), over 24 hours in five patients with nocturnal asthma. Reproduced with permission.[81]

Table 4

BAL Histamine and Plasma Histamine in Subjects with and without Nocturnal Asthma

Histamine (pg/mL)	Nocturnal Asthma (n = 5)		No Nocturnal Asthma (n = 10)	
	0400	*1600*	*0400*	*1600*
BAL	2122 ± 1800*	281 ± 75	191 ± 38	209 ± 58
Plasma	288 ± 64	194 ± 56	235 ± 48	173 ± 36

*P < 0.05 versus no nocturnal asthma (0400)

asthma. Our studies do not indicate if the source of histamine was the basophil or mast cell.

Taken together, these studies suggest a role for mast cell participation in nocturnal asthma either by initiating airway obstruction (i.e., through allergen stimulation) and/or by perpetuation of the inflammatory cascade by further release of mediators. This latter event may occur as a consequence of diminished concentrations of epinephrine at night. The end result of these events is nocturnal worsening of airway obstruction. Why differences in mast cell function are found only in patients who experience nocturnal exacerbations of asthma is not established.

The Role of Eosinophil in Nocturnal Asthma

The eosinophil is a principal effector cell in airway inflammation, and peripheral blood eosinophil counts correlate with the severity of airway obstruction. Of recent interest has been the identification of eosinophil heterogeneity which is reflected in cell density, possibly representing a subpopulation with greater inflammatory potential. The importance of eosinophil subpopulations, as reflected by cell density, to asthma severity has been addressed in a number of ways. For example, asthma patients have been found to have a higher proportion of circulating hypodense eosinophils (HE), and the proportion of these cells correlates directly with the severity of their airway obstruction.[53] Since low-density eosinophils may have greater inflammatory potential, it was thought that a greater proportion of HE represented, or caused, more airway obstruction in asthma. Such findings led us to investigate the density profile of peripheral blood eosinophils in relationship in nocturnal asthma.

Calhoun et al.[83] studied fifteen asthmatic subjects on two occasions, at 0400 hours and at 1600 hours, each separated by 36 hours. Five subjects had nocturnal asthma as defined by a fall in FEV_1 of $\geq 15\%$ from 1600 hours to 0400 hours. The remaining 10 subjects had little circadian change in FEV_1 (FEV_1 AM to PM variability of less than 15%). In each subject, spirometry was done and a blood sample drawn at 0400 and 1600 hours. Eosinophils were then separated on multiple discontinuous density Percoll gradients. The most dramatic finding was a significant increase in the percentage of low-density eosinophil (≤ 1.090 gm/mL) at 0400 hours versus 1600 hours only in subjects with nocturnal asthma (Fig. 5). Furthermore, there was a significant correlation between the percentage of low-density eosinophils at 0400 hours and the overnight change in

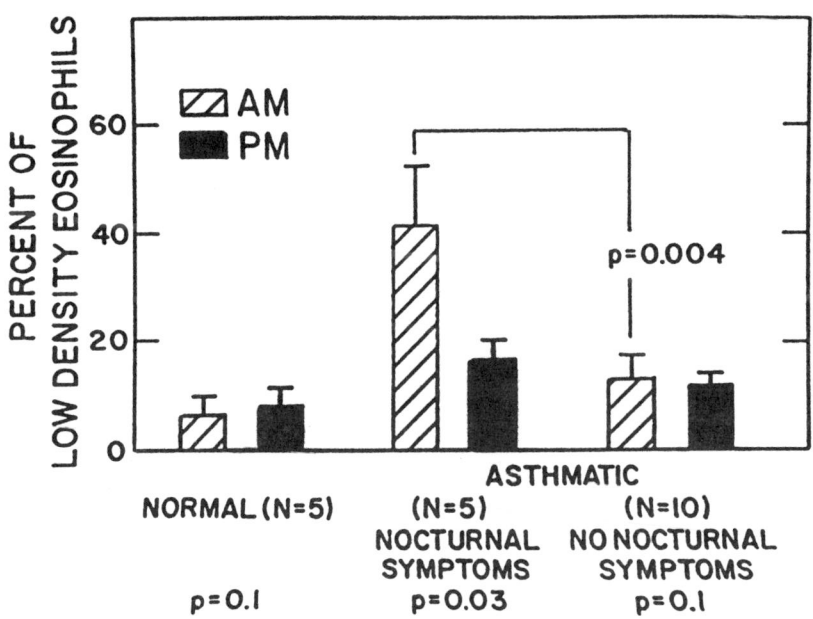

Figure 5. The percentage (mean ± SEM) of low-density eosinophils (≤1.090 g/mL) at 0400 (AM) and 1600 (PM) in normal and asthmatic patients. The difference between AM and PM values is indicated for each subject group. Reproduced with permission.[83]

FEV_1 ($r = 0.66$, $P = 0.002$). Finally, eosinophils at 0400 hours from subjects with nocturnal asthma had a tendency toward longer survival in culture than those recovered at 1600 hours; the longer survival represents a tendency towards greater functional inflammatory action. Both the eosinophil counts and the *in vitro* survival correlated significantly with the fall in FEV_1 ($P<0.03$ for both). From these observations it appears at this time that eosinophil upregulation occurs at night in selected patients, and these changes may contribute, or be associated with, increased airway obstruction.

Taken together, these studies suggest that there may be an overlap between the mechanisms of nocturnal asthma and those of LAR, both of which involve recruitment, upregulation, and activation of eosinophils. However, the peripheral blood is an insensitive representative of events in the lung, and direct evaluation of airway cells is needed to add greater relevance to our understanding of the mechanism in asthma.

To determine whether nocturnal asthma is associated with an increase in airway inflammatory cells at night, Martin et al.[84] obtained BAL fluid at 1600 and 0400 in asthmatic patients with ($n = 7$) and without ($n = 7$) nocturnal asthma. Nocturnal asthma was defined as a $\geq 20\%$ fall in PEFR from bedtime to morning awakening on three repeated nights. The cellular components of BAL fluid in the non-nocturnal asthma group did not change between 1600 and 0400. However, in patients with nocturnal asthma, there was a significant increase in the total leukocyte count, neutrophils, and eosinophils from 1600 to 0400 (Fig. 6). In addition, the overnight fall in PEFR was correlated with changes in airway eosinophils ($r = 0.77$, $P<0.05$). These findings further support the concept that nocturnal worsening of asthma is associated with a cellular inflammatory response and that the intensity of inflammatory changes at night determines the likelihood and intensity of bronchial obstruction.

To determine whether changes in BAL cellular components and their functions were related to nocturnal airway obstruction, we performed BAL at 0400 and 1600 in asthma subjects with ($n = 5$) and without ($n = 10$) nocturnal asthma.[82] No significant changes were observed in absolute total cells, percentage, or concentration of eosinophils or neutrophils from 1600 to 0400 in either group. Our differences with the observation of Martin et al.[84] may relate to patient selection of underlying severity of asthma. Importantly, spontaneous and stimulated superoxide anion generation by air-

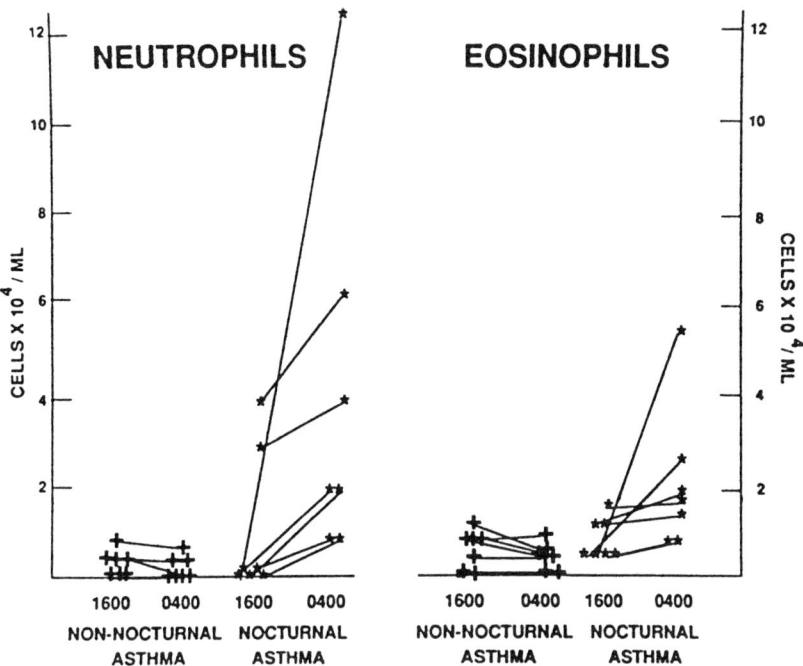

Figure 6. The neutrophil and eosinophil counts for each subject are shown. Between 1600 and 0400, the nocturnal asthma group had a significant increase in neutrophils ($P < 0.05$) and eosinophils ($P < 0.05$). Similarly, between groups at 0400 hours, the nocturnal asthmatics had a significant increase in neutrophils ($P < 0.01$) and eosinophils ($P < 0.05$). Reprinted with permission.[84]

space cells were significantly greater at 0400 compared to 1600 only in subjects with nocturnal asthma. Moreover, superoxide production of 0400 was greater in subjects with nocturnal asthma compared to individuals without nighttime airway obstruction (Fig. 7). In these subjects, the development of nocturnal airway obstruction in asthma was thus characterized by enhanced generation of oxygen radicals by airspace cells. Because oxygen radicals can influence airway function,[85] it is possible that the enhanced release of reactive oxygen compounds is causally associated with airway obstruction in nocturnal asthma. The mechanisms that regulate and determine cell recruitment and activation at night, particularly in asthma, are poorly defined and need clarification.

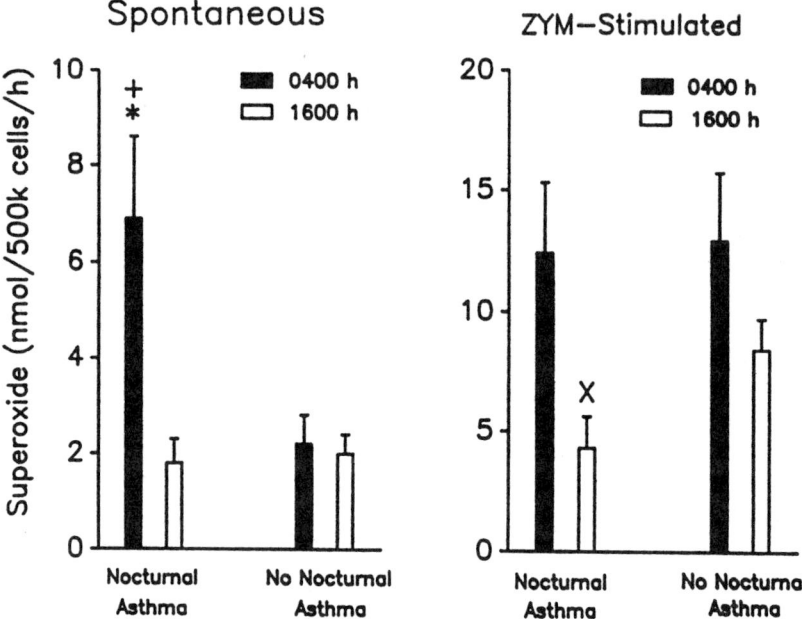

Figure 7. Superoxide anion release by airspace cells, spontaneous and zymosan (zym)-stimulated. In the nocturnal asthma group, spontaneous SO release was higher at 0400 hours compared to 1600 hours. Spontaneous SO release at 0400 hours was higher in subjects with, compared to those without, nocturnal asthma. ZYM-stimulated release at 1600 hours was lower in the group with nocturnal asthma. Subjects without nocturnal asthma showed no significant changes from 1600 hours to 0400 hours. Data are expressed as mean ± SEM. + = P<0.02, NA + vs. NA − group at 0400 hours; * = P<0.05, 0400 hours vs. 1600 hours in NA + group; X = P<0.02, NA + vs. NA − group at 1600 hours. Reprinted with permission.[82]

The Role of Neutrophil in Nocturnal Asthma

As discussed, activated neutrophils can produce a number of products that may damage the airway, produce bronchoconstriction, and enhance bronchial responsiveness. These products include leukotrienes, thromboxane B_2, oxygen radicals, and lysosomal enzymes.

In a study by Martin et al.,[84] the nocturnal asthma group had a threefold increase in the average neutrophil concentration in BAL fluid from 1600 to 0400; no changes in neutrophil numbers were seen in the nonnocturnal asthma group (Fig. 6). Furthermore, both 0400 neutrophil count and the 1600 to 0400 percent change in neutrophils correlated with the overnight fall in PEFR for the entire group (r = 0.54, P<0.05). This study suggested that overnight change in neutrophil counts may be an important factor in the pathogenesis of nocturnal asthma. Alternatively, these changes may simply reflect a heightened airway inflammatory response.

We sought to evaluate the importance of the changes in BAL cellular components and their function in a group of asthma patients in a study similar to that done by Martin et al.[84] BAL fluid was analyzed in asthma patients with (n = 5) and without (n = 10) nocturnal asthma at 1600 and 0400.[82] We observed a trend toward an increase in BAL neutrophils and eosinophils counts from 1600 for 0400; this occurred both in those with and without nocturnal asthma. As mentioned above, airspace oxidative cell metabolism was enhanced at 0400 only in those with nocturnal asthma. Furthermore, the change in generated superoxide from 1600 to 0400 was significantly correlated with the change in FEV_1 (r = -0.71, P<0.01) for the entire group (Fig. 8). Since neutrophils are such rich sources of superoxide and other oxygen radicals, it is possible that their

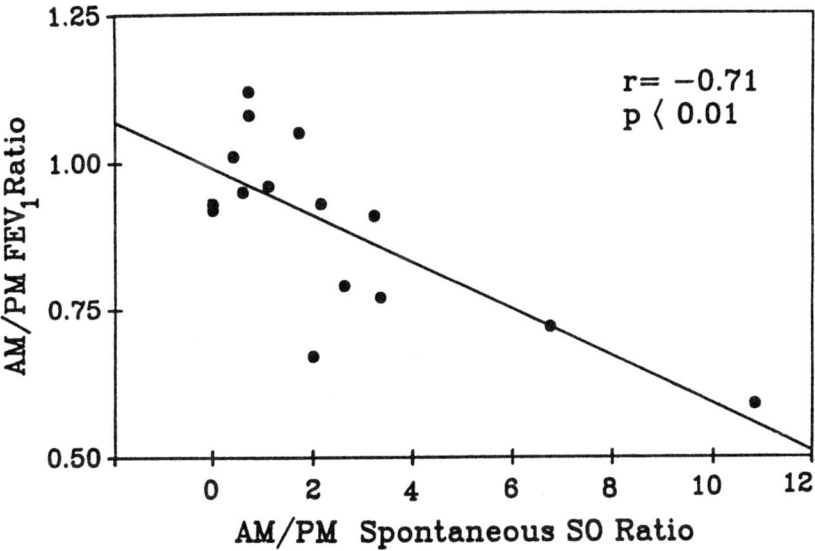

Figure 8. Correlation of overnight fall in FEV_1 and increase in spontaneous superoxide release at 0400 hours. A significant negative correlation was noted between changes in FEV_1 and spontaneous SO release from 1600 hours to 0400 hours in the entire group. Reprinted with permission.[82]

oxidative metabolism was enhanced at night in this group. However, other phagocytic cells like eosinophils and macrophages may be contributing to the observed increase in generated superoxide by airspace cells at night. The increase in superoxide may be relevant to the increased bronchoconstriction in asthma in view of the correlation between the overnight superoxide generation increase and the decline in FEV_1. Airspace metabolism of oxygen radicals is also enhanced in symptomatic asthma and correlates with clinical severity of disease,[85] and superoxide directly causes bronchoconstriction[86] lending further credibility to the possibility that this factor is involved in nocturnal worsening of asthma. It is possible that other functions of airspace cells, including those of neutrophils, are activated at night; thus, the combination of increased cell counts and function may enhance airway damage and constriction in those asthma patients with nocturnal asthma.

Circadian Changes in Endogenous Hormones and Their Relationship to Nocturnal Wheezing

Circadian variations in endogenous hormones have been linked to the development of nocturnal asthma and may be pivotal events determining this response. Barnes et al.[81] found a strong correlation between the circulating level of epinephrine and PEFR in five patients with nocturnal asthma (r = 0.97). PEFR reached trough values at 0400 and were maximum at 1600; the same was true for epinephrine (Fig. 4).

In view of the critical role that eosinophils have in asthma, we further evaluated their participation in nocturnal asthma with particular attention to an interaction with epinephrine and cortisol.[87] Ten patients with asthma were hospitalized for three days to monitor spirometry and blood eosinophil counts. Epinephrine and cortisol levels were obtained at 0400, 1000, 1600, and 2200 hours on each of the three consecutive days. As noted in previous studies, FEV_1 reached a nadir at 0400 hours and peaked at 1600 hours in subjects when nocturnal asthma occurred. Peripheral blood eosinophil count correlated with the average change in FEV_1 (%) over the three days (r = 0.67, P = 0.03), and with the frequency of days with nocturnal asthma (FEV_1 fall of 15% or more) (r = 0.74, P = 0.02). Although eosinophil counts were significantly higher in those with nocturnal asthma, there was not clear circadian variation in their values. Plasma epinephrine levels demonstrated a circadian rhythm in all patients, but a unique pattern emerged in patients when nocturnal asthma developed. Trough levels of epinephrine were noted at 2200 and 0400 and peak levels at 1000 hours in patients when nocturnal asthma occurred; in contrast, trough values occurred at 0400 and peak at 1000 on the nonnocturnal asthma days. The earlier fall (2200) in epinephrine, in patients experiencing nocturnal asthma, suggests that in addition to its direct smooth muscle regulatory effects, the decline in epinephrine may also promote inflammatory cell function a few hours later. This may occur through a number of mechanisms including regulation of inflammatory events including T-cell cytokine production, eosinophil migration activation, or a host of other factors.

A number of studies have begun to evaluate how circadian patterns in catecholamines can be important to the regulation of cytokine production by T-cells and the possible role that these changes may have on the occurrence of nocturnal asthma. Lymphocytes express beta$_2$-adrenergic receptors,[88] which suggest that beta-adrenergic agonists may regulate the function of lymphocytes (and possibly other leukocytes), in addition to their direct effect on airway smooth muscle. Moreover, stimulation of beta-receptors on lymphocytes markedly inhibits tritiated thymidine uptake in response to phytohemagglutinin (PHA).[89] Beta-adrenergic effects on cell functions are achieved by activation of adenylate cyclase which increases intracellular cAMP. It has been shown that the simultaneous stimulation of T-cells with isoproterenol and mitogen (PHA) results in a two- to fourfold increase in cAMP production compared to isoproterenol alone,[90] and IL-2 production by cultured human T-cells is markedly inhibited by agonists that increase intracellular cAMP concentration (stimulation of adenylate cyclase, inhibition of phosphodiesterase, etc.).[91] Thus these data suggest that stimulation of lymphocyte beta-receptors by endogenous catecholamines could increase intracellular cAMP levels, which then leads to reduced lymphocyte production of cytokines. Nocturnal falls in catecholamines can translate into a loss of this lymphocyte regulation, and enhanced cytokine production and release. The end result of these events should be heightened asthma.

Glucocorticoids, which are also pivotal in the management of chronic asthma, have profound inhibitory effects on T-cell function. Peripheral blood mononuclear cells from asthma patients, sensitive to corticosteroid therapy, showed significant inhibition of PHA-driven growth in the presence of methylprednisolone, whereas cells from corticosteroid resistant patients showed much less growth inhibition by this drug.[92] Clinically, the effects of corticosteroids on lymphocyte functions are usually seen after several hours. Thus, the contribution of circadian changes in cortisol levels to changes in airway function might not be as important as the contribution of catecholamine changes, since the latter can influence cell function over the course of minutes.

A Unifying Hypothesis for Mechanisms of Nocturnal Asthma

The specific roles of eosinophils, mast cells, and neutrophils in nocturnal asthma have begun to emerge. Other cells like lymphocytes and macrophages are also potentially important in this process and, no doubt, interact with other inflammatory cells. While definitive conclusions as to the cause of nocturnal asthma cannot be made from available data, a plausible hypothesis has evolved that centers on activation and recruitment of eosinophils, neutrophils, and mast cells to the airway. From emerging evidence, this process is most likely influenced by cytokines.

During the late phase skin reaction, there is evidence for recruitment and activation of T-lymphocytes as well as infiltration of eosinophils; a strong correlation exists between the numbers of recruited CD4 T-cells and activated eosinophils found 24 hours postintradermal antigen challenge.[93] It is tempting to extend the information in the skin and suggest that a similar sequence of events occurs following antigen challenge in asthma: T-cell activation leading to eosinophil recruitment and activation. Thus, if correct, the T-cell becomes a, or the, central factor in regulating and expressing inflammation of the airways, and may be pivotal in processing the initiating factor for asthma exacerbations.

Since T-cell is capable, through the release of various cytokines, of recruiting and activating other inflammatory cells including mast cell, eosinophil, macrophage, and neutrophil, and because of the known inhibitory effects of catecholamines and cortisol on T-cells, we propose that changes in the levels of these circulating hormones influence airway inflammation in asthma by modulating pulmonary T-cell release of proinflammatory cytokines, thus causing a circadian change in function of eosinophils, neutrophils, mast cells, and macrophages. Decreased levels of catecholamines and corticosteroids at night allow for enhanced pulmonary T-cell function which, in turn, leads to activation of airway inflammatory cells at night. The effect of these changes is translated into bronchospasm, airway edema, and increased secretion (Fig. 2).

Conclusion

Circadian variations in circulating catecholamines and cortisol are well established. In asthma, these hormonal changes correlate with circadian changes in airway obstruction, responsiveness, and inflammation. Several inflammatory cells participate in airway inflammation in asthma. Much remains to be discovered about the exact contribution of each cell to nocturnal asthma. However, currently available data suggest that eosinophils and neutrophils are recruited to the airway at night and activated in patients with nocturnal asthma. In addition, evidence suggests that histamine release from mast cells and/or basophils is increased at night in association with nocturnal exacerbation of asthma. The regulation of these changes and the exact contribution of circadian changes in endogenous hormones and other factors to changes at the level of the airway tone remain to be established. In view of the considerable importance of T-cell in directing airway inflammation, it is plausible that T-cell directs enhanced airway inflammation in nocturnal asthma as a consequence of diminished endogenous catecholamines at night, which in turn leads to increased T-cell production of proinflammatory cytokines which upregulate the functions of other airway inflammatory cells. With increased concentrations of proinflammatory cytokines, cell recruitment and activation, nocturnal asthma is more likely.

References

1. Turner-Warwick M. Epidemiology of nocturnal asthma. Am J Med 1988; 85:6S–8S.
2. Barnes PJ. A new approach to the treatment of asthma. N Engl J Med 1989; 321:1517–1527.
3. Rothenberg ME, Petersen J, Stevens RL. IL-5 dependent conversion of normodense human eosinophils to the hypodense phenotype uses 3T3 fibroblasts for enhanced viability, accelerated hypodensity, and sustained antibody-dependent cytotoxicity. J Immunol 1989; 144:1484–1487.

4. Lopez AF, Sanderson CJ, Gamble JR, Campbell HD, Young IG, Vadas MA. Recombinant human interleukin-5 is a selective activator of human eosinophil function. J Exp Med 1988; 167:219–224.
5. Teale JM, Abraham KM. The regulation of antibody class expression. Immunol Today 1987; 8:122–126.
6. Bousquet J, Chanez P, Lacoste JY, Barneon G, Ghavanian N, Enander I, Venge P, et al. Eosinophilic inflammation in asthma. N Engl J Med 1990; 323:1033–1039.
7. Jeffery PK, Wardlaw AJ, Nelson FC, Collins JV, Kay AB. An ultrastructural, quantitative study and correlation with hyperreactivity. Am Rev Respir Dis 1989; 140:1745–1753.
8. Azzawi M, Bradley B, Jeffrey PK, Frew AJ, Wardlaw AJ, Knowles G, Assoufi B, et al. Identification of activated T-lymphocytes and eosinophils in bronchial biopsies in stable atopic asthma. Am Rev Respir Dis 1990; 142:1407–1413.
9. Gleich GJ. The eosinophil and bronchial asthma: current understanding. J Allergy Clin Immunol 1990; 85:422–436.
10. White SR, Ohno S, Munoz NM, Gleich GJ, Abrahams C, Solway J, Leff AR. Epithelium-dependent contraction of airway smooth muscle caused by eosinophil MBP. Am J Physiol 1990; 259:L294–303.
11. Gundel RH, Letts LG, Gleich GJ. Human eosinophil major basic protein induces airway constriction and airway hyperresponsiveness in primates. J Clin Invest 1991; 87:1470–1473.
12. Rubin AH, Smith LJ, Patterson R. The bronchoconstrictor properties of platelet-activating factor in humans. Am Rev Respir Dis 1987; 136:1145–1151.
13. Drazen JM, Austen KF. Leukotrienes and airway responses. Am Rev Respir Dis 1987; 136:985–998.
14. Sedgwick JB, Calhoun WJ, Gleich GJ, Kita H, Abrams JS, Schwartz LB, Volovitz B, et al. Immediate and late airway response of allergic rhinitis patients to segmental antigen challenge: characterization of eosinophil and mast cell mediators. Am Rev Respir Dis 1991; 144:1274–1281.
15. Wenzel SE, Fowler AA III, Schwartz LB. Activation of pulmonary mast cells by bronchoalveolar allergen challenge. Am Rev Respir Dis 1988; 137:1002–1008.
16. Holtzman MJ, Fabbri LM, O'Byrne PM, et al. Importance of airway inflammation for hyperresponsiveness induced by ozone. Am Rev Respir Dis 1983; 127:686–690.
17. Kelly C, Ward C, Stenton CS, Bird G, Hendrick DJ, Walters EH. Number and activity of inflammatory cells in bronchoalveolar lavage fluid in asthma and their relation to airway responsiveness. Thorax 1988; 43:684–692.
18. Caughey GH. Roles of mast cell tryptase and chymase in airway function. Am J Physiol 1989; 257:L39–L46.
19. Caughey GH, Lazarus SC, Viro NF, Gold WM, Nadel JA. Tryptase and chymase: comparison of extraction and release in two dog mastocytoma lines. Immunology 1988; 63:339–344.

20. Drazen JM, Austen KF. Leukotrienes and airway responses. Am Rev Respir Dis 1987; 136:985–998.

21. Plaut M, Pierce JH, Watson CJ, Hanley-Hyde J, Nordon RP, Paul WE. Mast cell lines produce lymphokines in response to cross-linkage of Fc_E RI or to calcium ionophores. Nature 1989; 339:64–67.

22. Burd PR, Rogers HW, Gordon JR, Martin CA, Jayaraman S, Wilson SD, Dvorak AM, et al. Interleukin-3-dependent and independent mast cells stimulated with IgE and antigen express multiple cytokines. J Exp Med 1989; 170:245–257.

23. Wodnar-Filspowicz A, Heusser CH, Moroni C. Production of haemopoietic growth factors GM-CSF and interleukin-3 by mast cells in response to IgE-receptor-mediated activation. Nature 1989; 339:150–152.

24. Busse WW, Swenson CA. The relationship between plasma histamine concentrations and bronchial obstruction to antigen challenge in allergic rhinitis. J Allergy Clin Immunol 1989; 84:658–666.

25. Casale TB, Wood D, Richerson HB, Zehr B, Zavala D, Hunninghake GW. Direct evidence of a role for mast cells in the pathogenesis of antigen-induced bronchoconstriction. J Clin Invest 1987; 80:1507–1511.

26. Jarjour NN, Calhoun WJ, Schwartz LB, Busse WW. Elevated bronchoalveolar lavage fluid histamine levels in allergic asthmatics are associated with increased airway obstruction. Am Rev Respir Dis 1991; 144:83–87.

27. Manning PJ, Watson RM, Margolskee DJ, Williams VC, Schwartz JI, O'Byrne PM. Inhibition of exercise-induced bronchoconstriction by MK-571, a potent leukotriene D_4-receptor antagonist. N Engl J Med 1990; 323:1736–1739.

28. Knapp HR. Reduced allergen-induced nasal congestion and leukotriene synthesis with an orally active 5-lipoxygenase inhibitor. N Engl J Med 1990; 323:1745–1748.

29. Isreal E, Dermarkarian R, Rosenberg M, Sperling R, Taylor G, Rubin P, Drazen JM. The effects of a 5 lipoxygenase inhibitor on asthma induced by cold, dry air. N Engl J Med 1990; 323:1740–1744.

30. Hui KP, Barnes WC. Lung function improvement in asthma with a cysteinyl-leukotriene receptor antagonist. Lancet 1991; 337:1062–1063.

31. Gaddy JN, Bush RK, Margolskee D, Williams VC, Busse WW. The effects of a leukotriene D_4 (LTD_4) antagonist (MK-571) in mild to moderate asthma. J Allergy Clin Immunol 1990; 85:197(A).

32. Dolovich J, Hargreave FE, Chalmers R, Shier KJ, Gauldie J, Bienenstock J. Late cutaneous allergic responses in isolated IgE-dependent reactions. J Allergy Clin Immunol 1973; 52:38–46.

33. Melewicz FM, Spiegelberg HL. Fc receptors for IgE on a sub-population of human peripheral blood monocytes. J Immunol 1980; 125:1026–1031.

34. Joseph M, Tonnel AB, Topier G, Capron A, Arnoux B, Benveniste J. Involvement of immunoglobulin E in the secretory process of alveolar macrophages from asthmatic patients. J Clin Invest 1983; 71:221–230.

35. Dvorak AM, Ackerman SJ, Weller PF. Subcellular morphology and biochemistry of eosinophils. In: Blood Cell Biochemistry. Vol. 2. Megakaryocytes, platelets, macrophages and eosinophils, Harris JR (ed). Plenum Publishing, London, 1990 2, pp 37–344.
36. Holgate ST, Roche WR, Church MK. The role of the eosinophil in asthma. Am Rev Respir Dis 1991; 143:566–570.
37. Weller PF. The immunobiology of eosinophils. N Engl J Med 1991; 324:1110–1118.
38. Wasmoen TL, Bell MP, Loegering DA, Gleich GJ, Prendergast FG, McKean DJ. Biochemical and amino acid sequence analysis of human eosinophil granule major basic protein. J Biol Chem 1988; 263:12559–12563.
39. Butterworth AE, Wassom DL, Gleich GJ, Loegering DA, David JR. Damage to schistosomula of *Schistosoma mansoni* induced directly by eosinophil major basic protein. J Immunol 1979; 122:221–229.
40. Gleich GJ, Adolphson CR. The eosinophilic leukocyte: structure and function. Adv Immunol 1986; 39:177–253.
41. Frigas E, Loegering DA, Gleich GJ. Cytotoxic effects of the guinea pig eosinophil major basic protein on tracheal epithelium. Lab Invest 1980; 42:35–43.
42. Frigas E, Loegering DA, Solley GO, Farrow GM, Gleich GJ. Elevated levels of eosinophil granule major basic protein in the sputum of patients with bronchial asthma. Mayo Clin Proc 1981; 56:345–353.
43. Jacoby DB, Viki IF, Widdicombe JM, Loegering DA, Gleich GJ, Nadel JA. Effect of human eosinophil major basic protein on ion transport in dog tracheal epithelium. Am Rev Respir Dis 1988; 137:13–16.
44. Gundel RM, Letts LG, Gleich GJ. Human eosinophil major basic protein induces airway constriction and airway hyperresponsiveness in primates. J Clin Invest 1991; 87:1470–1473.
45. Olsson I, Persson AM, Winqvist I. Biochemical properties of the eosinophil cationic protein and demonstration of its biosynthesis *in vitro* in marrow cells from patients with an eosinophilia. Blood 1986; 67:498–503.
46. McLaren DJ, McKean JR, Olsson I, Venge P, Kay AB. Morphological studies on the killing of schistosomula of *Schistosoma mansoni* by human eosinophil and neutrophil cationic proteins *in vitro*. Parasite Immunol 1981; 3:359–373.
47. Peterson CGB, Skoog V, Venge P. Human eosinophil cationic proteins (ECP and EPX) and their suppressive effect on lymphocyte proliferation. Immunobiology 1986; 171:1–13.
48. Fredens K, Dahl R, Venge P. The Gordon phenomenon induced by the eosinophil cationic protein and eosinophil protein X. J Allergy Clin Immunol 1982; 70:361–366.
49. Henderson WR, Jony EC, Klebanoff SJ. Binding of eosinophil peroxidase to mast cell granules with retention of peroxidase activity. J Immunol 1980; 124:1383–1388.
50. Owen WF Jr, Soberman RJ, Yoshimoto T, Sheffer AC, Lewis RA, Austen KF. Synthesis and release of leukotriene C4 by human eosinophils. J Immunol 1987; 138:532–538.

51. Weller PF, Lee CW, Foster DW, Corey EJ, Austen KF, Lewis RA. Generation and metabolism of 5-lipoxygenase pathway leukotrienes by human eosinophils: predominant production of leukotriene C_4. Proc Natl Acad Sci USA 1983; 80:7626–7630.

52. de Monchy JGR, Kauffman HF, Venge P, Koeter GH, Jansen HM, Sluiter HJ, de Vries K. Bronchoalveolar eosinophilia during allergen-induced late asthmatic reactions. Am Rev Respir Dis 1985; 181:373–376.

53. Frick WE, Sedgwick JB, Busse WW. The appearance of hypodense eosinophils in antigen-dependent late phase asthma. Am Rev Respir Dis 1989; 139:1401–1406.

54. Prin L, Charon M, Capron M, Gosset PL, Taelman H, Tonnel AB, Capron A. Heterogeneity of human eosinophils. Variability of respiratory burst activity related to cell density. Clin Exp Immunol 1984; 57:735–742.

55. Kajita A, Yui Y, Mita H, Tanaguchi N, Saito H, Mischima T, Shida T. Release of leukotriene C4 from human eosinophils and its relationship to cell density. Int Arch Allergy Appl Immunol 1985; 78:406–410.

56. Fujisawa T, Abu-Ghazaleh R, Kita H, Sanderson CJ, Gleich GJ. Regulatory effect of cytokines on eosinophil degranulation. J Immunol 1990; 144:642–646.

57. Abu-Ghazaleh, Fujisawa T, Mestecky J, Kyle RA, Gleich GJ. IgA-induced eosinophil degranulation. J Immunol 1989; 142:2393–2400.

58. Wegner CD, Gundel RH, Reilly P, Haynes N, Letts LG, Rothlein R. Intercellular adhesion molecule-1 (ICAM-1) in the pathogenesis of asthma. Science 1990; 247:456–459.

59. Zoratti EM, Sedgwick JB, Bates ME, Vertis RF, Geiger K, Busse WW. Platelet activating factor primes human eosinophil generation of superoxide. Am J Resp Cell Mol Biol 1992;6:100–106.

60. Kroegel C. Yukawa T, Dent G, Venge P, Chung KF, Barnes PJ. Stimulation of degranulation from human eosinophils by platelet-activating factor. J Immunol 1989; 142:3518–3526.

61. Clutterbuck EJ, Hurst EMA, Sanderson CJ. Human interleukin-5 (IL-5) regulates the production of eosinophils in human bone marrow cultures: comparison and interaction with IL-1, IL-3, IL-6 and GMCSF. Blood 1989; 73:1504–1512.

63. Yamaguchi Y, Hayashi Y, Sugama Y, Miura Y, Kasahara T, Kitamura S, Torisu M, et al. Highly purified murine interleukin 5 (IL-5) stimulates eosinophil function and prolongs in vitro survival. J Exp Med 1988; 167:1737–1742.

64. Rothenberg ME, Petersen J, Stevens RL, Silberstein DS, McKenzie DT, Austen KF, Owens WF Jr. IL-5 dependent conversion of normodense human eosinophils to the hypodense phenotype uses 3T3 fibroblasts for enhanced viability: accelerated hypodensity and sustained antibody-dependent cytotoxicity. J Immunol 1989; 143:2311–2316.

65. Hamid Q, Azzawi MA, Ying S, Moqbel R, Wardlaw AJ, Corrigan CJ, Bradley B, et al. Expression of mRNA for interleukin-5 in mucosal bronchial biopsies from asthma. J Clin Invest 1991; 87:1541–1546.

66. Wodnar-Filspowicz A, Heusser CH, Moroni C. Production of haemopoi-

etic growth factors GM-CSF and interleukin-3 by mast cells in response to IgE-receptor-mediated activation. Nature 1989; 339:150–152.

67. Lopez AF, Sanderson CJ, Gamble JR, Campbell HD, Young IG, Vadas MA. Recombinant human interleukin-5 is a selective activator of human eosinophil function. J Exp Med 1988; 167:219–224.

67. Moqbel R, Hamid Q, Ying S, Barkens J, Hartnell A, Tsicopoulos A, Wardlaw AJ, et al. Expression of mRNA and immunoreactivity for the granulocyte/macrophage colony-stimulating factor in activated human eosinophils. J Exp Med 1991; 174:749–752.

68. Kita H, Ohnishi T, Okubo Y, Wecler D, Abrams JS, Gleich GJ. Granulocyte/macrophage colony-stimulating factor and interleukin-3 release from human peripheral blood eosinophils and neutrophils. J Exp Med 1991; 174:745–748.

69. Malech HL, Gallin JI. Neutrophils in human disease. N Engl J Med 1987; 317:687–694.

70. Babior BM. The respiratory burst of phagocytes. J Clin Invest 1984; 73:599–601.

71. Cross CE, Halliwell B, Borish ET, Pryor WA, Ames BN, Saul RL, McCord JM, et al. Oxygen radicals and human disease. Ann Int Med 1987; 107:526–545.

72. Stelzer J, Bigby BG, Stulbarg M, Holzman MJ, Nadel JA, Ueki IF, Leikauf GD, et al. O_3-induced change in bronchial reactivity to methacholine and airway inflammation in humans. J Appl Physiol 1986; 60:1321–1326.

73. Boschetto P, Fabbri LM, Zocca E, Milani G, Pivirotto F, DalVecchio A, Plevani M, et al. Prednisone inhibits late asthmatic reactions and airway inflammation induced by toluene diisocynate in sensitized subjects. J Allergy Clin Immunol 1987; 80:261–267.

74. O'Bryne PM, Walters EH, Gold BP, Aizawa HA, Fabbri LM, Alput SE, Nadel JA, et al. Neutrophil depletion inhibits airway hyperresponsiveness induced by ozone exposure. Am Rev Respir Dis 1984; 130:214–219.

75. Wardlaw AJ, Dunnette S, Gleich GJ, Collins JV, Kay AB. Eosinophils and mast cells in bronchoalveolar lavage in mild asthma. Relationship to bronchial hyperreactivity. Am Rev Respir Dis 1988; 137:62–69.

76. Busse WW. Pathogenesis and pathophysiology of nocturnal asthma. Am J Med 1988; 85:24–29.

77. Davies KJ, Green M, Schofield NM. Recurrent nocturnal asthma after exposure to grain dust. Am Rev Dis 1976; 114:1011–1019

78. Cockcroft DW, Hoeppner VH, Werner GD. Recurrent nocturnal asthma after bronchoprovocation with western red cedar sawdust: association with acute increase in non-allergic bronchial responsiveness. Clin Allergy 1984; 14:61–68.

79. Gervais D, Reinberg A, Gervais C, et al. Twenty-four-hour rhythm in the bronchial hyperreactivity to house dust in asthmatics. J Allergy Clin Immunol 1977; 59:207–213.

80. Mohiuddin AA, Martin RJ. Circadian basis of the late asthmatic response. Am Rev Respir Dis 1990; 142:1153–1157.

81. Barnes P, Fitzgerald G, Brown M, Dollery C. Nocturnal asthma and

changes in circulating epinephrine, histamine, and cortisol. N Engl J Med 1980; 303:263–267.

81a. Szefler SJ, Ando R, Cicuto LC, Surs W, Hill MR, Martin RJ. Plasma histamine, epinephrine, cortisol, and leukocyte β-adrenergic receptors in nocturnal asthma. Clin Pharmacol Ther 1991; 49:59–68.

82. Jarjour NN, Calhoun WJ, Busse WW. Enhanced metabolism of oxygen radicals in nocturnal asthma. Am Rev Respir Dis 1992;146:905–911.

83. Calhoun WJ, Bates ME, Schrader L, Sedgwick JB, Busse WW. Characteristics of peripheral blood eosinophils in patients with nocturnal asthma. Am Rev Respir Dis 1992;145:577–581.

84. Martin RJ, Cicutto LC, Smith HR, Ballard RD, Szefler SJ. Airways inflammation in nocturnal asthma. Am Rev Respir Dis 1991; 143:351–357.

85. Cluzel M, Damon M, Chanez P, Bousquet J, dePaulet AC, Michel FB, Godard P. Enhanced alveolar cell luminol dependent chemiluminescence in asthma. J Allergy Clin Immunol 1987; 80:195–201.

86. Katsumata U, Miura M, Ichinose M, Kimura K, Takahashi T, Inoue H, Takishima T. Oxygen radicals produce airway constriction and hyperresponsiveness in anesthetized cats. Am Rev Respir Dis 1990; 141:1158–1161.

87. Geiger K, Bates ME, Schrader L, Shultz T, Calhoun WJ, Busse W. Mechanisms of nocturnal asthma: The effect of diurnal concentrations of circulating epinephrine. J Allergy Clin Immunol 1991; 87:211.

88. Conolly ME, Greenacre JK. The β-adrenoreceptor of the human lymphocyte and human lung parenchyma. Br J Pharmac 1977; 59:17–23.

89. Hadden JW, Hadden EM, Middleton E. Lymphocyte blast transformation. Cell Immunol 1970; 1:583–595.

90. Carlson SL, Brooks WH, Roszman TL. Neurotransmitter-lymphocyte interactions: dual receptor modulation of lymphocyte proliferation and cAMP production. J Neuroimmunol 1989; 24:155–162.

91. Mary D, Aussel C, Ferrua B, Fehlmann M. Regulation of Interleukin-2 by cAMP in human T-cells. J Immunol 1987; 139:1179–1184.

92. Poznansky MC, Gordon ACH, Douglas JG, Karjewski AS, Whyllie AH, Grant IWB. Resistance to methylprednisone in cultures of blood mononuclear cells from glucocorticoid-resistant asthmatic patients. Clin Sci 1987; 639–645.

93. Frew AJ, Kay B. The relationship between infiltrating CD_4 lymphocytes, activated eosinophils, and the magnitude of the allergen-induced late phase cutaneous reaction in man. J Immunol 1988; 141:4158–4164.

6

The Macrophage and Lymphocyte in Nocturnal Asthma: Potential Roles and Importance

Sally E. Wenzel, M.D., and Richard J. Martin, M.D.

Introduction

As more knowledge is gained about the inflammatory cellular lung process occurring in asthma, we will be able to better understand which cell or cells initiate and/or control this phenomenon. We hope that by better understanding the control of inflammation in asthma, improved therapeutic regimens can be developed and drug toxicity minimized. For these reasons we have dedicated a chapter in this text to the alveolar macrophage and lymphocyte (alveolar and blood), as either one or both of these may be the "captain(s) of the ship." We apologize to the reader for the brevity of this chapter, as our knowledge of these cells and their role in asthma is just beginning to be understood.

For each cell, we will first discuss in general terms what is known about its role in asthma. Then we will focus on the area of

Martin RJ (editor): *Nocturnal Asthma: Mechanisms and Treatment,* © Futura Publishing Co., Inc., Mount Kisco, NY, 1993.

nocturnal asthma and circadian variation for that cell. Much of this later section will contain preliminary but interesting research information. We will then conclude with speculation about a potential scenario for control of the nocturnal (sleep related) inflammatory process.

The Macrophage in Asthma

The macrophage is an extremely complicated cell with the potential to function in a variety of capacities (Table 1). It can function as an accessory cell for the presentation of antigen to T-cells for immune processing and recognition, as well as control the influx of inflammatory cells through its multiple chemotactic factors. Additionally, it can play a role in the modulation of an inflammatory response through release of cytokines and growth factors, and it can act directly as an effector cell through its production of oxygen radicals and eicosanoids. Somewhat uniquely, the macrophage is a

---

Table 1

The Multiple Potential Roles for the Alveolar Macrophage in Asthma

---

I.   Accessory cell for immune activation
    A. Processes and presents antigen to T-lymphocytes for initiation of lymphocyte response
    B. Bears IgE receptors on surface for activation by specific allergen

II. Controls influx of inflammatory cells through production of chemotactic factors
    A. Leukotriene B4
    B. Interleukin-8
    C. Platelet-activating factor
    D. Platelet-derived growth factor

III. Modulates immune response through cytokines and growth factors
    A. Interleukins 1, 6, 8
    B. Interferon α, β, and γ
    C. Tumor necrosis factor
    D. Transforming growth factor $-\beta$
    E. Granulocyte macrophage colony-stimulating factor
    F. Monocyte colony-stimulating factor

IV. Direct effector cell
    A. Superoxide production
    B. Lysosomal enzymes
    C. Eicosanoids such as leukotriene C_4 and thromboxane

pleuripotential cell capable of undergoing differentiation to various functionally different forms or phenotypes (Fig. 1). These specific phenotypes may be present under specific stages of inflammatory or quiescent conditions, modulating the inflammation at each stage. At its most simple interpretation, these cells can be divided into inflammatory or "elicited" cells and "resident" cells, i.e., those found at a given tissue location in the normal steady state. The role that the various functions and phenotypes play in asthma in general, or in nocturnal asthma in particular, remains largely unknown.

It is believed by many investigators that the macrophage plays a

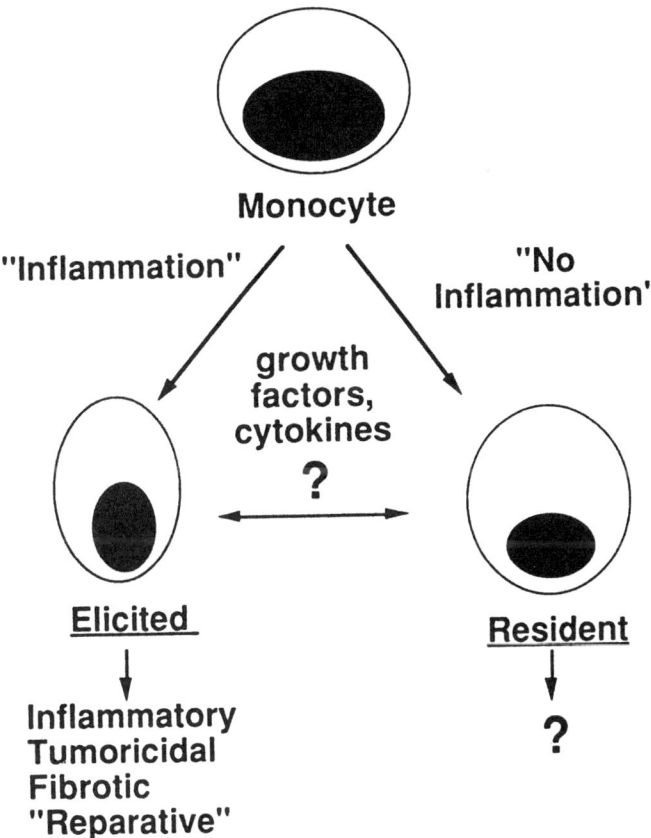

Figure 1. Effect of the state of inflammation on the differentiation (or phenotype) of the macrophage.

major modulatory role in asthma. This is particularly true for allergen induced asthma (Fig. 2). Macrophages and the presumed precursor cell, the blood monocyte, bear low affinity IgE receptors (FεRεII, CD-23) on their surface which are present in higher numbers in atopic individuals than in normals.[1,2] Exposure to allergen appears to further upregulate the CD-23 receptor in susceptible individuals.[3] These cells can be primarily activated following exposure to allergen. However, unlike the high affinity IgE receptor found on mast cells and basophils, it appears that the macrophage/monocyte must be activated by immune complexes of both allergen and antibody.[4] In vitro sensitization and activation of macrophages with IgE and specific allergen leads to the release of lipid mediators, superoxides, and lysosomal enzymes.[5,6] Certainly, the macrophage may be activated by other mechanisms that are not IgE dependent. These mechanisms may include the active phagocytosis of particles, or activation through receptors for endotoxin or IgG.[7,8]

Once activated, the macrophage may produce and/or release a variety of substances, as already mentioned (Table 1). Studies to evaluate the activation state of the macrophage in asthma have been done exclusively during the day. Most studies that have suggested macrophage activation in asthmatics have been based on response to an ex vivo stimulus, either calcium ionophore, the phagocytic stimulus zymosan, or allergen/IgE. Therefore the studies may not be evidence of in vivo "activation," but rather a primary or secondary index of "stimulatability" to an ex vivo stimulus. Thus, the evidence to suggest that macrophages may be primarily activated during the day is limited. In support of activation are the observations that the degree of chemoluminescence (as a marker for the presence of reactive toxic oxygen species) at baseline prior to any further stimulation is increased in macrophages from asthmatics and that the production of interleukin-1 at baseline is also increased.[9,10] Perhaps most interesting is the very recent finding that the alveolar macrophage population from asthmatics is made up of a higher percentage of hypodense cells than is normally seen.[11] Hypodense cells are considered to be a marker for an "activated" cell, as first described in eosinophils.[12] This suggests that a different population of macrophages is present in asthmatics. In contrast, when evaluating response to an ex vivo stimulus (rather than baseline production), a variety of results have been obtained. Most of the studies have suggested that macrophages from allergic asthmatics respond

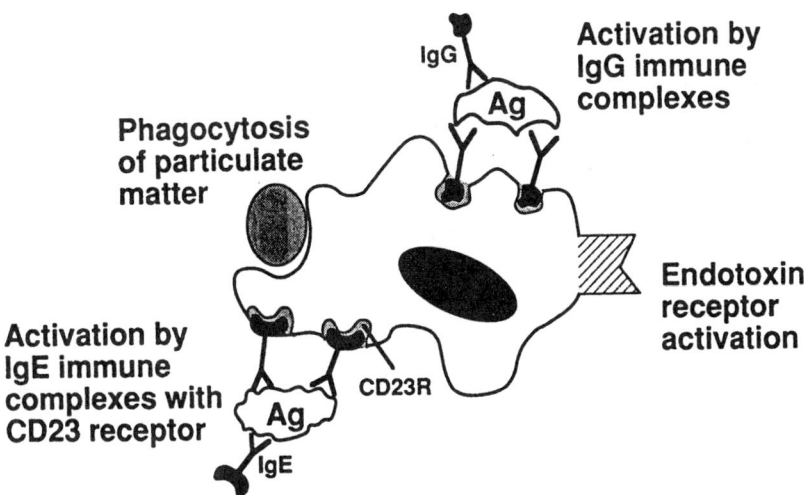

Figure 2. Macrophages may be activated by a number of different stimuli including phagocytosis of particles, activation by immune complexes, and through the endotoxin receptor.

to stimulus to a greater degree than normals.[13,14] This enhanced response in asthmatic subjects' macrophages was seen when evaluating superoxide and leukotriene production. In contrast, however, prostaglandin production appeared to be reduced compared to normals.[5] Similarly, alveolar macrophages from normal subjects respond to lipopolysaccharide with a nearly twentyfold increase in interleukin-1, while stimulation of alveolar macrophages from asthmatics failed to increase interleukin-1 production. These two findings implied a greater stimulatability for alveolar macrophages from normal subjects during daytime hours.[10] In contrast, Balter et al. evaluated eicosanoid production from asthmatics, allergic subjects without asthma, and normals, and could not detect any difference in measured amounts either at baseline or following stimulation.[15] Therefore the evidence regarding stimulatability is inconclusive for daytime hours. Some evidence suggests that the macrophage is more stimulatable in asthmatics than normals, others that it is less stimulatable, while some evidence suggests that there is no difference.

The data regarding alveolar macrophage function following allergen challenge, either endobronchially or inhaled, are sparse.

Activation of alveolar macrophages following endobronchial allergen challenge in atopic asthmatics has been suggested by demonstrating the release of the lysosomal enzyme β-glucuronidase into bronchoalveolar lavage fluid within several minutes of the challenge.[16] However, β-glucuronidase is not specific to macrophages, and is contained in neutrophils and epithelial cells as well. Direct study of the macrophage during the time of the immediate asthmatic response to allergen (IAR) has not been performed. Preliminary evidence evaluating the alveolar macrophage during and after the time of the late asthmatic response (LAR) suggests that the population or phenotype of the macrophage changes towards that of a newly migrated inflammatory cell.[17] That is, monocytic cells that contain peroxidase, an enzyme seen only in monocytes and immature monocytic phagocytes, appear to be present in increased quantities. In addition, a very recent study indicates that the alveolar macrophage may be the partial source for the cytokine, granulocyte macrophage colony-stimulating factor (GM-CSF) following allergen challenge.[18] In this study, allergic asthmatics were challenged endobronchially with allergen, and bronchoalveolar lavage (BAL) was performed 24 hours later. The cells obtained by BAL following allergen challenge demonstrated an increase in macrophages expressing mRNA for GM-CSF. This increase in positive cells was also seen among a subclass of lymphocytes, those bearing a receptor known as UCHL-1. Despite the greater percent positive lymphocytes than macrophages, the greater numbers of macrophages found in BAL fluid may make the alveolar macrophage output just as important. Further interactions between these cytokines, inflammation, and pulmonary physiological changes clearly need to be elucidated.

Other aspects of the alveolar macrophage function include its potential to secrete a variety of chemotactic factors, including leukotriene B_4 (LTB_4), platelet activating factor and interleukin-8 (IL-8), as well as one or more low molecular weight peptides (Fig. 3). While the macrophage production of IL-8 has been confirmed in interstitial lung disease,[19] the chemotactic properties of macrophage function have not been adequately explored in asthma. It does appear that the alveolar macrophage makes chemotactic substances, most likely LTB_4, following in vitro allergen challenge. This form of activation has not yet been demonstrated in vivo. However, there is some indirect pharmacological evidence to suggest the importance of macrophage LTB_4 production, based on the presence of the chemo-

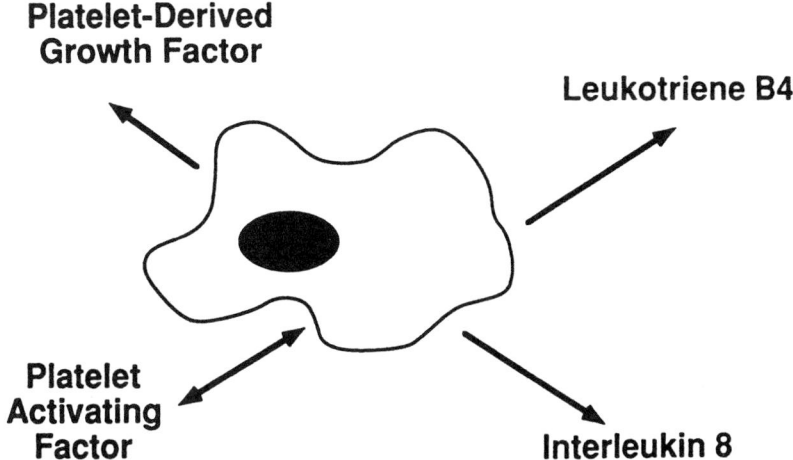

Figure 3. Macrophages are capable of generating and releasing a variety of chemotactic stimuli, some of which are shown here.

tactic substances in serum postallergen challenge, the very rough purification of these substances, and the sensitivity of these substances to steroid manipulation.[20]

Alveolar macrophages can also function as accessory cells or antigen presenting cells. They can process specific allergen/antigen to a form where it can be presented to the T-lymphocyte for activation of the T-cell response. Again, nothing is known about this role in asthmatics, but the potential for initiating and upregulating an inflammatory response is certainly there.

Lastly, and somewhat obtusely, the alveolar macrophage is known to be a highly steroid sensitive cell. To date, steroids are certainly the most effective drug available for the treatment of asthma. While steroids certainly may be affecting a wide variety of cells, the alveolar macrophage must be considered to be a prime target.

The Macrophage in Nocturnal Asthma

Very little data exists on the activity, function and/or phenotype of the alveolar macrophage in nocturnal asthma. Most data are highly preliminary. In our laboratory, we have been interested in determining the phenotype of the alveolar macrophage at night in nocturnal asthmatics. Our hypothesis is that newly migrated and/or inflammatory macrophages migrate to the lungs sometime during the night to initiate and/or participate in the inflammatory and physiological changes occurring during this time. This is certainly what is seen with neutrophils and eosinophils, as their numbers in BAL fluid increase at 4 AM in nocturnal asthmatics.[21] Macrophages increase in a similar fashion; however, the possibility exists that the macrophages are arising from stem cells in the lungs themselves, and are not necessarily part of the influxing inflammatory process.

In addition to the differentiation of phenotypes, we have further been interested in the activation state of these macrophages. Have these cells been activated in vivo by any of a variety of cytokines, growth factors, or combinations thereof, so that they are now producing and releasing more of the same or, in addition, eicosanoids, superoxides, etc? Or are they "primed," such that a small secondary stimulus will activate them?

In order to address these issues, we obtained alveolar macrophages from asthmatics with and without nocturnal symptoms, and normal controls at 4 AM.[22] To evaluate the inflammatory nature of these cells, we stained them for peroxidase content. It is known that monocytes and newly migrated monocytic phagocytes contain peroxidase in intracellular granules, while resident, noninflammatory cells, do not contain (or no longer contain) that enzyme. In BAL fluid from nocturnal asthmatics obtained at 4 AM, there appears to be a greatly increased percentage of peroxidase positive cells when compared to normal controls and asthmatics without nocturnal exacerbations, suggesting that these are newly migrated macrophages.

These macrophages were also evaluated for LTB_4 (as an example of a chemotactic substance) and thromboxane B_2 (TX B_2) (as an example of a macrophage derived bronchoconstrictor and inducer of hyperreactivity) production. This was done in the baseline, unstimulated state, and following stimulation with the calcium ionophore A23187. The baseline production of both LTB_4 and TX B_2 did not differ between nocturnal asthmatics and normal controls. However, following stimulation, there was a significant difference between the controls and the nocturnal asthmatics such that the nocturnal asthmatics produced less LTB_4 following stimulation with A23187 than did the controls. Given that there are also more "inflammatory" macrophages in the lungs of nocturnal asthmatics at 4 AM than controls, it implies either that these cells are less capable of generating LTB_4 or that these cells have already been activated and are continuously producing LTB_4 so that they are "exhausted" and not able to respond appropriately to further stimulus. No such differences existed for thromboxane.

In addition to eicosanoids, IL-1 has been measured in significant quantities in BAL fluid from nocturnal asthmatics, whereas it was not detectable in normal controls. While the cellular source for the IL-1 is not definitively known, in situ hybridization for IL-1 mRNA seems to demonstrate that the majority of the IL-1 is being produced by the alveolar macrophage.[23]

Effect of Prednisone on the Macrophage in Nocturnal Asthma

As noted earlier, macrophages (and lymphocytes) are highly steroid sensitive cells in vitro, and probably in vivo as well. In studies by Beam et al., it was determined that a single 3 PM dose of prednisone (50 mg) was capable of significantly blunting the fall in FEV_1 normally seen in nocturnal asthmatics at 4 AM, as well as decreasing the cellular influx.[24] Given the sensitivity of the macrophages and lymphocytes to steroids, we evaluated the effect of prednisone on the LTB_4 production of the alveolar macrophage in vivo. In the nocturnal asthmatics, there was a significant fall in stimulated LTB_4 production. In eight of nine controls and nonnocturnal asthmatics, there was actually an increase in stimulated LTB_4 production following 3 PM dosing of prednisone. These data would suggest that the macrophages are differentially susceptible to steroids depending on their state of activation and/or phenotype. This observation supports the peroxidase data and suggests that the macrophage population in BAL at 4 AM in nocturnal asthmatics must be different from non-nocturnal asthmatics and normals. Of note, the change in stimulated production of LTB_4 in asthmatics on prednisone versus placebo correlates significantly with the change in the FEV_1 on prednisone versus placebo, as well as with the change in cellular influx.[22]

In addition to eicosanoids, steroids also appear to blunt the production of interleukin-1 at 4 AM in nocturnal asthmatics. This was confirmed by measuring interleukin-1 levels directly in BAL fluid and by evaluating the cells for the presence of mRNA transcripts.[23] No comparisons of changes in interleukin-1 levels or mRNA with changes in pulmonary function were reported.

Speculation

Based on the data available from allergen-induced asthma and the preliminary data on nocturnal asthma, it is clear macrophages are contributing to the changes in the airways. However, the precise mechanisms and interrelationships with other cells are as yet unclear. One can speculate, however, that the alveolar macrophage and the lymphocyte (as well as the mast cell in allergic asthma) sit atop an inflammatory cascade, with the alveolar macrophage initiating, directing, and regulating the inflammatory response (Fig. 4). A circadian gene expression cycle, phagocytosis of aspirated particles or a form of late asthmatic response are only some of the possible mechanisms for triggering the alveolar macrophage either alone or in conjunction with lymphocytes. Additionally, the asthmatic macrophage may be chronically "primed" by a variety of cytokines (perhaps derived from lymphocytes) which may make it more susceptible to activation during the night. Once activated, the resident alveolar macrophage may then begin releasing chemotactic factors, such as LTB_4 and interleukin-8 which call in a host of inflammatory cells, including more monocytic phagocytes. In addition, it may release a variety of cytokines, including tumor necrosis factor-α, transforming growth factor-β and GM-CSF, each being generated and released, perhaps at different times, in response to a different combination of stimuli. The combination of stimuli will then determine the differentiation pathway that the alveolar macrophage will take which, in turn, controls the type of inflammatory response which occurs. The influx of newly migrated monocytic phagocytes from the blood brings in a new population of cells, which can further contribute to the inflammation, either directly or through release of more eicosanoids, growth factors, etc. In contrast, it is conceivable that the migration of monocytes serves to downregulate the inflammation through differentiation to a new "inhibitory" macrophage phenotype, thereby preventing the nocturnally-induced asthma from continuing into the daytime hours. Certainly, all of these hypotheses will need verification to further our understanding of the alveolar macrophage in nocturnal asthma.

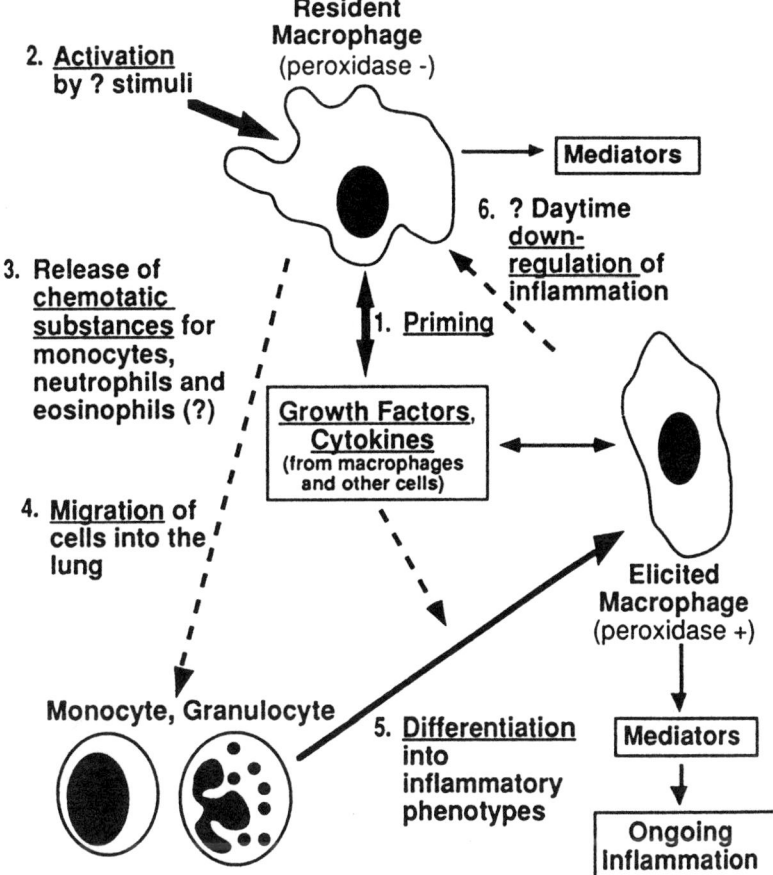

Figure 4. Speculation regarding the potential role for the alveolar macrophage in regulating the inflammatory influx into the lungs of nocturnal asthmatics. The figure should be read starting at point 1 and continuing counterclockwise through point 6.

The Lymphocyte in Asthma

From autopsy studies of asthma deaths, lymphocytes and eosinophils appear to be the prominent inflammatory cells found.[25] The lymphocyte is a fascinating cell that has the potential to control other cells as well as produce mediators of bronchoconstriction (Table 2). Over the last several years, the contribution of the

Table 2

Potential Role for the T-Lymphocyte in Asthma

I. Attract and activate effector cells shown to be involved in the asthma process
   A. Neutrophils
   B. Eosinophils
   C. Monocytes
   D. Basophils
   This is done via lymphokine secretion.
II. Lymphokines
   A. Further lymphocyte activation IL-2, IL-4
   B. Interaction with macrophages IL-3, GM-CSF
   C. Activate and recruit effector cells IL-3, IL-5, GM-CSF

lymphocyte to the pathogenesis of asthma has just started to be investigated. The lymphocyte can be divided into three distinct types. These include: the thymus derived or T-lymphocyte, the B-lymphocyte (those lymphocytes from the bone marrow that do not pass through the thymus), and those lymphocytes with natural killer activity (NK cells). NK activity is an inducible property of T-cells and of non B non T-lymphocytes. It appears that the T-lymphocyte is most prominently involved in the asthmatic process.

Separating the lymphocytes into these abovementioned subgroups can be done using flow cytometric techniques. Simply stated, flow cytometry involves passing a laser beam over a column of cells thus producing forward and side scattering of rays. Different size cells produce different scatter patterns, thereby grouping cells according to size. Once this is done, a "gate" around the cell population to be studied is made. Further differentiation of the specific population of cells is accomplished by monoclonal antibody staining for specific subsets. Thus, a T-lymphocyte can be differentiated from a granulocyte and further differentiated from a B-lymphocyte. The CD (cluster density) nomenclature from antibody staining shows that the T-lymphocytes are CD-3 positive cells while the B-lymphocytes are CD-19 positive. The T-lymphocytes can be further subtyped to "helper" or CD-4 positive cells and "cytotoxic/suppressor" or CD-8 positive cells. Additionally, specific receptor activation can be determined on these cells.

In studies of asthma and airways hyperreactivity, several investigations have suggested that T-lymphocytes play a role in this process. Lundgren et al. performed bronchial biopsies on patients with asthma and found a high number of inflammatory cells that consisted mostly of lymphocytes.[26] Additionally, Jeffrey and colleagues observed irregularly shaped lymphocytes in the lamina propria and submucosa of bronchial biopsies from symptomatic asthma patients.[27] From bronchoalveolar lavage (BAL) studies, CD-4 positive T-lymphocytes were markedly increased 48 hours after an inhaled antigen trial.[28,29] Other studies have shown a variability in the numbers and types of T-cell recovered in BAL fluid. These may relate to timing of the BAL postantigen challenge, the severity of the asthma, and/or the patient population. However, the T-lymphocyte still appears to play a prominent role in asthma.

If any cell is truly involved in a disease process, then it should be activated out of the resting state and producing mediators. The T-lymphocytes are capable of attracting and activating nonspecific effector cells or neutrophils, eosinophils, monocytes, and basophils,[30–33] all cells that are potentially involved in the inflammatory aspect of asthma. One marker of activation on lymphocytes associated with acute and atopic asthma patients is the interleukin (IL)-2 receptor, also called CD-25 positive lymphocyte.[34–36] Other activation markers on lymphocytes that have also been noted in association with acute asthma and increased bronchial responsiveness include HLA-DR and the very late activation antigen-1 (VLA-1).[36–37]

As stated above, autopsy studies of asthma deaths have a large proportion of lymphocytes and eosinophils. Additionally, a prominent feature of both intrinsic (nonallergic) and extrinsic (allergic) asthma is the presence of eosinophilia.[38] This eosinophilia appears to be related to decreased respiratory function and increased bronchial responsiveness. The eosinophil and the neutrophil are probably the end cells that produce the epithelial tissue damage associated with asthma. Lymphokines derived from T-cells may be instrumental in controlling immature and mature eosinophils and neutrophils.

Walker and colleagues[39] investigated whether peripheral blood eosinophilia correlated with T-lymphocyte activation and lymphokine production in asthmatic patients. They found that the increased IL-2 receptor (CD-25 positive) bearing T-cell subsets is a characteristic of asthmatic patients. The absolute number of IL-2 receptor positive T-cells correlated with the eosinophilia observed. In addition, the CD-4 and CD-8 positive T-cells from both intrinsic and extrinsic individuals spontaneously secreted factors that extend the lifespan of the eosinophil in vitro. These factors were identified as interleukin-5 and granulocyte macrophage colony-stimulating factor (GM-CSF). Their findings strongly support the role of T-lymphocytes and lymphokine secretion in the etiology of asthma.

The Lymphocyte and Nocturnal Asthma

As reviewed in detail in other chapters in this text, the nocturnal worsening of asthma is indeed common. Turner-Warwick has shown that in an outpatient population the worsening of asthma occurs between 39% for every night of the week and 74% for at least one night per week.[40] Additionally, when symptoms are mapped over a 24-hour period, independent of any medication, the vast majority of symptoms occur between 10 PM and 7 AM.[41] Martin and colleagues have shown that airways reactivity has a circadian rhythm to it whether the asthmatic subjects have nocturnal worsening of asthma or not.[42] The increase in bronchial reactivity is much greater at night in those asthmatics with nocturnal decrements in lung function.

The inflammatory aspect of nocturnal asthma is interesting. Marked increases in neutrophils and eosinophils can be seen in the BAL fluid at 4 AM compared to 4 PM in nocturnal asthma subjects.[21] In those asthmatics without this worsening of asthma at night, there is no increase in the cell populations. Both the increase in neutrophils and eosinophils in the 4 AM BAL fluid are significantly correlated to the overnight decrement in lung function. The BAL lymphocytes, although not as markedly increased between 4 PM and 4 AM, are also significantly higher at 4 AM than the control non-nocturnal asthma population.

Taking the circadian rhythms that occur and the potential for inflammation to be worse at night, we have also shown that the late asthmatic response is both more frequent and severe when an evening challenge is given.[43] That is, in mild asthmatics without nocturnal worsening of their asthma and in a controlled randomized fashion, the frequency of the late asthmatic response increased from the typical 40% when challenged during the daytime to 90% when challenged in the evening. Again, this points to the importance of the potential for inflammation to play a marked role in asthma on a circadian basis.

Figure 5 shows an electron micrograph of an endobronchial biopsy from a nocturnal asthma patient at 4 AM. Several lymphocytes can be seen even with the small endobronchial biopsy obtained

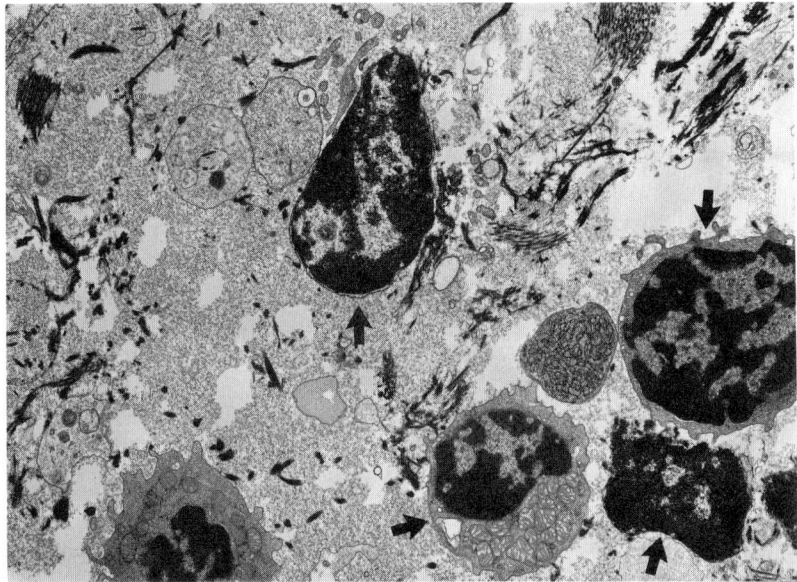

Figure 5. Electron micrograph from an endobronchial biopsy of the subcarina of the right lower lobe from a patient with nocturnal asthma. The biopsy was done via a fiberoptic bronchoscope at 4 AM. Arrows (4) point to a cluster of lymphocytes.

via the fiberoptic bronchoscope. Figure 6 is a representation of how the lymphocytes move from the peripheral or intravascular lymphocyte pool to the various organ systems. The majority of lymphocytes are recirculating and can be found in the bone marrow, spleen, lymph nodes, thoracic duct, and lung. The interruption by corticosteroids probably occurs at the area of the recirculating intravascular pool heading towards the tissues. This will be important in discussions below on the alteration of lymphocytes with steroids in individuals with nocturnal asthma. Figure 7 shows that the percentage of lymphocytes in the bronchoalveolar lavage fluid does correlate with the overnight fall in lung function. The higher the percentage of the BAL lymphocytes, the greater the decrement in lung function.

As discussed in greater detail in Chapter 8, a time-dependent dosing of oral corticosteroids may be very important in disrupting the inflammatory process and improving the overnight decrement in

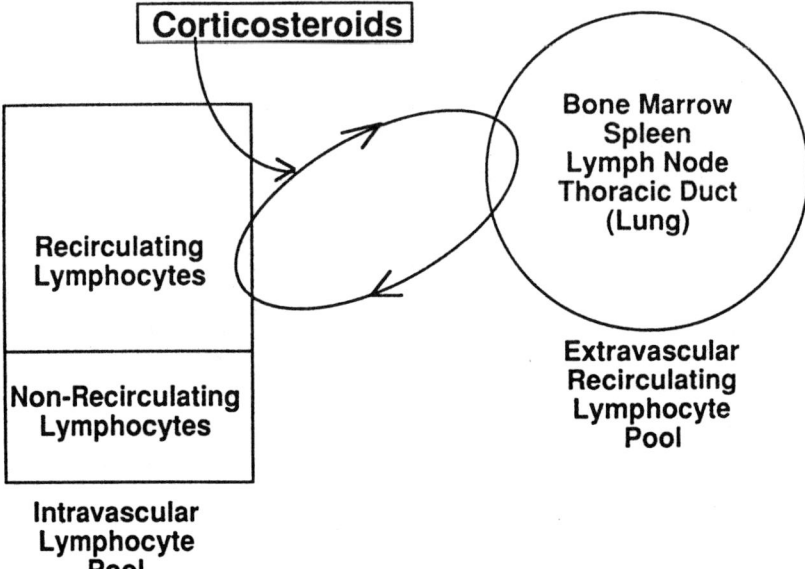

Figure 6. This figure demonstrates the possible trafficking of the blood lymphocyte to the lung and other organs. Corticosteroids probably interrupt this process where indicated.

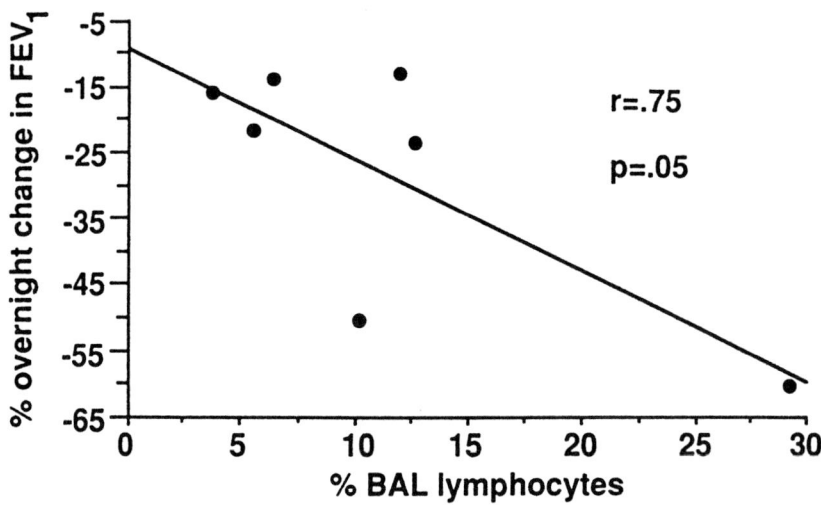

Figure 7. The percent of lymphocytes in bronchoalveolar lavage fluid at 4 AM from patients with nocturnal asthma correlates with the overnight fall in lung function.

FEV_1. Figure 8 shows that the 3 PM dosing compared to the 8 AM or 8 PM dosing is the only time when the overnight function is improved. This correlates very nicely with a pancellular decrement in BAL fluid at 4 AM from the 3 PM steroid dosing. The 8 AM or 8 PM dosing did not affect the inflammatory process in the lung. This certainly suggests that the 3 PM steroid dosing is probably interrupting the start of a cascade of events somewhere around 11 PM to midnight and preventing the nadir lung function at approximately 4 AM. It is of interest that in the peripheral blood the peak lymphocyte level is at 11 PM.[44]

The phenotypic lymphocyte subsets seen prominently at 4 AM in nocturnal asthma individuals are the T-lymphocytes. That is, there was no difference from placebo to steroid in regard to alteration of B-lymphocytes. The T-lymphocyte (CD-3 positive) had a significant decrement from placebo to steroid dosing as lung function improved. Of the T-lymphocytes the CD-4 positive (helper) cell showed a marked reduction with steroids, whereas the cytotoxic suppressor (CD-8 positive) lymphocytes were not altered. Looking at markers of

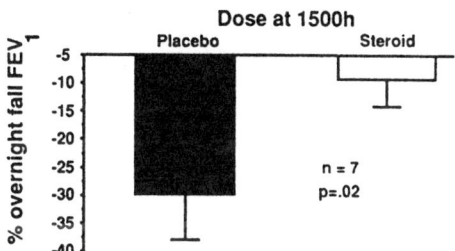

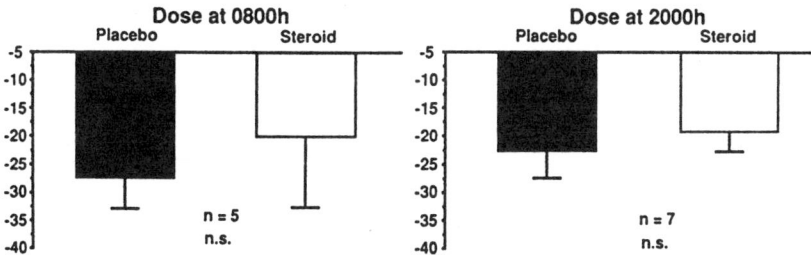

Figure 8. A 3 PM dosing of prednisone produces a significant reduction in the overnight decrement in lung function, whereas the 8 AM and 8 PM doses do not. Reproduced with permission from reference 24.

activation, it appears that the most prominent receptor present in nocturnal asthma was the IL-2 (CD-25 positive) receptor. This is similar to the findings for asthma in general, as discussed above. In fact, the IL-2 receptor markedly fell from placebo to steroid dosing. The percent alteration of the T-lymphocyte with the IL-2 receptor marker was significantly correlated (r = .82) with the percent improvement in overnight lung function on the steroids.

Although we are still in our infancy regarding our knowledge of the inflammatory process in nocturnal asthma, it appears that the lymphocyte is an important cell.

Speculation

Figure 4, discussed above, shows a detailed series of events that may be involved in asthma. Figure 9 is a simplified speculation of events that may initiate or play an important role in the nocturnal inflammatory process in asthma. This figure stresses the interactions between the macrophage and lymphocyte. We do know that IL-1 specific to the macrophage is increased at night and IL-1 will stimulate resting lymphocytes to become active. The IL-2 receptor marker appears to be increased at night in those individuals with nocturnal worsening of their asthma. IL-2 and IL-4 produced by the active lymphocyte can recycle to further stimulate lymphocytes. Interleukin-3 and GM-CSF can go back and forth between macrophage and lymphocyte for further activation. IL-3 and IL-5 plus GM-CSF can activate and recruit eosinophils and neutrophils and other asthma related effector cells. Again, the eosinophil and neutrophil are probably the cells that cause epithelial disruption and enhance bronchoconstriction.

IL-1 has multiple proinflammatory properties. It can cause leukocyte adhesion to endothelial cells via intercellular adhesion molecules (ICAM) and endothelial leukocyte adhesion molecules (ELAM-1). IL-1 also can cause neutrophil chemotaxis and degranulation. As noted above, we have shown the neutrophils to be increased in quantity at 4 AM. Finally, IL-1 can cause microvascular leakage and edema formation. With these multiple ongoing effects at night, airways hyperreactivity and lung function are at a vulnerable point for worsening.

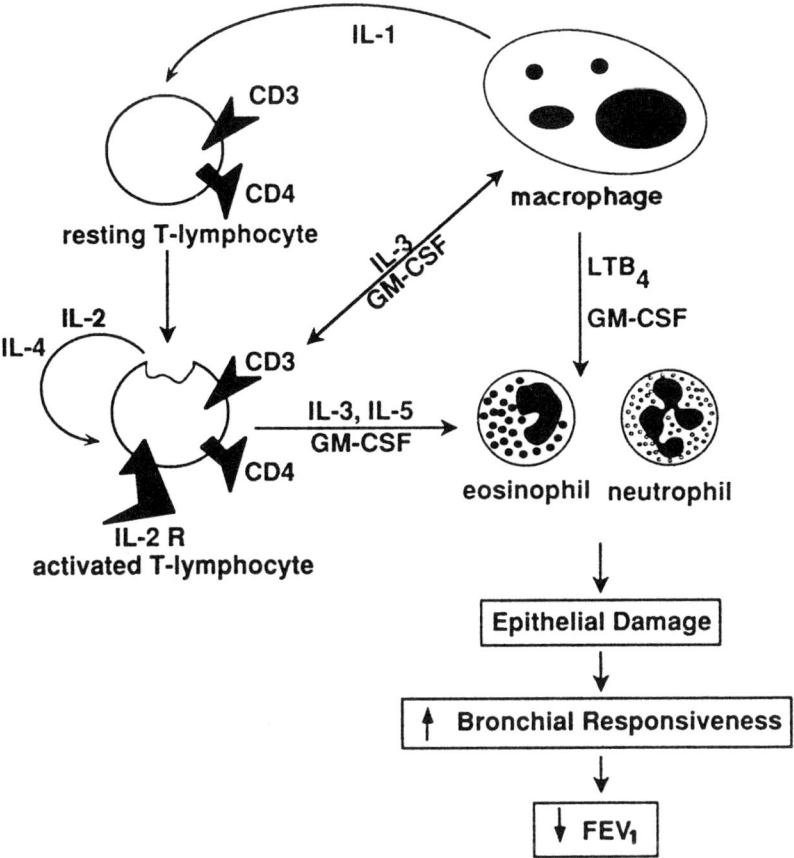

Figure 9. This figure represents a possible scenario of macrophage IL-1 production stimulating resting lymphocytes with resultant influx of inflammatory cells and nocturnal asthma.

References

1. Joseph M, Tonnel AB, Capron A, Dessaint JP. The interaction of IgE antibody with human alveolar macrophages and its participation in the inflammatory process of drug allergy. Agents Actions 1981;11:619–622.
2. Melewicz FM, Zeijer RS, Mellon MH, O'Connor RD, Speigelberg HL. Increased peripheral blood monocytes with Fc receptors for IgE in

patients with severe allergic disorders. J Immunol 1981;126:1592–1595.

3. Williams J, Borish L. Expression and regulation of low affinity receptors on human mononuclear phagocytes. J Allergy Clin Immunol 1991; 87:245A.

4. Rankin JA, Askenase PW. The potential role of alveolar macrophages as a source of pathogenic mediators in allergic asthma. In: Asthma: Physiology, Immunopharmacology and Treatment. Academic Press, London, 1984, pp 157–168.

5. Godard P, Chaintreuil J, Damon M, et al. Functional assessment of alveolar macrophages: Comparison of cells from asthmatic and normal subjects. J Allergy Clin Immunol 1982; 70:88–94.

6. Joseph M, Tonnel AB, Capron A, Voisin C. Enzyme release and superoxide anion production by human alveolar macrophages stimulated with immunoglobulin E. Clin Exp Immunol 1980; 40:416–422.

7. Cohn ZA, Wiener E. The particulate hydrolases of macrophages. II. Biochemical and morphological response to particle ingestion. J Exp Med 1963; 118:1009.

8. Rankin JA, Schrader CE, Smith JM, Lewis RA. Recombinant interferon-gamma primes alveolar macrophages cultured in vitro for the release of leukotriene B_4 in response to IgG stimulation. J Clin Invest 1989; 83:1691–1700.

9. Cluzel M, Damon M, Chung P, et al. Enhanced alveolar cell luminol-dependent chemiluminescence in asthma. J Allergy Clin Immunol 1987; 80:195–201.

10. Pujol JL, Cosso B, Pauves J–P, et al. Interleukin-1 release by alveolar macrophages in asthmatic patients and healthy subjects. Int Arch Allergy Appl Immunol 1990; 91:207–210.

11. Chanez P, Bousquet J, Couret I, et al. Increased numbers of hypodense alveolar macrophages in patients with bronchial asthma. Am Rev Respir Dis 1991; 144:923–930.

12. Fukuda T, Dunnett SL, Reed CE, et al. Increased numbers of hypodense eosinophils in the blood of patients with bronchial asthma. Am Rev Respir Dis 1987; 132:981–985.

13. Damon M, Chavis C, Paures JP. Increased generation of the arachidonic metabolites LTB_4 and 5-HETE by human alveolar macrophages in patients with asthma: effect in vitro of nedocromil sodium. Eur Respir J 1989; 2:202–209.

14. Damon M, Chavis C, Crastes de Paulet A, et al. Arachidonic acid metabolism in alveolar macrophages. A comparison of cells from healthy subjects, allergic asthmatics and chronic bronchitis patients. Prostaglandins 1987; 34:291–306.

15. Balter MS, Eschenbacher WL, Peters-Golden M, et al. Arachidonic acid metabolism in cultured alveolar macrophages from normal, atopic and asthmatic subjects. Am Rev Respir Dis 1988; 138:1134–1142.

16. Tonnel AB, Gosset P, Joseph M, et al. Stimulation of alveolar macrophages in asthmatic patients after local provocation test. Lancet 1983; 1:1406–1409.

17. Metzger WJ, Zvala D, Richardson HB, et al. Local allergen challenge

and bronchoalveolar lavage of allergic asthmatic lungs. Am Rev Respir Dis 1987; 135:403–440.

18. Broide DH, Firestein GS. Endobronchial allergen challenge in asthma. Demonstration of cellular source of granulocyte macrophage colony-stimulating factor by in situ hybridization. J Clin Invest 1991; 88:1048–1053.

19. Carre PC, Mortenson RL, King RE, et al. Increased expression of Interleukin-8 gene by alveolar macrophages in diopathic pulmonary fibrosis. J Clin Invest 1991;88:1802–1810.

20. Venge P, Dahl R, Håkansson L. Heat-labile neutrophil chemotactic activity in subjects with asthma after allergen inhalation: Relation to the late asthmatic reaction and effects of asthma medication. J Allergy Clin Immunol 1987; 80:679–681.

21. Martin RJ, Cicutto LC, Smith HR, et al. Airway inflammation in nocturnal asthma. Am Rev Respir Dis 1991; 143:351–357.

22. Wenzel SE, Trudeau JB, Beam WR, et al. Alteration of the alveolar macrophage population and eicosanoid production following oral prednisone in nocturnal asthmatics. Presented, American Thoracic Society, 1992.

23. Borish L, Mascali JJ, Dishuck J, et al. Detection of alveolar macrophage derived Interleukin 1 β in asthma inhibition with corticosteroids. Accepted for publication. J. Immunol.

24. Beam WR, Torvik JA, Martin RJ. Timing of prednisone and alterations of airway inflammation in nocturnal asthma. Am Rev Respir Dis 1991; 143:A626.

25. Dunnill MS. The pathology of asthma. In: Allergy, Principles and Practice, Middleton E Jr, Reed CE, Elto EF (eds). CV Mosby, St. Louis, MO, 1978, pp 678–686.

26. Lundgren R, Sodarberg M, Horstedt P, Stenling R. Morphological studies of bronchial mucosal biopsies from asthmatics before and after ten years of treatment with inhaled steroids. Eur Respir J 1988; 1:833–889.

27. Jeffrey PK, Wardlaw AJ, Nelson FC, Collins JV, Kay AB. Bronchial biopsies in asthma: an ultrastructural, quantitative study of correlation with hyperreactivity. Am Rev Respir Dis 1989; 140:1745–1753.

28. Metzger J, Zavala H, Richerson HB, et al. Local allergen challenge and bronchoalveolar lavage of allergic asthmatic lungs: a description of the model and local airway inflammation. Am Rev Respir Dis 1987; 135:433–440.

29. Gerblich AA, Salik H, Schuyler MR. Dynamic T-cell changes in peripheral blood and bronchoalveolar lavage after antigen bronchoprovocation in asthmatics. Am Rev Respir Dis 1991; 143:533–537.

30. Miyajima A, Miyatake S, Schreurs J, de Vries J, Arai N, Yokota T, Arai K. Coordinate regulation of immune and inflammatory responses by T-cell derived lymphokines. FASEB J 1988; 2:2462.

31. Sustiel A, Rocklin R. T-cell responses in allergic rhinitis, asthma and atopic dermatitis. Clin Exp Allergy 1989; 19:11.

32. Silberstein DS, David JR. The regulation of human eosinophil function by cytokines. Immunol Today 1987; 8:380.

33. Schleimer RP. Effects of glucocorticosteroids on inflammatory cells relevant to their therapeutic applications in asthma. Am Rev Respir Dis 1990; 141:s59.

34. Azzawi M, Bradley B, Jeffrey PK, et al. Activated T-lymphocytes and eosinophils in bronchial biopsies in stable atopic asthma. Am Rev Respir Dis 1990; 142:1407–1413.

35. Mattoli S, Mattoso VL, Soloperto M, Allegra L, Fasoli A. Cellular and biochemical characteristics of bronchoalveolar lavage fluid in symptomatic non-allergic asthma. J Allergic Clin Immunol 1991; 87:794–802.

36. Corrigan CJ, Hartnell A, Kay AB. T-lymphocyte activation in acute severe asthma. Lancet 1988; 1:1129–1131.

37. Poulter LW, Power C, Burke C. The relationship between bronchial immunopathology and hyperresponsiveness in asthma. Eur Respir J 1990; 3:792–799.

38. Durham SR, Kay AB. Eosinophils, bronchial hyperreactivity and late-phase asthmatic reactions. Clin Allergy 1985; 15:411.

39. Walker C, Virchow Jr JC, Bruijnzeel PLB, Blaser K. T-cell subsets and their soluble products regulate eosinophilia in allergic and nonallergic asthma. J Immunol 1991; 146:1829–1835.

40. Turner-Warwick M. Epidemiology of nocturnal asthma. Am J Med 1988; 85(1B):6–8.

41. Dethlefsen U, Repgas R. Ein neues therapieprinzip bei nachtilchen asthma. Klin Med 1985; 80:44–47.

42. Martin RJ, Cicutto LC, Ballard RD. Factors related to the nocturnal worsening of asthma. Am Rev Resp Dis 1990; 141:33–38.

43. Mohiuddin AA, Martin RJ. Circadian basis of the late asthmatic response. Am Rev Respir Dis 1990; 142:1153–1157.

44. Signore A, Lugini P, Letizia C, et al. Study of the diurnal variation of human lymphocyte subsets. J Clin Lab Immunol 1985; 17:25–28.

7

Chronopharmacology of Theophylline and Beta-2 Sympathomimetic Therapies

Gilbert E. D'Alonzo, D.O., and Michael H. Smolensky, Ph.D.

Introduction

It has been recognized for centuries that many patients with asthma experience their worst symptoms at night, during the early morning hours, and that these symptoms disturb sleep and often disrupt daily job and school performance. The symptoms of most patients with asthma tend to worsen at night with the majority of dyspneic episodes occurring between 2:00 AM and 7:00 AM (Fig. 1).[1] We have learned from survey data that as many as 90% of asthmatic patients, both adults and children, experience nocturnal symptoms severe enough to awaken them from sleep, at least periodically.[2] More alarmingly, in one large survey it was found that almost 40% of the patients had nightly awakenings and over 60% were awakened at least three nights a week. Furthermore, a more important consequence of nocturnal asthma is death. Douglas[3] reviewed several studies which investigated asthma death and

Martin RJ (editor): *Nocturnal Asthma: Mechanisms and Treatment,* © Futura Publishing Co., Inc., Mount Kisco, NY, 1993.

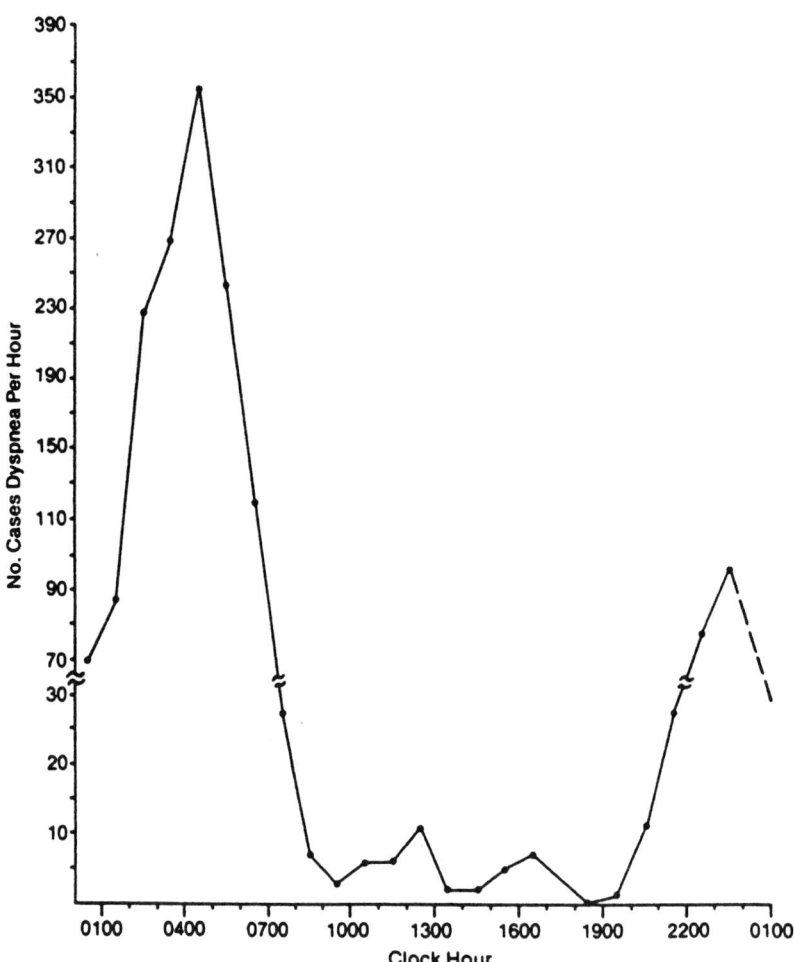

Figure 1. Day-night variation in the occurrence of 1,631 attacks of dyspnea in 3,121 untreated, mainly asthmatic patients. Most dyspnea, likely asthma, occur between 2:00 AM and 7:00 AM with very few occurring during the daytime. From Dethlefsen and Repges.[1]

found that 93 of 219 deaths occurred between midnight and 8:00 AM Deaths often occur outside the hospital, at home, when asthma is poorly controlled or during recovery from an acute exacerbation. These findings are supported by the presence of large circadian variations in airflow which have been recognized as a risk factor for fatal asthma.[4–6] It is important to recognize that this unacceptably high level of mortality and morbidity is occurring despite the availability of numerous therapeutic options for the control of asthma. One possible solution is to concentrate on directing more intense therapy at a time when the disease is known to be more unstable. The standard asthma therapeutic interventions with theophylline and beta-sympathomimetics will not be discussed, but how to therapeutically intervene with these agents on a chronopharmacological basis will be reviewed.

Circadian rhythm adapted chronotherapy for nocturnal asthma involves the scheduling of bronchodilator and, possibly, other medications once daily in the morning or evening, or several times daily using an unequal dosing strategy. The goal of this therapeutic approach is to control nighttime symptoms without a worsening or breakthrough of daytime symptoms, while minimizing or averting the adverse effects of the medications. Chronotherapy aims to increase both drug effectiveness and tolerance by selecting an experimentally validated dosing schedule in accordance with the biological rhythm time structure of the disease that is being treated. Therefore, the principal objective of the treatment plan for nocturnal asthma would be a good night's sleep, uninterrupted by the need for medication, and a reduction in the circadian variability in airflow and in bronchial responsiveness, all of which should improve daytime productivity and the overall quality of life of patients.

Certain treatment strategies must be considered if nocturnal asthma is to be treated effectively. Perhaps an ideal bronchodilator medication should increase in effect at a time when bronchoconstriction is known to occur. For asthma, bronchodilator therapy taken after dinner or before bedtime would have to reach a peak effect after approximately 6 to 8 hours, respectively, and the duration of drug activity would have to be considerable. Therefore, the pharmacokinetics and pharmacodynamics of the bronchodilator during sleep and the patient's daily schedule would have to be known in order to recommend an optimum dosing plan.

The duration of action of many of the currently available beta-adrenergic and anticholinergic bronchodilator aerosols is likely to be too brief to protect against nocturnal bronchoconstriction. Thus far, primarily theophylline and oral beta$_2$-agonist sustained-release medications have been evaluated as chronotherapies for asthma. Newer long-acting beta-agonist inhalation therapy appears promising.

Theophylline Therapy

Recently, there have been numerous articles on the superiority of inhaled beta-adrenergic agonists over theophylline as bronchodilators in treating asthma. However, there are data to suggest that theophylline therapy has substantial value as a form of maintenance therapy for patients with nocturnal asthma,[7,8] including those who are steroid-dependent.[9] Joad et al.[7] studied the relative benefit of maintenance therapy with theophylline, inhaled albuterol, and combination therapy in a cohort of adolescents and adults with chronic asthma during a 3-month trial. Twice-daily sustained-release theophylline alone or in combination was associated with significantly fewer days with symptoms (52% and 55%) in comparison to treatment with albuterol only (72%). The greater frequency of symptoms during the albuterol regimen was increasingly apparent more than 4 hours after dosing and was greatest between 4:00 and 8:00 AM (Fig. 2). Zwillich et al.[8] demonstrated the superiority of a twice-daily sustained-release theophylline preparation as compared to 8-hour duration inhaled beta$_2$-agonist (bitolterol) in treating patients with mild to moderate nocturnal asthma (Fig. 3). In this study, sleep quality and architecture were unchanged between the two medications, showing that the presence of theophylline did not alter sleep as compared to an inhaled beta-agonist. Moreover, not only was early morning airflow improved, but there was less nocturnal oxygen desaturation while the patients were on theophylline (Fig. 4).

In patients with more severe nocturnal asthma, higher serum theophylline concentrations (STC) during the night and lower STC during the daytime, when there may be less or easier to control bronchoconstriction, has been shown to be successful.[10–12]

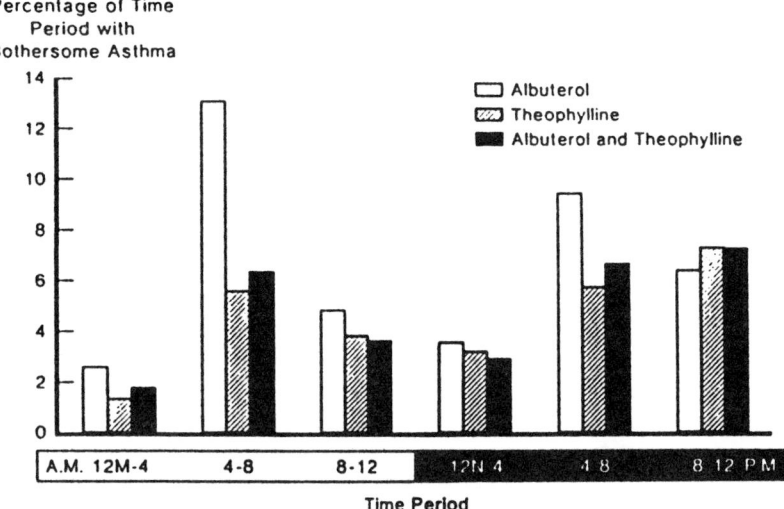

Figure 2. Percentage of time with asthma as recorded in the daily diaries by 18 patients during month-long treatment periods using regimes of inhaled albuterol, sustained-release theophylline, or the combination of both medications. Differences between the albuterol and the theophylline-containing regimens were greatest at night between 4:00 and 8:00 AM. From Joad et al.[7]

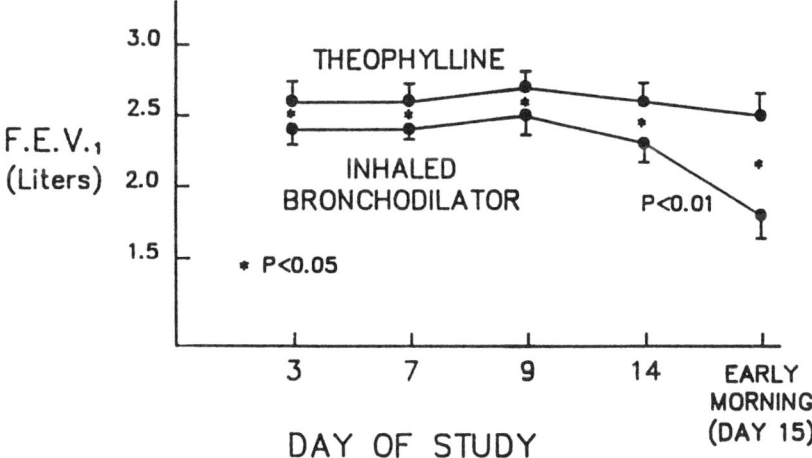

Figure 3. Sustained-release theophylline therapy was associated with a consistently higher mean forced expiratory volume in one second (FEV_1) than the inhaled bronchodilator, bitolterol, during each phase of study. A significant decrease ($P < 0.01$) in airflow occurred in the early morning during inhaled bronchodilator therapy, but not during theophylline therapy. From Zwillich et al.[8]

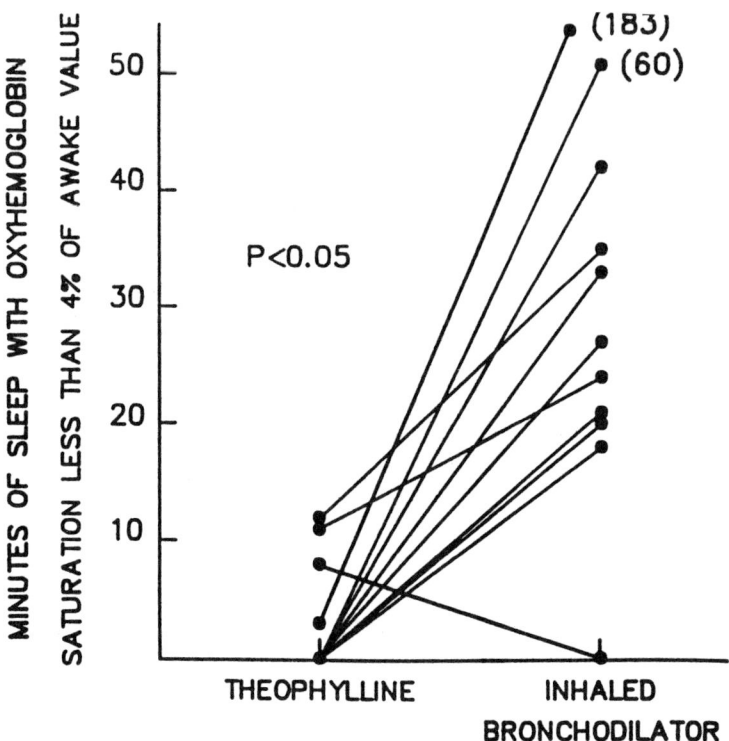

Figure 4. Oxyhemoglobin desaturation episodes occurred in 10 of the 26 asthmatic patients during inhaled bronchodilator therapy and in only 4 patients during sustained-release theophylline therapy. From Zwillich et al.[8]

Unlike aerosol therapy, where the medication effect is likely to be waning as nocturnal bronchoconstriction is increasing, certain slow-release theophylline preparations can be administered in a way that provides a rising STC during the time that airflow is deteriorating (Fig. 5). For this reason theophylline chronotherapy has been employed. Two strategies for theophylline chronotherapy have been explored, first in Europe and then in the United States. However, before discussing these therapeutic strategies it is important to review a phenomenon which has been recognized by several investigators during the past decade,

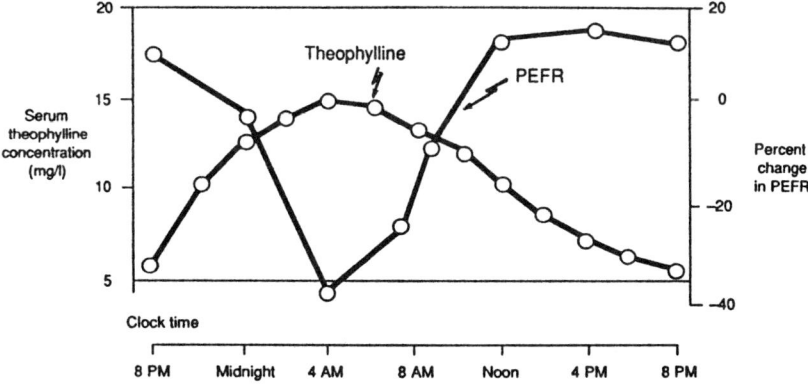

Figure 5. A chronopharmacological approach to sustained-release (SR) theophylline therapy. Theoretically, an ideal SR-theophylline should provide a rising drug blood level with a sustained plateau during the time that airway obstruction is likely to be increasing.

that is, dose-time-dependent variability in theophylline kinetics (chronokinetics).

Chronokinetics of Theophylline

The pharmacokinetics of a variety of theophylline formulations are dosing-time-dependent.[13] Immediate, sustained-release, and ultra slow-release theophylline preparations have all demonstrated certain chronokinetics using a variety of different experimental protocols. The term chronokinetics specifically refers to administration-time-dependent variability in drug disposition; including aspects of absorption, distribution, metabolism, and elimination, resulting from biological rhythm influences. In investigations where such factors as the timing and content of meals and body positional effects have been controlled and chronokinetic phenomena have been recognized, the former cannot be used as an explanation for the latter. When these and other factors are not controlled, it becomes more difficult to uncover chronokinetic influences. Some studies have been done on subjects or patients who were dosed with theophylline until a steady-state was achieved, while others were performed after acute administration. At times, the timing and content of meals and posture were regulated. Some studies were conducted on asthmatic patients, while others involved healthy young adults. Even though the type of theophylline formulation studied and certain features of the experimental protocols differed, chronokinetic phenomena have been identified.[13]

Immediate-Release Theophyllines

Only a few studies have investigated the chronopharmaco-kinetics of elixir[14,15] and immediate-release aminophylline and theophylline tablet medications.[15-19] Kyle et al.[19] investigated administration-time kinetic differences of an immediate-release theophylline (IRT), Theolair (3M Riker, USA). Both acute time-specific, once daily, and chronic studies during an 8-day administration period were performed. Meal timing and content were standardized during each day when the drug was ingested at a given clock time in the acute study and throughout the 24-hour sampling periods of the first and last days of the chronic study. Single doses of IRT (12 mg/kg/day) were administered to 13 diurnally active healthy adults at four different clock times in separate trials, each performed at weekly intervals (Fig. 6). The time to peak or maximum serum theophylline concentration (T_{max}) and the maximum serum theophylline concentration (C_{max}) achieved following oral dosing at 7:00 AM and 1:00 PM was quicker and higher than after the 7:00 PM or 1:00 AM administration. A similar chronokinetic pattern was found during the chronic studies performed on both day one and day eight.

Table 1 summarizes the results of several studies which also demonstrate, for the most part, that certain tablet forms of IRT have a shorter T_{max} and greater C_{max} when administered in the morning versus the evening. As with tablet forms of therapy, elixirs when administered in the morning have a more rapid T_{max} than after an evening administration.[14,15]

Sustained-Release Theophyllines

In 1950, Kelly and Murphy[20] reported that there was an absence of 24-hour variation in serum theophylline concentrations (STC) in children treated with a sustained-release theophylline (SRT), Theo-Dur (Key, USA), administered twice daily at 8:00 AM and 8:00 PM.

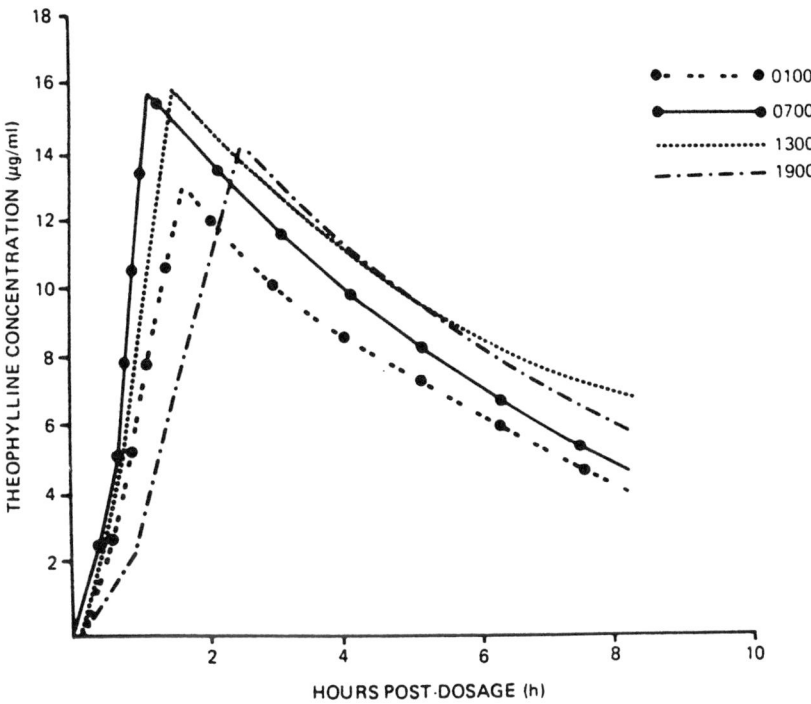

Figure 6. Administration-time-dependent kinetics of an immediate-release theophylline in 13 healthy adult subjects synchronized to diurnal activity and nocturnal rest and with meal timing and composition controlled. Single-dose studies done at different times (military time) on different days. From Kyle et al.[19]

Table 1

Review of Circadian Chronopharmacokinetics of Immediate-Release Theophyllines

Author (yr) [ref]	Type of theophylline	Clock hour of Rx (military time)	Number of subjects (age range)	Chronopharmacokinetics Major findings
Kyle (1980) [19]	Theolair (Riker)	Acute single dose: 0100, 0700, 1300, 1900	12 ♂ healthy (21–35 yr)	T_{max} shortest for 0700 R_x, longest for 1900 R_x; $t_{1/2}$ shortest for 0700 R_x, longest for 1900 R_x
Kyle	Theolair (Riker)	Steady-state: 0100, 0700, 1300, 1900	8 ♀, 4 ♂ healthy (17–35 yr)	T_{max} as above; C_{max} greatest for 0700 R_x; $t_{1/2}$ shortest for 0700 and 1300 R_x, longest for 1300 R_x
Lesko (1980) [16]	Oral aminophylline	Steady-state: 0100, 1300 1900, 0100	7 ♂, 7 ♀ healthy (21–40 yr)	STC greater at 0700 vs. 1300
Nakano (1982) [17]	Oral aminophylline	Acute single dose: 0900, 2100	8 ♂ healthy (23–27 yr)	C_{max} greater after 0900 R_x
Decourt (1982) [18]	Oral theophylline	Acute studies: 0900, 2100	7 ♂, 1 ♀ nightworkers (22–39 yr)	Absorption faster for 0900 R_x; $t_{1/2}$ shorter after 2100 R_x
Taylor (1983) [14]	Theophylline elixir	Acute studies: 0900, 2100	4 ♂, 4 ♀ healthy (19–25 yr)	T_{max} shorter after 0900 R_x; no Δ for C_{max}, $t_{1/2}$
Reed (1986) [15]	Elixophyllin elixir (Berlax), 150 mg	Acute studies: 0800, 1400, 2000, 0200 (5 days)	13 healthy subjects	0800 STC trough < 1400 trough by 14% (P < 0.005); 1400 STC trough >2000 trough by 10% (P < 0.05); 0800 STC trough = 2000 trough

Abbreviations: STC = serum theophylline concentration; T_{max} = time to maximum drug concentration; $t_{1/2}$ = half-life of drug; C_{max} = maximum concentration of drug

Since then, several studies have demonstrated substantial chronokinetic differences in STC following twice daily treatment. Theo-Dur has been most studied, but chronokinetic differences have been found for several other SRTs (Table 2). The studies by Scott et al.,[21,22] which were performed on children, best demonstrate the distinct administration time-dependent differences in drug kinetics of a SRT preparation (Fig. 7). The upper portion of Figure 7 demonstrates the 12-hour dosing interval differences in STC following morning versus evening Theo-Dur administration in an initial study performed on 13 diurnally active asthmatic children followed after treatment at 8:00 AM and 8:00 PM daily for at least one week. The "steady-state" STC following the evening dosing was statistically lower than that following the morning administration.[21] Essentially identical results were found in a second study performed by the same investigators, who evaluated 12 children and adolescents with Theo-Dur dosed at 7:00 AM or 7:00 PM.[22] In both studies, asthmatic children were evaluated over 24-hour periods after having attained steady-state conditions for theophylline after at least 7 days of twice daily dosing. The studies were controlled for the timing and content of meals. Of the 25 total children studied, 22 (88%) demonstrated comparable temporal patterns in STC. T_{max} occurred earlier and C_{max} was greater after the morning theophylline administration as compared to the evening dosing. In both studies, the differences between C_{max} and minimal serum theophylline concentration (C_{min}) over the entire 24 hours was not trivial, averaging nearly 7 mg/L for the total group (Fig. 7).

The temporal occurrence of C_{max} and C_{min} for theophylline during each 12-hour dosing interval for the 25 asthmatic children who were evaluated in both studies are shown in Figure 8. If C_{max} and C_{min} are determined without regard to each 12-hour dosing interval, it appears they are randomly distributed over the 24 hours. However, when the morning and evening dosing intervals are analyzed separately, specific patterns are found. After evening administration of Theo-Dur, the C_{max} in 22 of 25 patients occurred 8 to 12 hours after dosing, while in a similar number of patients the C_{min} occurred during the initial 4 hours after the evening dosing. In comparison, after morning dosing in 21 of 25 patients, the C_{max} occurred within 4 hours after Theo-Dur administration, while the C_{min} occurred during the last 4 hours of the dosing interval. Thus, the time to maximum and minimum serum concentration, T_{max} and

Table 2

Review of Circadian Chronopharmacokinetics of Sustained-Release Theophyllines

Author (yr) [ref]	Type of theophylline	Clock hour of Rx (military time)	Number of subjects (age range)	Chronopharmacokinetics Major findings
Lesko (1980) [16]	Somophyllin-CRT (Fisons)	Steady-state: 0800, 2000	7 ♂ 7 ♀ healthy (21–40 yr)	STC slightly higher at 0800 vs. 2000
Kelly (1980) [20]	Theodur (Key)	Steady-state: 0800, 2000	20 asthmatics (6–18 yr)	STC essentially nonvarying as a group phenomenon
Scott (1981) [21]	Theodur (Key)	Steady-state: 0800, 2000	13 asthmatic children (6–12 yr)	Large circadian change in STC; STC higher at 0800 vs. 2000
Simons (1982) [121]	Theodur (Key)	Chronic study: 0800, 2000	8 ♂ 2 ♀ asthmatic children (3–7 yr)	STC at 0800 vs. 2000 greater ($\bar{x}$ change between STC peak and trough = 6.5 µg/ml)
Primrose (1984) [122]	Theodur (Key)	Once daily for 3 days at 1000 only and 2200 only	3 asthmatics (x = 53 yr)	$\dfrac{R_x \text{ at } 1000}{5.3 \text{ hr}} \quad T_{max} \quad \dfrac{R_x \text{ at } 2200}{8.3 \text{ hr}}$ low $\quad C_{max} \quad$ high
Reed (1983) [123]	Theodur (Key)	0900, 2100 (48 hr)	9 asthmatics	More variability in STC after 2200 R_x; 0800 STC trough > 2100 trough by ~11–16% (P < 0.02); AM–PM difference in STC trough not affected by concurrent ingestion of antacid;
(1986) [15]	Bronkodyl-SR (Breon)	0800, 2000, (5 days)	14 healthy subjects	0800 STC trough >2000 trough by 11% (P ≈ 0.01)
Coulthard (1983) [124]	SR250 Nuelin (Riker)	Steady-state: 0900, 2100	10 asthmatics (5–13 yr)	The predose STC at 0900 2.6–2.9 times greater than at 2100

Study (year) [ref]	Product	Dosing times	Subjects	Results
Thompson (1983) [43]	SR250 Nuelin (Riker)	Steady-state: 0900, 2100	14 COPD including asthma (x = 50 yr)	The predose STC at 0900 greater than at 2100 (P < 0.01) studied under fasting and nonfasting conditions
Samaan (1984) [125]	Choledyl (Davis) Theodur (Key) Slophyllin (Rorer)	Steady-state: 0800, 2000	12 ♂ healthy (20–25 yr)	AUC_{0-12} for nighttime R_x reduced (~20%) for all three; STC at 0800 greater than at 2000
Jonkman (1984) [37]	Choline theophyllinate	Acute studies: 1000, 2200	6 ♂ 2 ♀ healthy (20–25 yr)	No change of AUC; STC at 1000 higher than at 2200
St. Pierre (1984) [126] (1985) [127]	Theophylline as compressed tablet and Theodure (Key)	Acute studies: 0800, 2000 / 0800, 2000	8 ♂ healthy / 8 ♂ healthy	T_{max} shortest, 0800 R_x; C_{max} highest, 2000 R_x / T_{max} shortest for 0800 R_x
Schultz (1984) [63]	Pulmidur Forte	Steady-state: 0800, 2000	12 ♂ healthy	C_{max} morning R_x greater than C_{max} night R_x
Segrestaa (1984) [128]	Armophylline (Armour)	Steady-state: 0900, 2100	8 healthy and 2 COPD (22–28 yr)	AUC post-0900 R_x greater than 2100 R_x; C_{max} morning R_x greater than C_{max} night R_x
Rogers (1985) [129]	Theodur (Key)	Chronic dosing: 0600, 1800	8 asthmatics (4–17 yr)	Circadian change in STC in 50% of patients
Straughn (1984) [25] (1985) [26]	Theodur (Key)	Steady-state: 0800, 2000	14 ♂ healthy (22–28 yr)	Predose STC AM > PM; trough STC R_x AM < PM R_x; T_{max} AM R_x < PM R_x; AUC AM R_x > PM R_x; C_{ss} AM R_x > PM R_x
Thuresson (1985) [27]	Theodur (Key) 300 mg b.i.d.	Steady-state: 0800, 2000	6 ♂ 6 ♀ healthy (26–40 yr)	$\bar{x}$ Peak-trough change over 24 hr 3.9 µg/mL; T_{max} PM R_x greater than T_{max} AM R_x
Kramer (1989) [130]	Theodur (Key) Somophyllin-CRT (Fisons)	Steady-state: 0700, 1900	12 asthmatics (6–17 yr)	T_{max} shortest for 0700 R_x; C_{max} highest for 0700 R_x for both

Table 2
Review of Circadian Chronopharmacokinetics of Sustained-Release Theophyllines *(continued)*

Author (yr) [ref]	Type of theophylline	Clock hour of Rx (military time)	Number of subjects (age range)	Chronopharmacokinetics Major findings
Fanta (1985) [81]	Theodur (Key)	Steady-state: 0800, 2000	8 asthmatics (24–46 yr)	T_{max} 8 hr after 2000 R_x vs. 5.3 hr after 0800 R_x; AUC slightly > for 0800–2000; no change in C_{max} by R_x time
Fairshter (1985) [69]	Theodur (Key)	Steady-state: 0800, 2000	12 nonsmoking adult asthmatics	T_{max} 5.2 hr after 0800 R_x, 8.2 hr after 2000 R_x; C_{max} 15 μg/mL after 0800 R_x, 12.7 after 2000 R_x; (P < 0.05); AUC and mean STC after 0800 R_x > 2000 R_x
Neuenkirchen (1985) [10]	Phyllotemp (Mundipharma)	Steady-state: 5 days at 0800 and 2000	9 patients with wheeze (25–62 yr)	For the 0800–2000 dosing intervals (DI): AUC, C_{SS}, C_{max}, $C_{max} - C_{min}$ greater (P ≤ 0.05) than 2000–0800 DI; K_{el} and Cl greater during 2000–0800 DI
Welsh (1986) [71]	Theodur (Key)	0800, 2000 (14 days)	12 asthmatics (14–65 yr)	T_{max} = 6 hr for AM R_x; T_{max} = 8 hr for PM R_x (P < 0.02)

Abbreviations: Same as Table 1. Additionally, AUC, area under time-drug concentration curve; C_{SS}, mean steady-state drug concentration; $C_{max} - C_{min}$, mean peak to trough difference in drug concentration; K_{el}, elimination rate constant; Cl, clearance

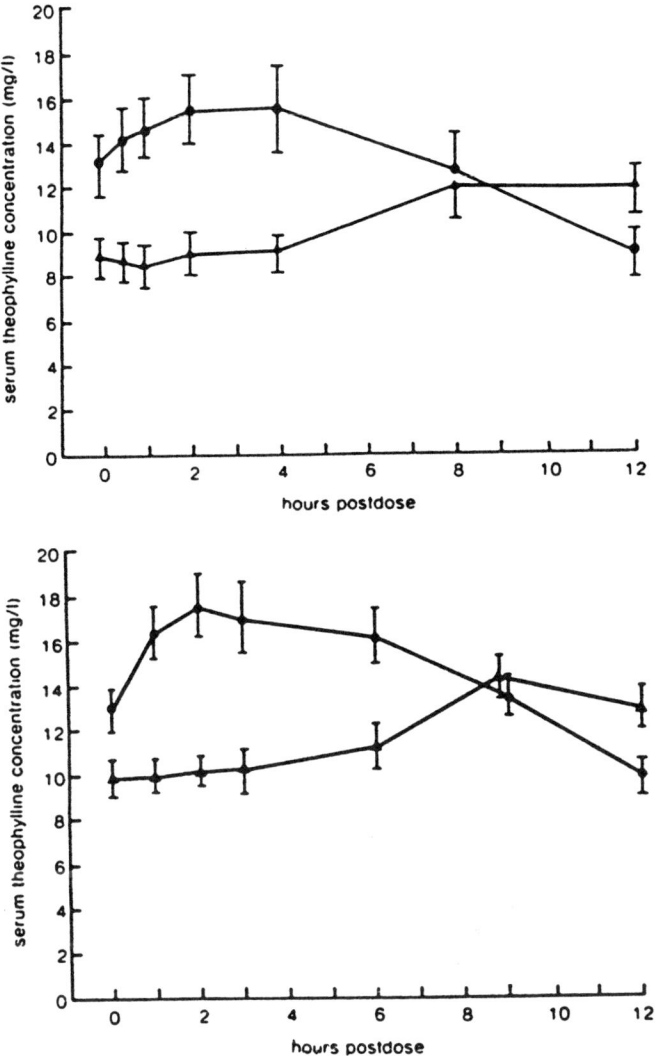

Figure 7. *Top panel,* comparison of steady-state serum theophylline concentrations (STC) after morning (8:00 AM; ●) and evening (8:00 PM; ▲) Theo-Dur administration to 13 young children with asthma. Note that the STCs immediately following the morning dosing were greater than those following the evening dosing. *Bottom panel,* in a separate study involving the administration of Theo-Dur at 7:00 AM; ● and 7:00 PM; ▲ to a different group of 12 young asthmatic patients, the findings were nearly identical. In both studies, the differences between the maximal and minimal serum theophylline concentration over the entire 24 hours was not small, averaging nearly 7 mg/L for the total group. From Scott et al.[21] and Smolensky et al.[22]

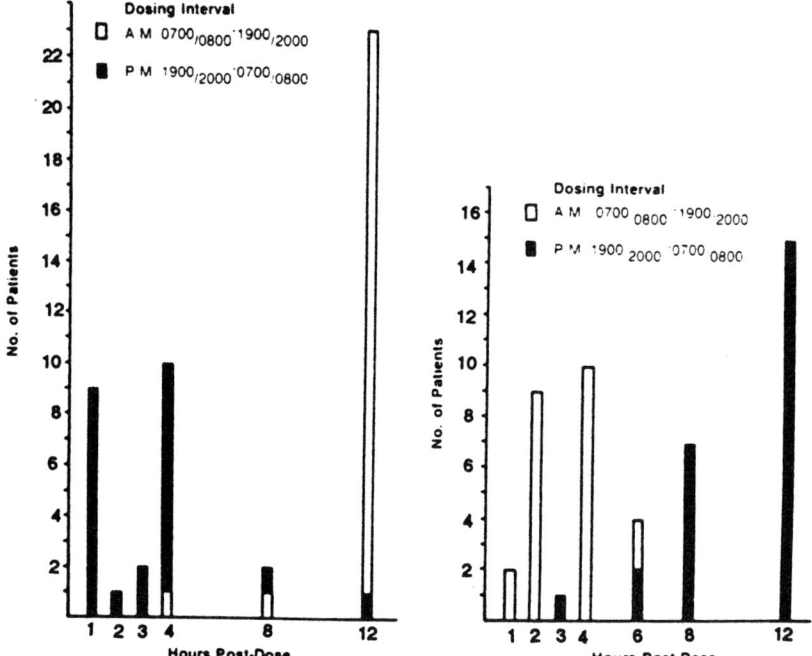

Figure 8. *Left:* Occurrence of trough STC after morning and evening administrations of Theo-Dur in 25 asthmatic children who were treated daily with equal doses either at 0700 and 1900 hours or 0800 and 2000 hours. When the SRT was taken in the morning, the trough STC occurred at the end of the dosing interval; when the SRT was taken in the evening, most patients exhibited the trough STC within the initial 4 hours after dosing. *Right:* Occurrence of peak STC after morning and evening administrations of Theo-Dur in 25 asthmatic children who were treated daily with equal doses either at 0700 and 1900 hours or at 0800 and 2000 hours. When the SRT was taken in the morning, the peak STC in most of the children occurred within the ensuing 4 hours; when taken in the evening the peak STC was likely during the last 4 hours of the dosing interval. From Smolensky et al.[13]

T_{min}, respectively, within each 12-hour interval was predictable in 22 of 25 asthmatic children.

In summary, these studies on children indicate that morning dosing of sustained-release theophylline such as Theo-Dur results in highest drug levels during the first half of the 12-hour dosing interval and trough levels just prior to the second daily dosing in the evening. However, evening dosing with Theo-Dur is associated with a trough (rather than a peak) level during the initial hours postdrug ingestion and peak (rather than a trough) at the end of the evening dosing interval. This finding of a difference in the pharmacokinetics of certain sustained-release theophyllines, such as Theo-Dur, has significant implications regarding the timing and interpretation of patient theophylline titration studies. Patients who exhibit therapeutic drug levels in the morning might experience subtherapeutic ones overnight, at a time when high theophylline levels are critical for patients prone to nocturnal asthma.

It is important to point out that the magnitude of these dosing-time related differences in theophylline kinetics seems to be greater in children than adults and varies by drug formulation (Table 2).

Ultrasustained-Release Theophyllines

Advances in drug formulation have led to the development of ultrasustained-release theophylline preparations that can be administered once daily for the treatment of asthma. Once daily therapy is expected to improve patient compliance and control of disease symptoms, especially the nocturnal manifestations of asthma.[23] For many asthma patients a once daily sustained-release theophylline (ODSRT) would be best administered in the evening, but for some patients whose asthma is more prevalent during the day, a morning administration would be better.[24] These differences in therapeutic strategy point out the importance of understanding any day-night differences in the chronokinetics of individual ODSRT formulations.[10,25–32] Several different ODSRT preparations have been evaluated for these day-night differences (Table 3).

Reinberg and colleagues,[28] in a study comparing the therapeutic effect of one ODSRT Armophyllin (Rorer, France) when given in the morning versus the evening, demonstrated rather substantial chronokinetic differences. For a group of eight diurnally active asthmatic

Table 3

Review of Circadian Chronopharmacokinetics and Chronoeffectiveness of Ultrasustained-release Theophyllines (Once Daily Sustained-releases Theophylline; ODSRT)

Author (yr) [ref]	Type of theophylline	Clock hour of Rx (military time)	Number of subjects (age range)	Chronopharmacokinetics Major findings
Thompson (1981) [131]	ODSRT 250 Nuelin (Riker)	Steady-state: 5 days ODSRT at 2100	12 patients with airways obstruction (50–82 yr)	Delayed absorption at night
Neuenkirchen (1985) [10]	ODSRT Uniphyllin vs. b.i.d. Phyllotemp (Mundipharma)	Steady-state: 5 days ODSRT at 2000 vs. b.i.d. R_x at 0800 and 2000	9 patients with wheeze (25–62 yr)	C_{max} and $C_{max} - C_{min}$ and therapeutic effect greatest following ODSRT at 2000
Reinberg (1985) [28]	ODSRT Armophyllin (Armour)	Steady-state: 8 days at 0800; 8 days at 2000	8 asthmatics (41–61 yr)	T_{max} shortest and STC fluctuation greatest for 0800 R_x; C_{max} greatest for 0800 R_x; therapeutic effect best following 2000 R_x
Busse (1985) [68]	ODSRT Uniphyl (Purdue Frederick)	Steady-state: 0700 vs. 1900	16 adult asthmatics (21–41 yr)	Suggestive higher nocturnal STC following 1900 R_x; 0700 "steady-state" STC after 1900 R_x about twice that after 0700 R_x; therapeutic effect reduction of "nocturnal dip" and 24-hr mean PEF better for PM dosing
Rivington (1985) [29]	ODSRT Uniphyl (Purdue Frederick)	Steady-state: 0800 vs. 2200	17 adult asthmatics ($x = 36$ yr)	Higher STC maintained during nocturnal hours when R_x at 2200; peak STC and T_{max} following 2200 R_x greater than that following 0800 R_x; peak-to-trough ratio greater after evening R_x; therapeutic effect —

Reference	Drug (formulation)	Dosing schedule	Subjects	Results
Straughn (1984) [25] (1985) [26]	ODSRT Theodur (Key)	0800 vs. 2000 single daily	6 healthy (22–27 yr)	reduction of "nocturnal dip" and 24-hr mean PEF better for PM dosing; T_{max} longer after PM dose (P = 0.035); fluctuation in STC greater after PM dose (P = 0.044); peak-trough difference greater for PM dose (P = 0.011); C_{max} greater for PM dose (P = 0.064)
Kramer (1986) [132]	ODSRT Theo-24 (Searle)	0600 vs. 1500 vs. 2100 steady-state	8 asthmatic children (8–13 yr)	% fluctuation, C_{max}, and AUC much greater when R_x at 1500 or 2100 (P < 0.01) than at 0600; T_{max} and C_{min} not dependent on R_x time
Smolensky (1987) [32]	ODSRT Theo-24 (Searle)	Steady-state: o.d.-0700, o.d.-1500, o.d.-2200	18 healthy adults	STC fluctuation increased from o.d.-1700 and o.d.-2200 vs. 0700 R_x; AUC least for o.d.-0700 and greatest for o.d.-2200
Frankoff (personal communication)	ODSRT Theodur (Key)	Steady-state: o.d.-0800 vs. b.i.d. 0800/2000	10 adult asthmatics with symptoms	STC fluctuation greater (P < 0.05) for o.d. schedules than b.i.d.; C_{max} for o.d. schedules greater than b.i.d.; other kinectic indices not R_x time or schedule affected
Van den Brande (1987) [30]	Theo-1 (Galephar)	8 day o.d. dosing at 0800 or 2000	11–16 COPD patients	STC fluctuation greater for AM vs. PM ingestion (P < 0.05); no difference in C_{max}, T_{max}, or AUC according to R_x time
Lamont (1987) [31]	ODSRT (Riker)	6 day o.d. dosing at 0800 or 2000	13 reversible airways obstruction	No R_x-time-dependent differences in STC fluctuation, C_{max}, T_{max} or AUC

Abbreviations: As in Tables 1 and 2. Additionally, o.d., once-daily; b.i.d., twice-daily; C_{min}, minimum drug concentration.

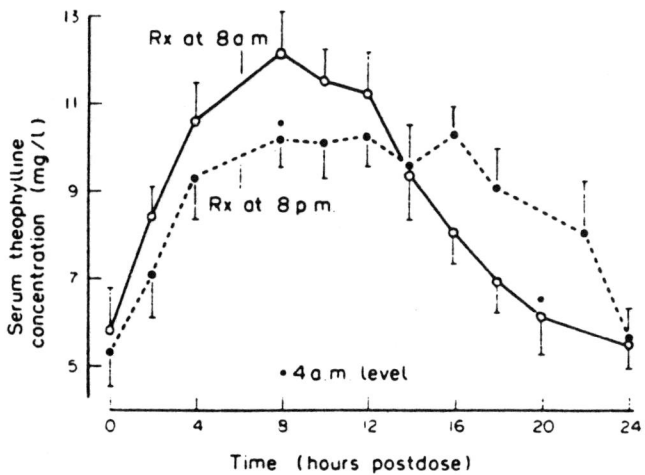

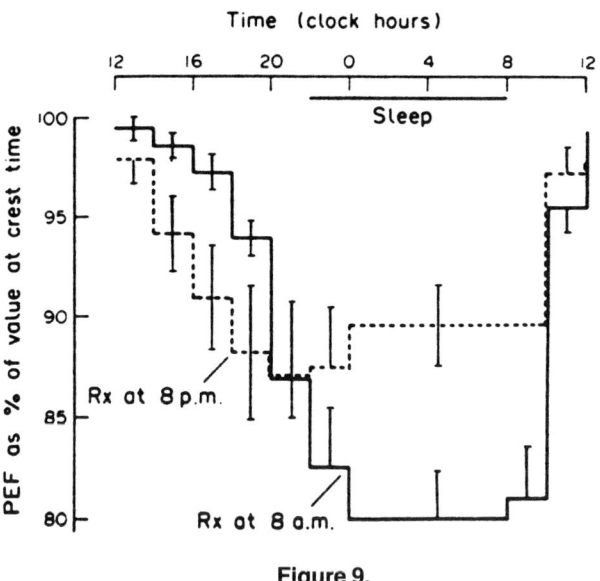

Figure 9.

adults, the T_{max} occurred earlier and the C_{max} was greater for Armophyllin following a 8:00 AM versus a 8:00 PM administration (Fig. 9). Treatment at 8:00 PM produced a more prolonged STC plateau between 4 and 18 hours postdose, while that at 8:00 AM resulted in a comparatively large swing in STC. Comparable dosing-time kinetic dependencies are suggested by studies performed with Theo-Dur when administered once daily.[25,26]

Of the ODSRT preparations, Uniphyl (Purdue Frederick) has been the most frequently studied. Rivington et al.[29] investigated the steady-state pharmacokinetics of Uniphyl administered either at 8:00 AM or 10:00 PM in 17 asthmatic adults using a crossover design protocol (Fig. 10). The 8:00 AM medication administration resulted in a slightly shortened T_{max}, but the evening dosing resulted in a slightly greater peak-to-trough STC ratio (3.3 for the evening dosing versus 2.7 for the morning dosing) over 24 hours.

For both Uniphyl and Armophyllin, the evening administration resulted in a more sustained elevated STC, well within the standard "therapeutic" range of 10 to 20 mg/L, during the crucial nighttime hours when asthma typically worsens. This higher level has been demonstrated to convey a beneficial and a protective effect against asthma.

Dramatic administration time-dependent pharmacokinetics have been demonstrated with Theo-24 (Searle) in children.[32] The most stable time versus STC curves were found when Theo-24 was administered at 6:00 AM (Fig. 11). The same dose given to the same patients for one week at either 5:00PM or 9:00 PM (Figs. 12 and 13,

---

Figure 9. Administration-time-dependent aspects of theophylline disposition in eight adult asthmatics treated with Armophylline. *Top:* Armophylline dosing at 0800 hours (solid line) was associated with a relatively short-lasting STC plateau during the afternoon, while dosing at 2000 hours (dashed line) was associated with a prolonged STC plateau lasting throughout the nocturnal sleep span. In the graph, time is shown as hours postdosing at 0800 or 2000 hours during different phases of the investigation. Each point plotted is the mean + SEM. The asterisk indicates the 0400 hour clock time for each theophylline dosing schedule. *Bottom:* Circadian changes in peak expiratory flow (PEF) of the same adult asthmatics. Armophylline dosing at 0800 hours (solid line) was associated with considerable nocturnal dip in airway patency, while such at 2000 hours (dashed line) was associated with a moderation of the nocturnal PEF dip. It is believed that the higher and prolonged STC during the nighttime is responsible for the better control of airway patency during this span. From Reinberg et al.[28]

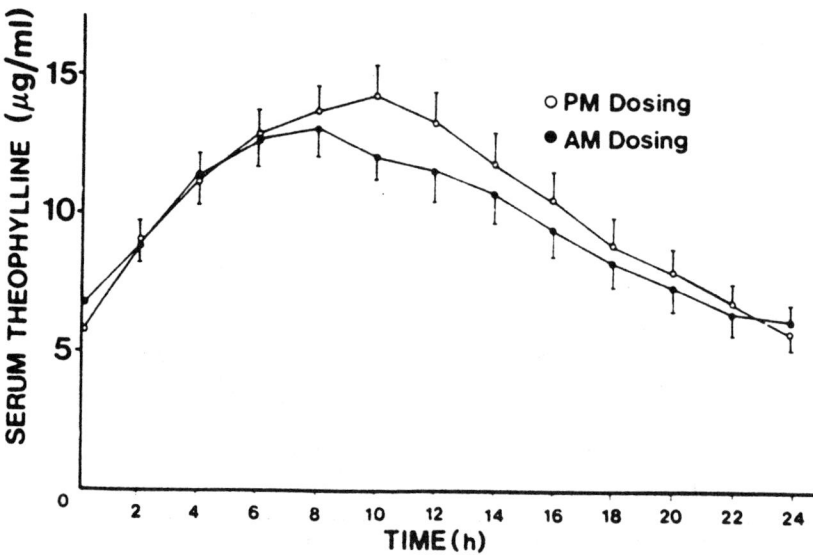

Figure 10. Comparison of STC profiles following dosing of 17 asthmatic patients in the morning (0800 hours) versus the evening (2200 hours) with Uniphyl. The effect of dosing time had only a small effect on the STC profile. Note that the STC data are plotted relative to the time after dosing at 0800 or 2200 hours. In consideration of the dosages administered, treatment at 2200 hours resulted in therapeutic levels between 0200 and 1400 hours. Treatment at 0800 hours resulted in the undesirable situation with respect to patients with nocturnal asthma of the STC being below 10 μg/mL between about 2200 and 0800 hours. From Rivington et al.[29]

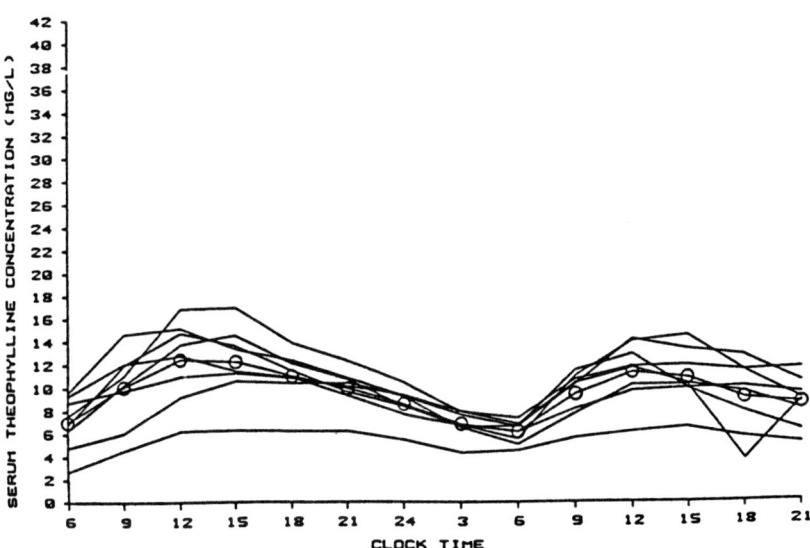

Figure 11. Steady-state 3-hour STC for each of eight asthmatic children and clock-hour means (0) for the group when Theo-24 (Searle, United States) was given daily at 0600 hours. Shown are two consecutive 0600 hour dosings over the 39-hour study period. For many patients, the STC varies between 8 and 14 mg/mL during most of each 24-hour dosing interval, but with low STC between 0000 and 0600 hours. From Smolensky et al.[32]

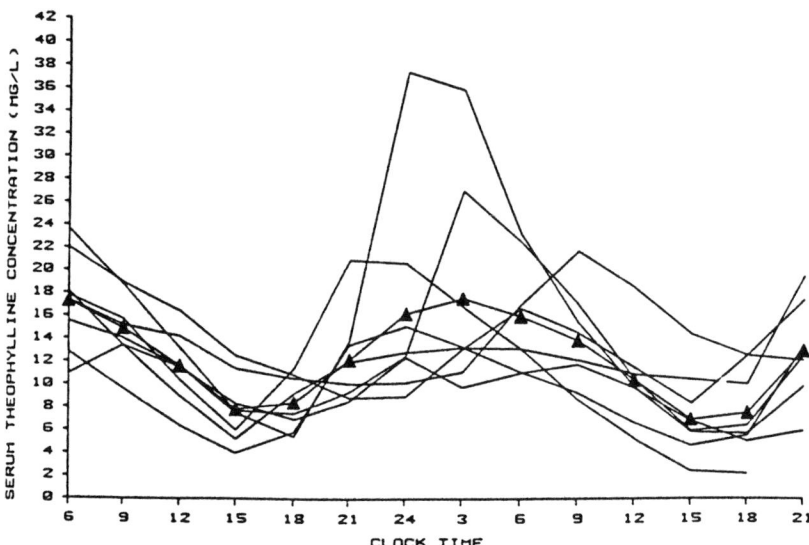

Figure 12. Steady-state 3-hour STC for each of eight asthmatic children and clock-hour means (△) for the group when Theo-24 was given daily at 1500 hours. Shown are two consecutive 1500 dosings over the 39-hour study period. Note the increases in STC fluctuation relative to the once daily 0600 hour dosing schedule. With the afternoon once daily dosing schedule, higher STC are achieved overnight than during daytime. From Smolensky et al.[32]

respectively) resulted in marked STC fluctuation. In fact, certain patients developed potentially toxic, and others subtherapeutic, STCs. These marked pharmacokinetic differences occurred despite adequate control for meal timing, since Theo-24 is known to be affected by both meal timing and composition.[33–36]

From these studies one can see how the degree of variability in the administration-time dependency of theophylline kinetics differs between formulations. Therefore, it cannot be assumed that the pharmacokinetics of every ODSRT is comparable when dosed at different times during the day and night.

The results of the ODSRT studies have shown wide fluctuations in STC over 24 hours, a finding which is in conflict with the belief that optimal control of asthma by theophylline is dependent on the achievement and maintenance of a stable STC over the 24-hour period. Nonetheless, from a chronotherapeutic perspective, certain

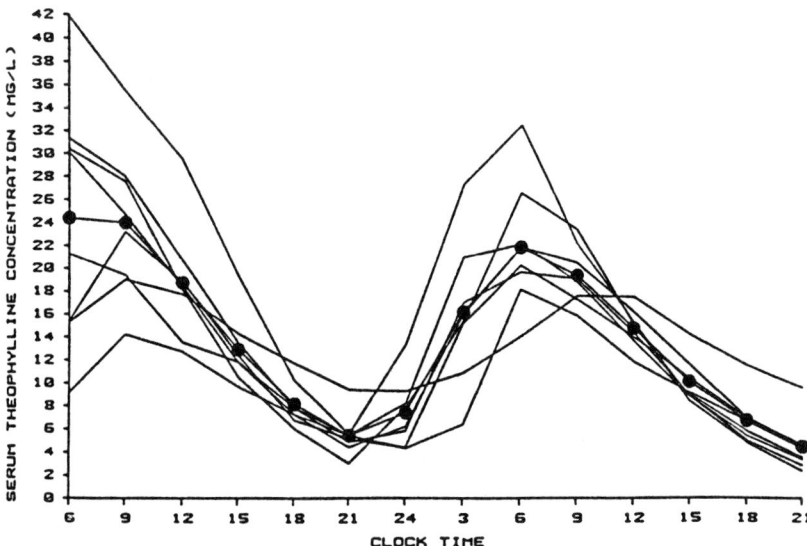

Figure 13. Steady-state 3-hour STC for each of eight asthmatic children and clock-hour means (●) for the group when Theo-24 was ingested once daily at 2100 hours. Shown are two consecutive 2100 hour dosings during the 39-hour study period. Note great fluctuation in STC with quite high levels overnight compared to the Theo-24 once daily 0600 or 1700 hour ingestion schedules. From Smolensky et al.[32]

attributes of the temporal pattern in the variation of STC are critical.[10,28] In asthmatic patients suffering mainly from nocturnal symptoms, an elevated STC, within therapeutic range, throughout the nighttime is crucial (Fig. 5).

Mechanisms of Dosing-time-dependent Variation in the Pharmacokinetics of Theophylline

Several consistent trends have been found in studies which have investigated day-night differences in the kinetics of theophylline.

1. In diurnally active persons who use NSRT, the morning predose STC is likely to be higher than it is at any other clock-hour predose level.

2. STC T_{max} is shortest and C_{max} is greatest following the administration of a morning NSRT.
3. In diurnally active children, and to a lesser extent adults, who are administered equal dosages of a SRT every 12 hours, the STC T_{max} is likely to be shorter and, depending on the formulation, the C_{max} will be slightly to moderately higher following a morning rather than an evening treatment.
4. For SRT therapy, the magnitude of the day-night STC difference is greater for children than adults, but additional studies are needed to establish this finding with certainty.
5. Intravenous infusions of aminophylline show that there is little to no day-night difference in STC,[37–39] thereby implying that the underlying explanation for the chronokinetics of theophylline with oral therapies is due mainly to circadian rhythms in gastrointestinal drug absorption rather than elimination.

Several factors have been implicated as playing a role in the mechanism responsible for the dosing-time-dependent variability in the pharmacokinetics of theophylline. These factors include the dosage form, the timing of meals and food's effect on drug absorption, posture, age and circadian rhythms.

Although it is known that gastric emptying, bowel transit time, pH and other activities of the gastrointestinal system are circadian rhythmic,[40–42] the chronophysiology of this system requires more intense study. The presence of food in the gastrointestinal tract can influence the absorption of certain medications. Food slows gastric emptying, alters intestinal transit time, and increases the secretion of hydrochloric acid, bicarbonate, bile, and certain digestive enzymes. Additionally, food or its components may interact chemically with medications. These factors can potentially enhance or suppress drug absorption. SRT preparations, especially ODSRT medications, remain in the gastrointestinal tract for long periods of time, increasing their susceptibility to food and its potential effects on drug absorption. The effect of food varies both qualitatively and quantitatively according to the theophylline preparation being studied.[36] Most data related to effects of food on theophylline disposition have been derived from research conducted on medication ingested in the morning using test breakfasts of varying size and content.[36] Absorption of certain SRT preparations is slowed by a standardized breakfast,[43,44] whereas the absorption of others is enhanced,[45] as compared to the situation following dosing after an overnight fast.

For Theo-Dur the effect of food on theophylline absorption appears minimal.[46–49] Sips et al.[47] sampled blood for STC measurement frequently, every 30 minutes for 7.5 hours postdose under three conditions: following Theo-Dur ingestion after a standardized breakfast, after an overnight fast, and 3 hours before breakfast. There was no evidence of a food effect on theophylline absorption. From these data it is impossible to imply that a food effect is responsible for the chronokinetics of Theo-Dur previously discussed.[20–22] However, this may not be the case for those theophylline preparations which seem to be affected by food content and timing.[36,50–56]

The pharmacokinetics of theophylline are affected by posture, but just how and to what extent body position influences dosing-time differences in the kinetics of certain theophylline preparations remains unknown. The alterations in posture during each 24-hour cycle are often programmed-in-time due to set patterns in lifestyle, which may result in certain physiological adjustments (e.g., organ blood flow, gastric emptying, plasma volume changes, etc.) affecting drug kinetics and dynamics. Warren et al.[57,58] suggested that the differences in theophylline kinetics following morning and evening dosings were due to a postural influence on plasma volume[58,59] and the rate of gastric emptying, thereby affecting medication absorption. They showed that there was a 1 to 1.5 mg/L difference in STC attributable to a change in posture from the supine versus the standing position. Additionally, the T_{max} was delayed when the subjects were tested while in the supine position. However, such small kinetic differences as are found with a change in posture do not fully explain some of the marked dosing-time differences in the theophylline kinetics reported by others (Tables 1–3).

Age is an important factor to consider when pharmacokinetics are being evaluated. Obviously there are differences in size, volume, and functional capacity of the gastrointestinal tract between children and adults. Studies have shown what appears to be dosing-time-dependent differences in the kinetics of theophylline in both children and adults.[32] However, this matter is complicated by the fact that not all theophylline preparations have been evaluated for age related differences in chronokinetics by comparable methods. Generally speaking, the results of studies done on children demonstrate substantial differences in theophylline kinetics following a morning versus an evening administration.[32] Comparatively, though with certain exceptions, studies performed on adults reveal less extensive chronokinetic differences.[32] At this time, the absence of data from standardized protocols that control for time and content of meals plus posture make the resolution of the clinical significance of age as a factor that influences the chronokinetics of theophylline difficult.

Finally, one has to consider the influence of large amplitude circadian rhythms in a variety of gastrointestinal functions on the pharmacokinetics of theophylline. Circadian rhythms have been reported in gastric emptying, fluid volume, and acid secretion;[40-42] for example, gastric emptying is 50% slower and acidity is higher in the evening and overnight than during the daytime.[40] For certain theophylline formulations whose dissolution rate is pH-dependent, the endogenous circadian rhythms of gastric emptying and acidity may play a role in the dosing-time-dependent differences in theophylline release and eventual rate and extent of drug absorption.

Theophylline Chronotherapeutic Strategies

The chronotherapy of nocturnal asthma with sustained-release theophyllines consists of circadian rhythm adapted dosing schedules using appropriately designed formulations.[60] As mentioned previously, two strategies exist. One consists of dosing theophylline at 12-hour intervals using unequal doses of medication. The daily dose is typically divided so that one-third is administered in the morning and two-thirds in the evening. The goal of this rather unconventional manner of treating asthma is to achieve a higher therapeutic STC throughout the night. Airway patency throughout the entire 24 hours is greater with this schedule, especially during sleep when bronchoconstriction is likely to be more intense, as compared to theophylline dosing schedules not adapted to meet the demands imposed by circadian rhythmic changes in airway caliber. A second dosing schedule strategy for nocturnal asthma involves a once daily evening administration schedule. Caution must be exercised in prescribing such once daily theophyllines. A formulation must be selected so that when administered in the evening it exhibits a predictable and safe pharmacokinetic profile. The kinetic profile of each theophylline preparation seems to be medication-delivery-system-dependent. Specially designed formulations for once daily dosing have been shown to be preferable to theophylline used twice daily in an equal-interval, equal-dose schedule. Both of these dosing strategies will be discussed in detail.

Unequal Twice Daily Theophylline Chronotherapy

Theophylline medication administered twice daily in unequal doses has been used to treat nocturnal asthma.[61–64] Euphyllin CR, a micropellet formulation of aminophylline (Byk Gulden, Germany), was at one time a popular medication for the treatment of asthma in Europe. One-third of the daily dose is taken in the morning and the remaining in the evening. Darow and Steinijans[61] treated 20 asthmatic patients by an unequal dosing schedule of Euphyllin CR, 500 mg aminophylline in the evening and 250 mg in the morning. Compared with a 24-hour peak expiratory flow rate reference profile taken prior to the onset of the theophylline testing period, Euphyllin CR resulted in a marked improvement in airflow during both the day and night. Rhind et al.[64] demonstrated that both daytime and nighttime symptoms were reduced when asthma patients were treated with a sustained-release choline theophyllinate in an unequal twice daily fashion. Although the subjective quality of sleep improved with theophylline therapy, the electroencephalographic determined sleep quality decreased, with an increase in wakefulness and drowsiness and a reduction in nonrapid eye movement sleep.

Once Daily Theophylline Chronotherapy

In recent years there has been great interest in the efficacy of once daily administered in the evening sustained-release theophylline therapy for the control of nocturnal asthma. The first studies with theophylline compounds using this treatment concept were performed by Cole and Al-Khadar[65] and by Fairfax and colleagues,[66] both using 450 mg doses of a slow-release aminophylline preparation. These investigators demonstrated that there was significant improvement in morning peak expiratory flow rates in some patients, but the doses were felt to be inadequate. Barnes et al.[23] placed 12 asthmatic patients with nocturnal wheezing on a larger dose of a slow-release aminophylline (Phyllocontin, Napp Laboratories, England), making sure that the dose provided an early morning STC of at least 12 mg/L. The mean dose for this study amounted to 10.4 mg/kg theophylline at bedtime. Peak expiratory flow rates were significantly higher in the morning on aminophylline as compared to placebo. Additionally, all patients reported symptomatic relief on aminophylline, and they used only one-fourth as much bronchodilator metered-dose inhaler medication as compared to placebo treatment. According to Barnes, theophylline-induced side effects are uncommon at night, possibly due to less pronounced cardiovascular effects during recumbency, when plasma epinephrine levels are known to decline. Since these early reports, numerous studies have been published, each demonstrating the efficacy of treating asthma patients, both children and adults, with nocturnal disease expression with once daily sustained-release theophylline chronotherapy.[10–12,24,28–30,67–79]

In a group of eight diurnally active adult asthmatic patients, the once daily sustained-release theophylline, Armophyllin (Armour, France) when taken at 8:00 PM for 8 days prevented the nocturnal decline in bronchial patency; the effect was considerably less when the identical dosage was administered to the same patients at 8:00 AM during another 8 days (Fig. 9)[28] Importantly, STC correlated highly with PEFR (r = .86; P < 0.01) (Fig. 14) when Armophyllin was administered in the evening but not in the morning. Subjective characteristics of sleep were also assessed and the effects of theophylline taken in the evening versus the morning were compared. Neither the sleep latency, duration, continuity and quality of sleep, nor the restfulness of sleep, self-rated upon awakening in the morning, was affected by the administration time of Armophyllin. The findings of Neuenkirchen et al.[10] have similarities to this previously discussed study of Reinberg et al.[28] Uniphyllin (Mundipharma, Germany), another once daily theophylline, was administered at 8:00 PM to nine patients with asthma. Nocturnal symptoms and the nighttime decline in airflow were markedly attenuated by evening Uniphyllin as compared to when the same patients were dosed with the same daily dose of a conventional twice daily sustained-release theophylline (Fig. 15). The PEFR circadian variability or amplitude was large, being equal to 47% of the 24-hour mean when the patients were treated with the twice daily theophylline, but only 17% when the treatment was Uniphyllin administered in the evening. The reduction in the amplitude of the circadian rhythm in PEFR following treatment with the once daily theophyl-

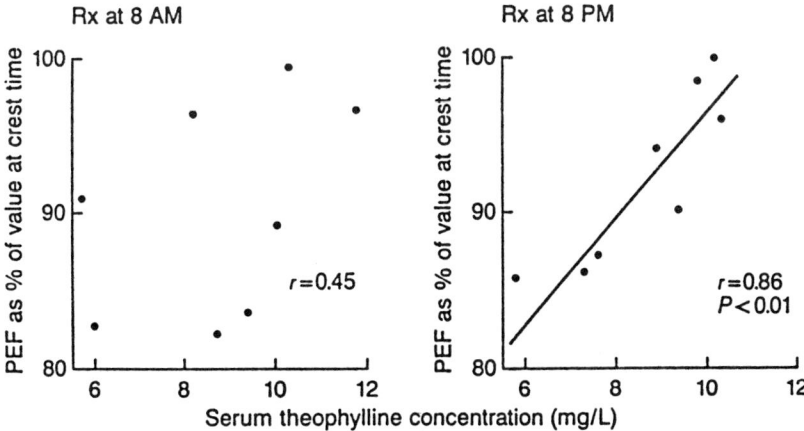

Figure 14. Correlation of time-qualified mean data between peak expiratory flow (PEF) [as a percentage of best value occurring between 12:00 PM and 5:30 PM in eight asthmatics] and serum theophylline concentration (STC). Note the high correlation found between STC and PEF when theophylline (Armophyllin) was administered in the evening. From Reinberg et al.[28]

line in the evening was due to the control and prevention of the decline in airway patency particularly between 2:00 and 6:00 AM.

A once daily theophylline formulation that has been found to be well-suited for evening administration is Uniphyl tablets. A large body of clinical evidence indicates that the chronotherapy of asthma with this medication delivers peak theophylline blood levels during the nighttime or early morning hours when airflow obstruction in many patients is likely to be most severe.[24,80] The use of the medication marks the first available antiasthmatic treatment in North America to incorporate dosing time as an important therapeutic element. Numerous studies have evaluated the once daily evening regime of Uniphyl. Placebo controlled, double-blind, randomized, crossover investigations include comparisons of morning versus evening administrations of this medication.[29,68] Evening administered Uniphyl, in similarly well-controlled studies, has been compared with a twice daily theophylline.[71,74] Uniphyl chronotherapy afforded better nocturnal control of asthma with significantly better early morning airflow than observed with the same daily doses of twice daily or once daily in the morning regimens; neverthe-

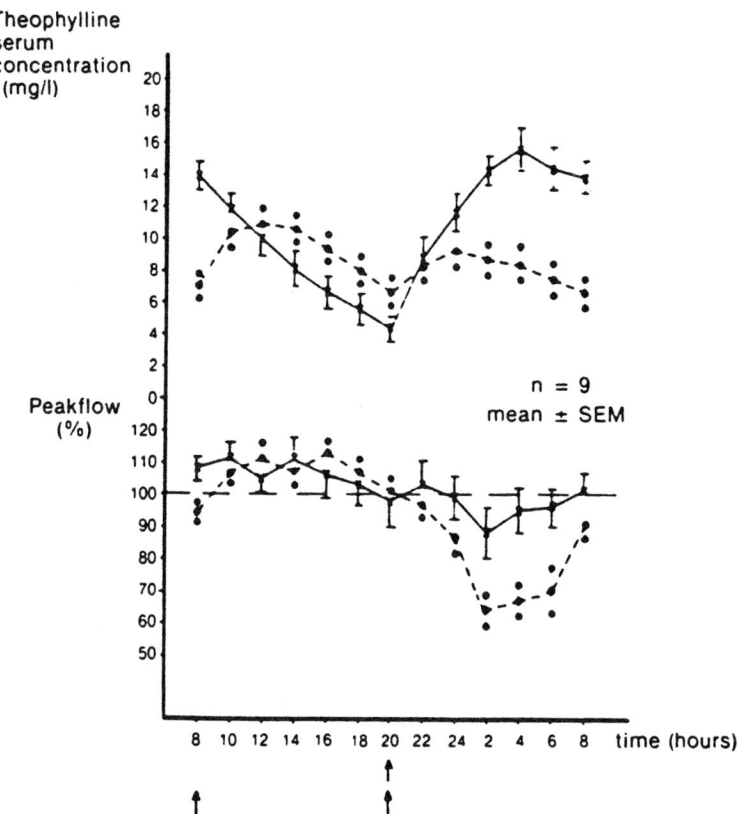

Figure 15. Comparison of STC and PEF of a 2000 hour ODSRT-Uniphyllin (solid line) and 0800/2000 hour twice daily ingested SRT (dashed line) formulation Phyllotemp in nine asthmatic patients studied while hospitalized. PEF expressed as a percentage change from the mean level while patients were managed on the twice daily or once daily theophylline schedule. Arrows indicate times of once or twice daily theophylline schedule. Note the fluctuation in the group mean STC greater for the ODSRT than the twice daily SRT schedule with higher STCs achieved during the nocturnal hours by the ODSRT schedule. The nocturnal decline in PEF was great when treatment was by the twice daily SRT; it was small when treatment was once daily in the evening. From Neuenkirchen et al.[10]

less, once daily evening therapy provided similar protection at other times of the day.[29,68,71,74]

Arkinstall et al.[74] treated a group of 22 patients with predominantly severe asthma and compared once daily in the evening Uniphyl with a conventional twice daily theophylline preparation. The comparison was based on symptom control and patient acceptance and preference with particular focus on early morning airflow parameters and the stability of daytime spirometry. This study was a placebo controlled, double-blind, randomized, crossover design investigation in which each phase consisted of a two-day washout period and seven days of theophylline dosing. On the last three days of each phase spirometry was performed and STC measured.

Both theophylline preparations maintained STCs within therapeutic range. The twice daily preparation was associated with more stable STCs than was the once daily theophylline which had a high peak-trough variation. In contrast, PEFR and FEV_1 were more stable when the patients were taking once in the evening therapy. The spirometry values were higher at 8:00AM with the once daily formulation than the twice daily theophylline (Fig. 16). With Uniphyl, there was no difference between 8:00 AM spirometry values and those obtained later in the afternoon and in the evening. Therefore, the once daily in the evening formulation improved early morning airflow and maintained satisfactory bronchodilation throughout the day. On the other hand, the PEFR and FEV_1 measured while the patients were treated with twice daily theophylline differed significantly between measurement times, with the higher values measured at 4:00 PM and the lowest ones at 8:00 AM. Both theophyllines were well tolerated; however, wheezing was reduced with once daily in the evening Uniphyl, the medication which showed a significant patient preference.

Consistent with the above findings, Welsh et al.[71] who performed spirometry at two early morning time points, found significantly higher airflow at 6:00 AM and 8:00 AM with Uniphyl. More recently, several investigators[11,75,77] have further substantiated this favorable effect of Uniphyl. Helm[77] reported the results of a large-scale, multi-investigator open outpatient evaluation comparing the effect of once daily, evening administered Uniphyl with previous twice or thrice daily methylxanthine therapies. Three hundred asthmatic patients, most (78%) prone to nocturnal disease, completed the study. Following a baseline period where the patients remained on their original theophylline and "concomitant" medica-

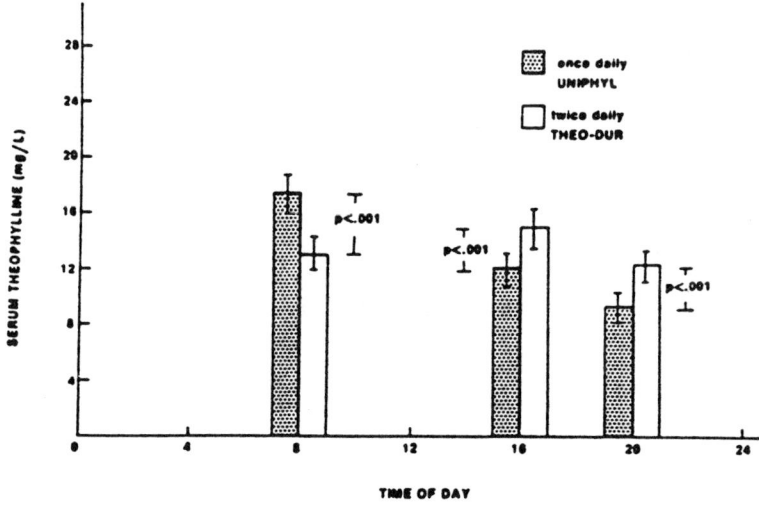

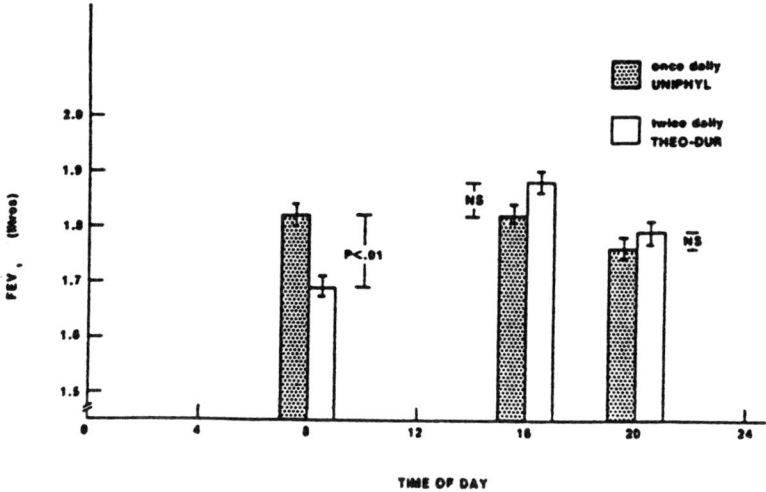

Figure 16. *Top panel* shows the mean steady-state serum theophylline concentrations (STC) in 22 asthmatics during once daily evening (8:00 PM) administered theophylline (Uniphyl) and twice daily (8:00 AM and 8:00 PM) theophylline (Theo-Dur). *Bottom panel* shows the mean FEV_1 values during once daily evening Uniphyl and twice daily Theo-Dur as described. Both theophyllines maintained STCs in therapeutic range. The twice daily therapy had a more stable STC profile over 24 hours than did the once daily formulation which had a considerable peak-trough variation. Nonetheless, airflow was more stable when the patients were taking once daily evening theophylline therapy. From Arkinstall et al.[74]

tions, the patients substituted their theophylline for evening administered Uniphyl on an approximately equivalent dose basis. Once daily therapy resulted in markedly fewer nocturnal asthma awakenings, less inhaler "dependency," and better control of morning wheeze, dyspnea, chest tightness and airflow which was measured at home or in the office. Similar results in 96 asthma patients were reported by Grossman.[75]

In subjects with very significant nocturnal asthma, a higher STC during the night and lower values during the daytime has been found to be successful. Martin et al.[11] showed that a higher STC, about 16 mg/L achieved with 7:00 PM administered Uniphyl produces a marked improvement in airflow overnight as compared to a level of about 11.5 mg/L accomplished with twice-daily Theo-Dur. Even though the twice daily theophylline had higher daytime theophylline levels, the FEV_1 measured every two hours during the daytime was not different. Sleep quality and architecture were not different between the two theophylline regimens, except with the once daily preparation there was a decreased number of hypopneas and fewer minutes below an oxygen saturation of 90 percent.

Not all studies have demonstrated once daily evening administered theophylline therapy superiority. Fanta and McFadden[81] and Faushter et al.[69] found evening therapy with Uniphyl tablets to be about equally effective as twice daily therapy with Theo-Dur. These studies involved few patients and did not restrict their recruitment to individuals with nocturnal asthma complaints.

Recently, D'Alonzo et al.[12] reported similar findings as Martin et al.[11] using another sustained-release theophylline formulation designed for once daily evening administration (TheoNite) for the control of asthma. TheoNite (Euphylong, Byk Gulden, Germany) is an encapsulated micro-osmotic system which allows pH-independent *in vitro* release of theophylline.[82] After a theophylline clearance study to substantiate normal or slow metabolism of the drug, a dose-tritration period and a 24-hour baseline (free of all medication) spirometric study, 25 asthmatics were randomized to 7-day treatment phases with either once daily in the evening TheoNite or twice daily Theo-Dur. Each outpatient segment of 6 days was followed by a 24-hour inpatient study on day 7 when STC and spirometry were obtained every 2 hours (Fig. 17). Twice daily theophylline therapy produced a nearly constant STC over the 24 hours, while the once daily in the evening therapy was associated with a larger peak-to-trough drug level fluctuation, with higher

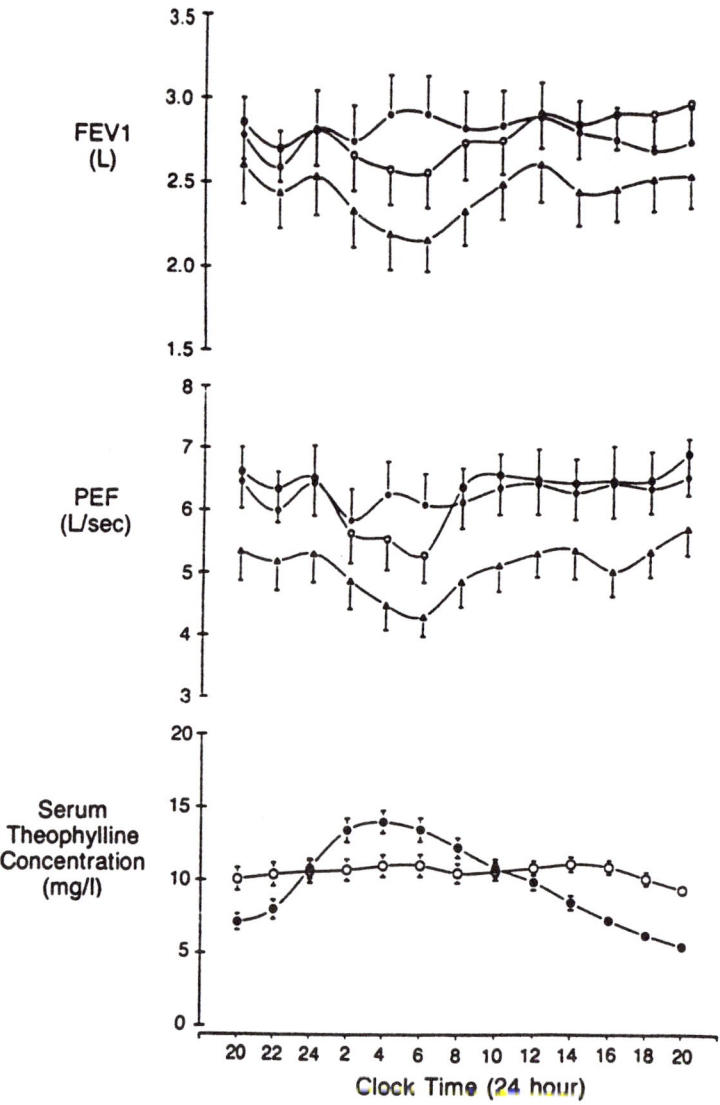

Figure 17. Mean FEV_1 peak expiratory flow (PEF), and serum theophylline concentration ($\triangle$ = baseline; $\bullet$ = once daily theophylline (TheoNite) administered at 8:00 PM; $\circ$ = twice-daily theophylline (Theo-Dur) dosed at 8:00 AM and 8:00 PM). Both theophylline medications improved airflow over the entire 24-hour period when compared to baseline values. A comparison between both medications shows a higher FEV_1 and PEF between 2:00 and 6:00 AM with the once daily regimen, at other times they were similar. From D'Alonzo et al.[12]

levels overnight and lower ones in the afternoon and evening, at the end of the dosing interval. Importantly, when compared to baseline both therapies improved airflow over the entire 24 hours. However, in the early morning hours, between 2:00 and 6:00 AM, both PEFR and FEV_1 were significantly greater with the once daily theophylline and correlated with the STC (Fig. 18). This was not the case for the twice daily theophylline. The improvement in airflow over baseline values between 2:00 and 6:00 PM was not correlated with STC of either treatment regime.

Beta-agonist Therapy

Plasma epinephrine concentration exhibits marked circadian rhythmicity in both normal subjects[83] and patients with asthma,[84,85] with lowest levels at 4:00 AM and highest at 4:00 PM. This is due to variability in catecholamine secretion from the adrenal medulla,[86] regulated by the hypothalamus,[87] the site of the central "clock,"[88] since plasma clearance seems to be unchanged.[86] This fall in plasma epinephrine concentration at night closely correlates with the fall in airflow, suggesting that the hormone is protective against asthma in that its withdrawal overnight induces bronchoconstriction. This protective effect may be due to direct beta-receptor stimulation on airway smooth muscles leading to bronchodilation or by an indirect stabilizing influence on airway inflammation. In support of this concept, plasma histamine concentrations, perhaps an indication of inflammatory cell secretion, rise at night in certain asthmatic patients, and its concentration inversely correlates with PEFR and plasma epinephrine levels.[85] An intravenous infusion of a low concentration of epinephrine not only reverses the fall in PEFR,[89] but also in the elevated plasma histamine concentration[85] at night. In fact, inhalation of epinephrine easily restores PEFR when administered during the circadian trough (Fig. 19).[89]

As mentioned previously, beta-agonist inhaler therapy administered routinely does not seem to be as effective as sustained-release theophylline therapy in the management of nocturnal asthma.[7,8,90] It would be gratifying if an inhaled medication could provide protection all night long. Unfortunately, the inhalation of the recommended dose of currently available (in the U.S.) $beta_2$-agonists from a metered-dose inhaler at bedtime provides bronchodilation for only 4 to 6 hours, and little protection for most cases against early

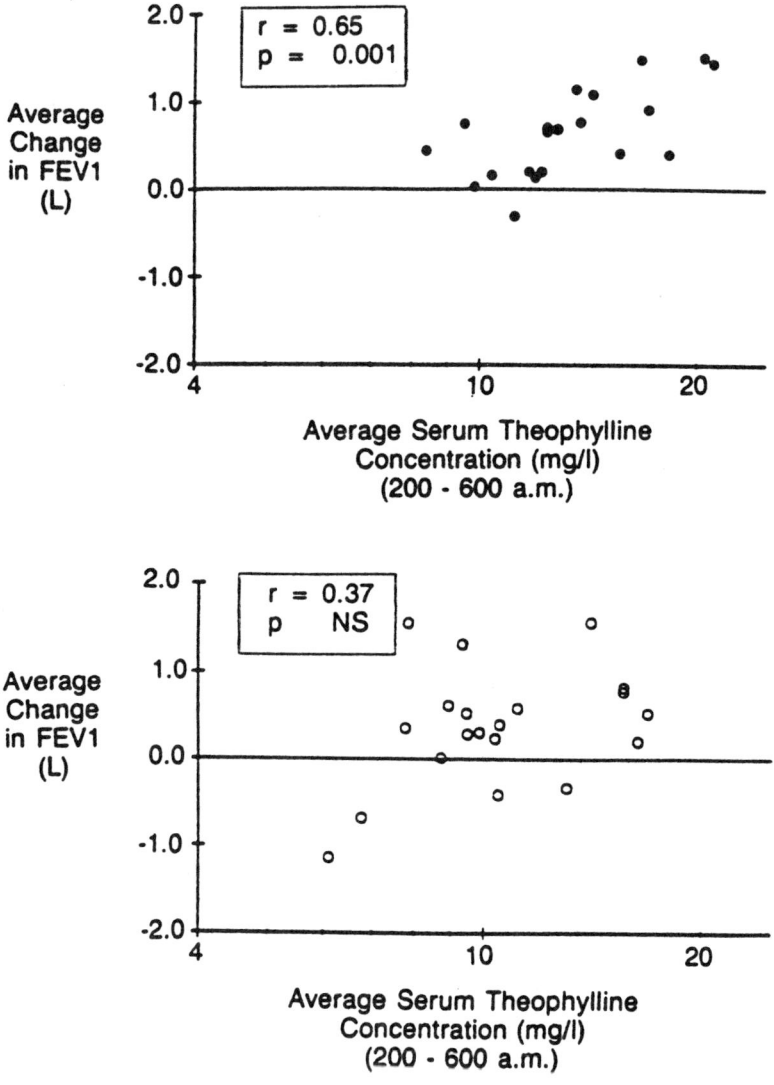

Figure 18. Time-averaged values of the change from baseline between 2:00 and 6:00 AM in FEV_1 in relation to the average serum theophylline concentration (STC) after repeated dosing with either a once daily in the evening theophylline (TheoNite) (top panel) or a twice daily theophylline (Theo-Dur) (bottom panel). When TheoNite was administered in the evening, a significant correlation was found between the nocturnal time-averaged change from baseline in FEV_1 and the STC. This relationship was not found with the twice daily theophylline preparation. From D'Alonzo et al.[12]

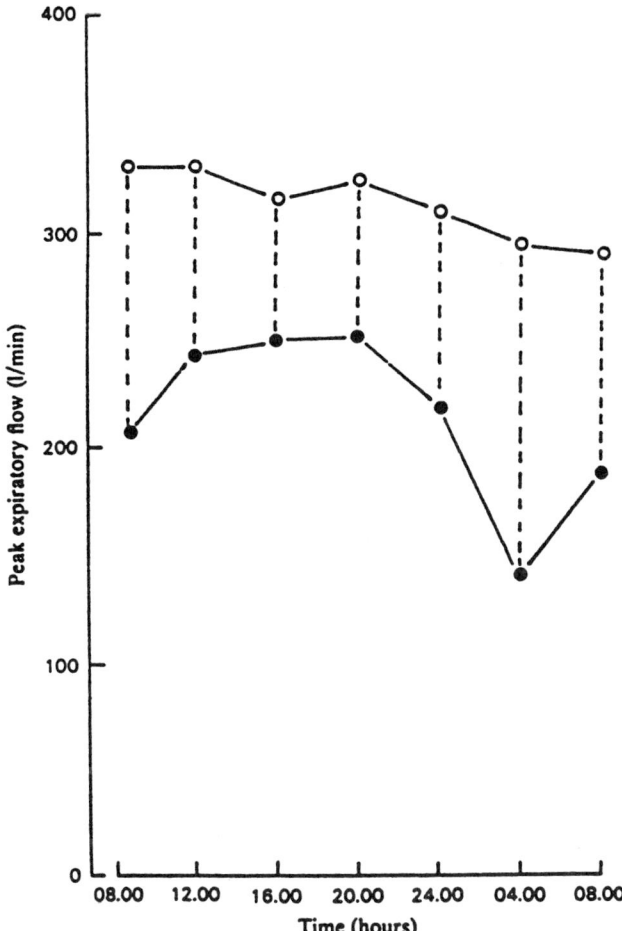

Figure 19. Mean circadian variation in peak expiratory flow (PEF) in five asthmatic patients before (●) and after (○) inhaled epinephrine (0.56 μg). Inhalation of epinephrine restored PEF when administered during the circadian trough. From Barnes et al.[82]

morning asthma. Longer acting beta-agonist inhalants appear promising, but oral sustained-release beta$_2$-agonist therapy is currently available for the treatment of nocturnal asthma.

Beta$_2$-agonist Inhalation Therapy

Day-night differences in the effect of metaproterenol (Alupent, Boehringer Ingelheim) were studied in diurnally active, healthy and asthmatic children.[91,92] Airways resistance and dynamic lung compliance were measured at 4 different times, 7:30 AM, 11:30 AM, 4:30 PM, and 10:30 PM, before and after the inhalation of metaproterenol. Metaproterenol was found to be minimally effective when inhaled around midday and in the afternoon, when airways resistance, i.e., airflow, was best during the 24 hours. Conversely, inhalation of the beta-agonist at the airflow circadian minimum, at 7:30 AM, resulted in greatest bronchodilation as evidenced by the largest reduction in airways resistance and an improvement in dynamic lung compliance. Similar findings were detected in studies on healthy, nonasthmatic adults. The bronchodilator effect of isoproterenol (a short-acting inhaled β-agonist), when inhaled during the night, caused statistically significantly greater bronchodilation than during the day.[93]

Inhaled beta-agonists have been studied alone and in combination with other inhalant medications for the treatment of nocturnal asthma. Horn and colleagues studied 14 asthmatic patients with nocturnal symptoms and morning reductions in peak expiratory flow rate.[94] In this uncontrolled, retrospective study each patient was treated with inhaled albuterol for 1 or 2 weeks, followed by inhaled beclomethasone dipropionate, in addition to albuterol, for 2 more weeks. PEFR was measured three times daily; on arising, in the early evening, and at bedtime. Albuterol (800 µg) was administered four times daily, and beclomethasone was then added in a dose of 400 µg four times daily. Mean PEFR improved 8% with Albuterol and a further 5% when beclomethasone was added. Eight patients demonstrated a striking reduction in their PEFR variability. Morning dips in PEFR were reduced 56% during Albuterol therapy, and the addition of the inhaled corticosteroid reduced the dips to only 31% of the baseline value. In three patients the morning falls in airflow were eliminated. More recently, in a study involving eleven asthmatics, the addition of an inhaled long-acting anticholinergic bronchodilator, atropine methylnitrate, to inhaled albuterol by nebulized mist treatment failed to show a statistically significant difference from albuterol alone for the treatment of nocturnal asthma.[95] Similarly, inhaled albuterol has been compared to a combination of another inhaled anticholinergic bronchodilator, ipratropium bromide, and a beta-agonist, fenoterol, which is available in Europe in a metered-dose inhaler (Duovent, Boehringer Ingelheim) for the control of nocturnal asthma.[96] Over the 10 weeks of the study there was no difference between Duovent and albuterol in any of the parameters measured, including morning and evening PEFR and diary recorded symptoms of nocturnal asthma. However, Carpenter and colleagues, using a higher daily dose of inhaled fenoterol for 3 consecutive days eliminated the circadian rhythm in airflow and they showed that fenoterol was most effective when administered upon arising.

New, long-acting inhaled beta-agonist therapies are now being evaluated for the control of asthma. To date, formoterol[97] and salmeterol[98] have been studied for the treatment of nocturnal asthma. Studies performed during the daytime have demonstrated that formoterol is a long-acting bronchodilator with an acceptable side effect profile.[99] Therefore, formoterol was more recently studied for its effect during the nighttime.[97] Formoterol was administered by metered-dose inhaler (12 μg) to 16 patients with nocturnal asthma and compared to 200 μg of albuterol in a double-blind, randomized fashion. Inhalations were given at 10:00 PM and the FEV_1 was measured at regular intervals up to 12 hours postadministration. Formoterol produced sustained bronchodilation throughout the night and minimized the requirement for "rescue" inhalations with albuterol as compared to treatment with albuterol. This study did not evaluate efficacy during sustained therapy; therefore, the influence of formoterol on sleep was not assessed.

Salmeterol is another new long-acting beta-agonist which has been shown to produce effective bronchodilation for 12 hours after an inhaled dose.[100] Like formoterol, salmeterol has been evaluated for its effect on nocturnal bronchospasm and sleep quality in patients with nocturnal asthma.[98] A randomized, double-blind, placebo-controlled, crossover study which involved 20 patients with nocturnal asthma showed that salmeterol improved the overnight PEFR using both 50 μg and 100 μg doses twice daily (Fig. 20).[98] These improvements were associated with a reduction in the mean 24-hour "rescue" use of albuterol by both doses of salmeterol. Additionally,

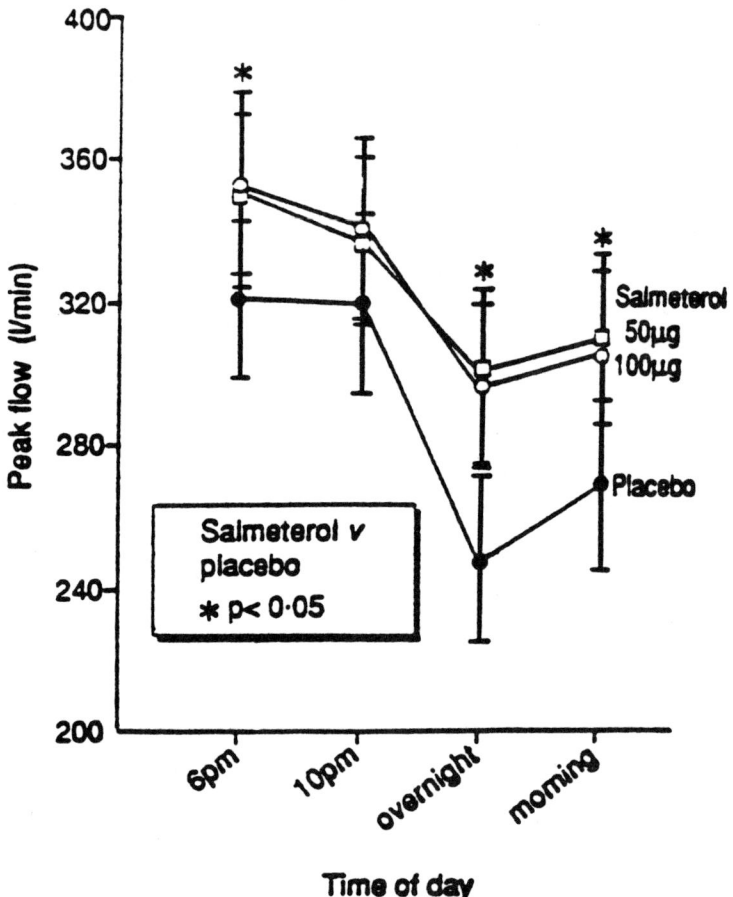

Figure 20. Mean peak expiratory flow in 18 asthmatic patients administered placebo or Salmeterol twice daily by inhalation. From Fitzpatrick et al.[98]

an objective improvement in sleep quality was found while the patients were taking salmeterol 50 μg twice daily. Less time spent awake or in light sleep and more stage 4 sleep were noted (Fig. 21). Although these changes in sleep quality appear small at first glance, they are likely significant from the standpoint of cognitive performance. Daytime cognitive performance was not measured in this study, but others have shown that even small amounts of sleep deprivation can lead to appreciable impairment of cognitive per-

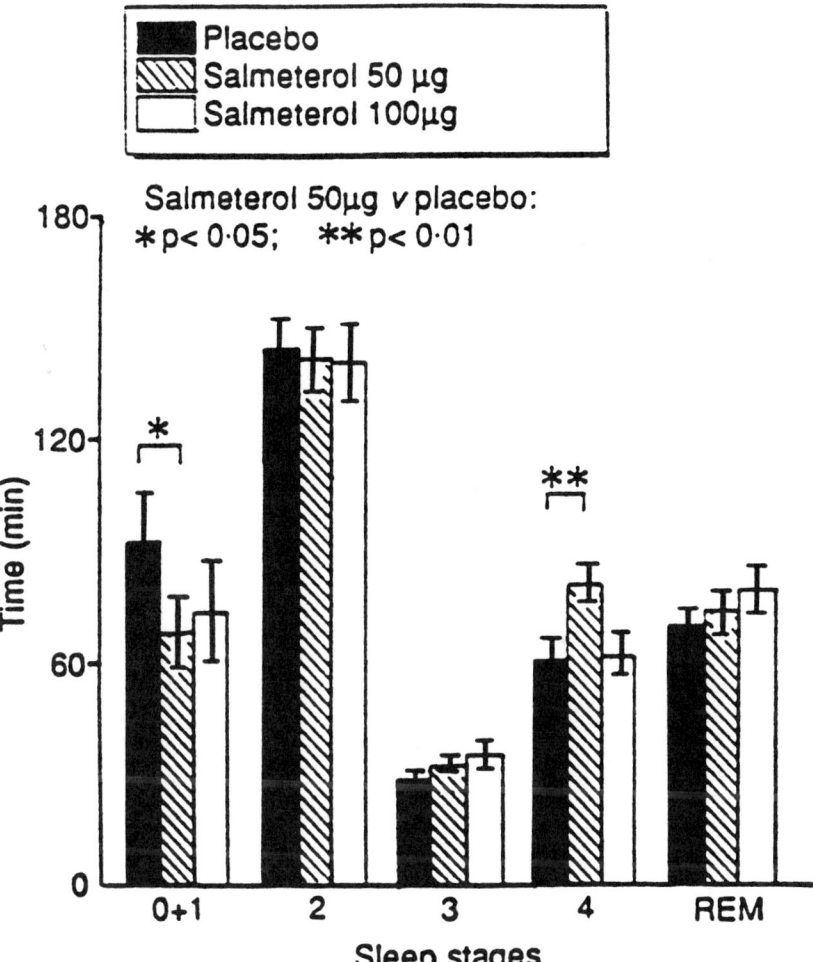

Figure 21. Mean time spent in each stage of sleep by 18 asthmatic patients receiving placebo or salmeterol twice daily by inhalation. Less time awake or in light sleep and more stage 4 sleep were noted during therapy with inhaled salmeterol. From Fitzpatrick et al.[98]

formance.[101,102] In fact, it has been recently shown that patients with nocturnal asthma who had minor problems with sleep quality do have impaired daytime cognitive performance to a degree which could affect work performance.[103]

Sustained-release Beta-agonists

Terbutaline

Compared to ordinary terbutaline tablets, treatment with sustained-release terbutaline tablets increases morning airflow in asthmatic patients.[104] Koeter et al.[105] and Postma et al.[106,107] further evaluated the chronotherapeutic advantage of sustained-release terbutaline tablets for the management of early morning dyspnea. Eight men with partially reversible airflow obstruction and who had a 20% circadian variation in PEFR were treated either with placebo or a slow-release terbutaline tablet (Bricanyl Depot, AB Draco, Sweden) using a dosing regime consisting of 5 mg at 8:00 AM and 10 mg at 8:00 PM. Throughout the study maintenance therapy, including theophylline, anticholinergic drugs, and inhaled corticosteroids, was withheld. On day 8 of each period, airway function and serum terbutaline concentration were measured at 4- and 2-hour intervals, respectively. PEFR and FEV_1 during terbutaline therapy showed a significant increase at all intervals throughout the day and night as compared to placebo (Fig. 22). The nocturnal decline in airflow was averted and the daytime level of PEFR and FEV_1 improved with a complete loss of circadian rhythmicity in bronchial patency. Even though twice the dose of terbutaline was given at 8:00 PM than at 8:00 AM, nearly comparable serum concentrations of terbutaline were found during the morning and evening 12-hour dosing intervals. Furthermore, nocturnal decreases in arterial oxygen tension and saturation were eliminated with unequal morning-evening terbutaline dosing.

Dahl and colleagues also performed a study to determine whether a higher evening dose of twice daily sustained-release terbutaline (Bricanyl Depot) would produce a further improvement in patients with a pronounced circadian variation in airflow.[108] In a randomized, double-blind, crossover study nine adult patients were treated during three periods with SR terbutaline 7.5 mg twice daily, 7.5 mg in the morning and 15 mg in the evening, and placebo twice

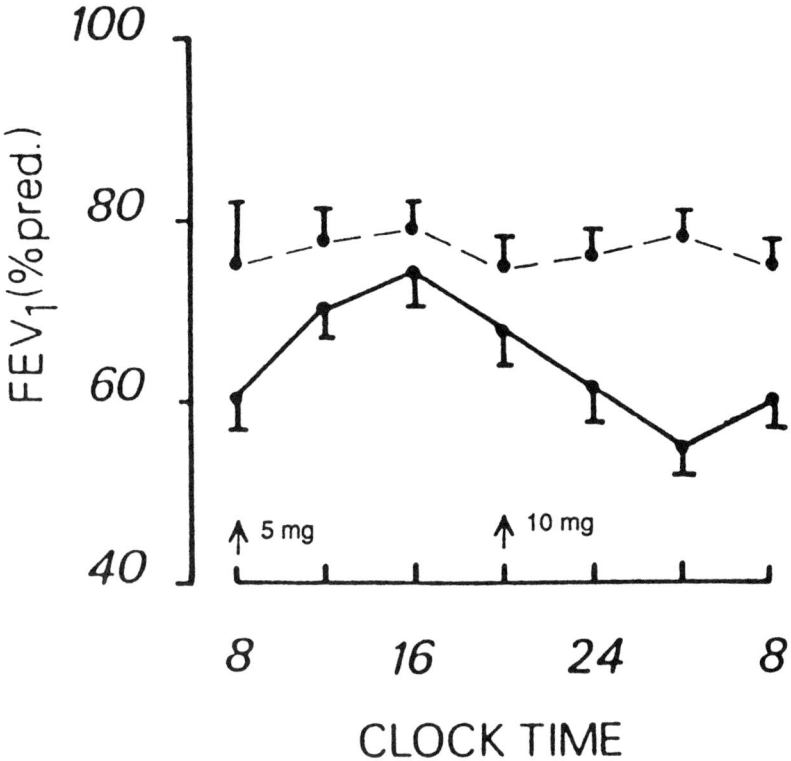

Figure 22. Terbutaline given as 5 mg at 0800 hours and 10 mg at 2000 hours induced a rise in the 24-hour mean level and averted the nocturnal decline of bronchial patency, as assessed by the 1-second forced expiratory flow (FEV_1), in eight patients suffering from chronic reversible airways obstructive disease. Darkened and dashed line plots represent before and during terbutaline treatment conditions. From Koëter et al.[105]

daily. Each treatment period was one week in duration. The nocturnal decline in PEFR was 45% with placebo, 27% after the lower, and 22% after the higher active drug evening doses; leading to a mean morning PEFR that was significantly higher with the high rather than the low evening dose. In another study, a single evening 15 mg dose of SR terbutaline (Bricanyl Depot) was administered at 10:00 PM to 14 adults with nocturnal asthma.[109] This double-blind, placebo-controlled, crossover study was divided into three 8-day periods, a run-in period and 2 study periods, one on active drug and the other on placebo. At the end of each period, an inpatient evaluation with serial spirometry was performed. There were no differences in FEV_1 between placebo and terbutaline at 4:00 and 10:00 PM, but after tablet administration FEV_1 steadily increased during the night with terbutaline, and decreased following placebo (Fig. 23). On request, the patients were permitted to administer a short-acting beta-agonist inhalation for relief of patient perceived asthma. Twenty-eight extra doses were taken during the placebo night, compared to 10 during the terbutaline night. Medication tolerance was good, but sleep quality was not assessed. However, Stewart and colleagues[110] assessed sleep quality by electroencephalography. In this placebo-controlled, double-blind, crossover study 7.5 mg twice daily terbutaline improved morning PEFR and reduced nocturnal inhaler usage without an impairment of sleep quality. As compared to placebo, the use of terbutaline did not affect total sleep time, sleep architecture, and the number of awakenings. However, at the dose tested, this medication was not ideal. A marked improvement in airflow did not occur, and awakenings related to asthma or the use of β-agonist inhalers at night continued to occur.

Sustained-release terbutaline has been compared to other antiasthma medication for the control of nocturnal asthma and related problems. SR terbutaline has been compared to sustained-release theophylline.[111-113] SR theophylline therapy seems to be at least as efficacious as SR terbutaline,[111,112] if not better.[113] Heins and colleagues[113] compared the effects of a twice daily SR terbutaline and a SR theophylline with one-third of the daily dose taken in the morning and two-thirds in the evening using a double-blind crossover protocol in 11 patients with nocturnal asthma. On day 7 of each treatment period PEFR was measured every 2 hours over a 24-hour period. In the night and early morning there was no significant difference in airflow between the two treatments. During theophylline treatment, fewer inhalations of beta-agonist therapy were required and there were fewer side-effects. Similar findings occurred when a once in the evening SR theophylline was used.[112] When nocturnal body movement and oxygen saturation were assessed in another study of comparable design, no differences were detected between treatments.[111] Finally, SR-terbutaline (10 mg b.i.d.) has been compared to an inhaled steroid (budesonide, Pulmicort) (400 μg b.i.d.) and combined treatment.[114] During a 1-week run-in period the 37 patients who completed the study were treated with inhaled terbutaline, and nocturnal awakenings and PEFR were measured.

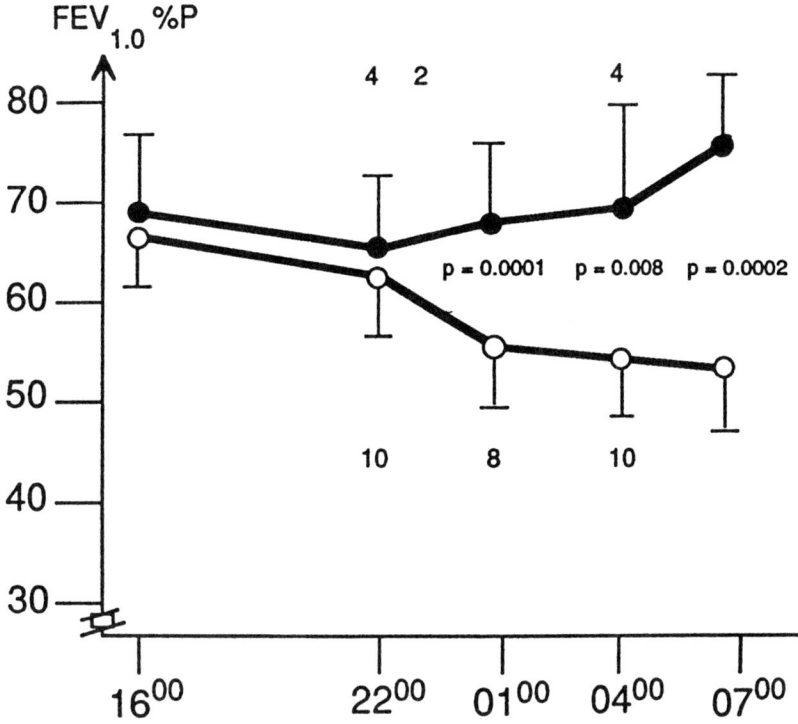

Figure 23. Mean percent predicted FEV_1 value during treatment with sustained-release terbutaline (●) and during placebo (○). The numbers above and below the curves indicated the number of extra doses of a short-acting beta-agonist inhalative for the relief of asthma. From Eriksson et al.[109]

At the completion of run-in the patients were randomly enrolled in a double-blind, multiple crossover study involving three treatment periods (mentioned above) of 3 weeks each. There were no clinically significant differences between the single treatments. There was a slightly lower overnight fall in PEFR with combined therapy; 7% compared to 9% during SR-terbutaline and 10% with budesonide (Fig. 24). Similarly, fewer nocturnal awakenings were found with combined therapy than either treatment alone. Finally, the concomitant daily number of puffs of inhaled terbutaline was significantly lower during treatment with budesonide and combination therapy compared to treatment with SR-terbutaline doses (Fig. 25).

Terbutaline is available in Europe in a prodrug form, bambut-

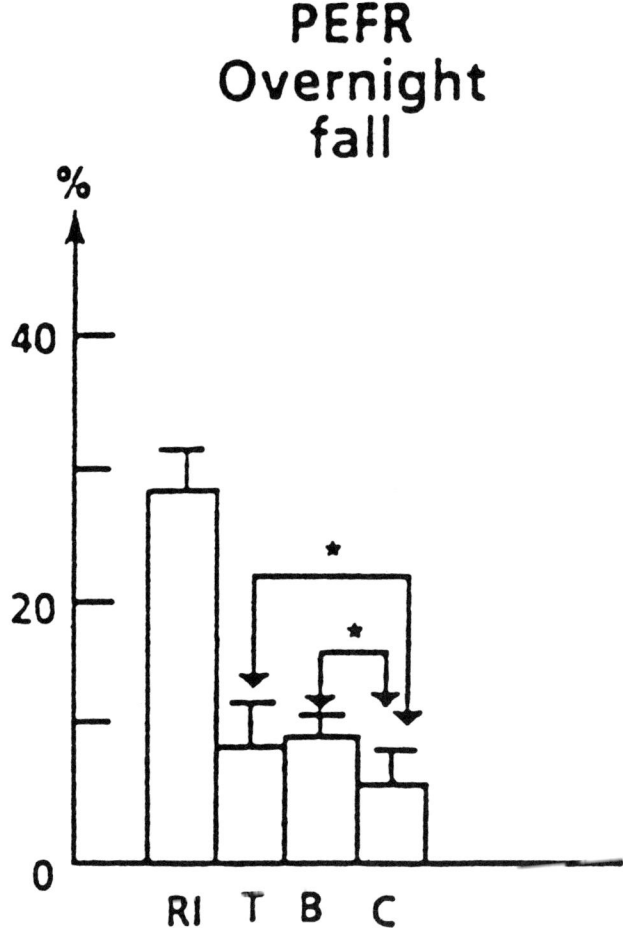

Figure 24. A comparison of the mean percentage fall in peak expiratory flow rate (PEFR) for several treatment periods. RI, run-in; T, sustained-release terbutaline; B, budesonide; C, combination treatment; *P< 0.05. From Dahl et al.[114]

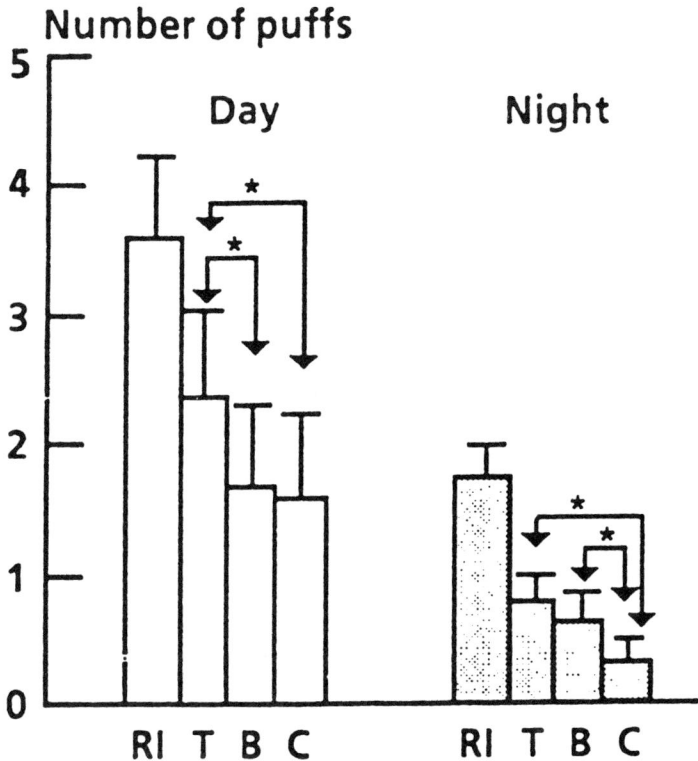

Figure 25. A comparison of the mean number of puffs of inhaled terbutaline taken during the day and night during several treatment periods previously described in Figure 24. From Dahl et al.[114]

erol (Bambec, Draco-Astra). Bambuterol is the bisdimethyl-carbonate of terbutaline. The terbutaline part of this prodrug is protected from being metabolized during absorption and its first pass through the liver, meaning that bambuterol acts as an inner depot from which terbutaline is slowly generated over the 24 hours.[115] Because of this, bambuterol has been found to be suitable for once daily administration, comparable to SR-terbutaline taken twice daily.[116] Recently a randomized, double-blind, crossover, placebo-controlled study of morning at 7:00 AM versus evening at 10:00 PM once daily bambuterol (20 mg) was completed by 29 diurnally active adult asthma patients.[117] Each of the three treatment periods (placebo plus once daily morning and evening bambuterol) lasted 1 week and on the sixth and seventh day an inpatient study was performed measuring bambuterol-derived terbutaline plasma concentration, spirometry, and drug tolerance every 3 hours. The 24-hour mean plasma terbutaline concentrations were comparable for morning and evening administered bambuterol. However, the maximum concentration of terbutaline was greater for the evening administration. The 24-hour mean FEV_1 was greater for the bambuterol treatments than for placebo (Fig. 26). The effectiveness of bambuterol was greater for the evening versus the morning administration schedule. Evening dosing resulted in a statistically significantly greater FEV_1 and PEFR in the morning upon arising in comparison to the morning dosing schedule and placebo treatment. Bambuterol had a low side effect profile regardless of when it was administered; however, evening dosing was best tolerated by patients (less tremor).

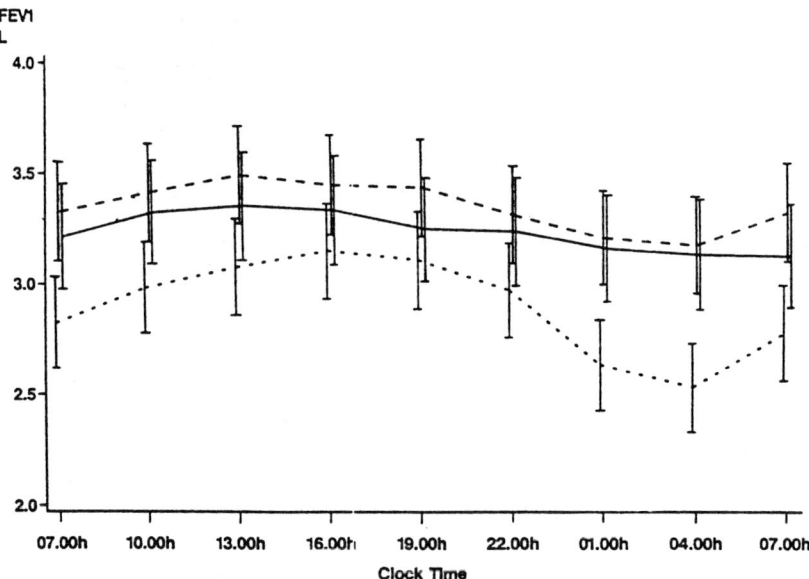

Figure 26. Mean FEV_1 values measured every 3 hours over 24 hours during three steady-state condition treatment periods in 29 asthmatic patients. The dotted line refers to a placebo period, the solid and dashed lines to treatment with bambuterol administered in the morning and evening, respectively. The 24-hour mean FEV_1 was greater for bambuterol than for placebo. From D'Alonzo et al.[117]

The chronotherapy of nocturnal asthma with tablet terbutaline and bambuterol may be achieved with either unequal morning-evening dosing of the former by means of a twice daily dosing regimen or an evening, once-a-day regimen of the latter.

Albuterol

Sustained-release albuterol[118] has been evaluated on a limited basis for the control of nocturnal asthma. Storms et al.[119] at four medical centers studied in a double-blind fashion 98 stable asthma patients who had histories of awakening three times weekly with asthma, nocturnal declines of PEFR of at least 15%, and who were not taking oral beta-agonist medications. These patients were randomly treated with either SR-albuterol (Proventil Repetabs, Schering Corp.), 4 mg in the morning and 4 to 16 mg at bedtime, or a placebo for 2 weeks. Diary data and PEFRs were recorded on a daily basis and spirometry was measured after a 1-week baseline period and at the end of the first and second weeks of medication. Forty-seven patients received SR-albuterol and 51 received placebo. Albuterol-treated patients had fewer nocturnal awakenings, fewer nights with a greater than 15% dip in PEFR between bedtime and morning arousal, and an improvement in mean FEV_1 over the entire study. Tremor was noted more frequently in the albuterol group (18 of 53 patients); however, only two patients in this group discontinued therapy because of side effects.

Another study compared a SR-albuterol (Ventolin Spandets, Glaxo) medication with a SR-theophylline in the form of aminophylline (Phyllocotin continuous tablets) for the treatment of nocturnal asthma.[66] Fourteen asthma patients with regular nocturnal exacerbations were studied under double-blind, randomized conditions taking either 16 mg of SR-albuterol, 450 mg of SR-aminophylline or placebo at midnight for 1-week periods each. Asthma diary data and PEFR were measured daily and on the last night of each week patients were admitted to the hospital for spirometry, and plasma drug concentrations were measured. Mean PEFR in the morning, upon awakening, was significantly higher on bronchodilator therapy but, in this study, neither medication abolished the morning dip. Those patients unresponsive to either oral aminophylline or albuterol all showed rapid increases in PEFR after a standard inhaled dose of albuterol upon awakening. Medication tolerance was not discussed. In another study,[120] which used the same medications and dosing regime in 26 asthmatics, there were appreciable side effects. Each drug produced an equal incidence of their chief side effect (gastrointestinal distress for aminophylline, and tremor for albuterol) in about 20% of patients, but both medications improved nocturnal asthma symptoms and peak expiratory flow rates.

Summary

Effective chronotherapy of asthma depends on an understanding of an individual's temporal pattern of disease expression and the chronokinetics and (chrono)effectiveness of each medication used for its treatment. Theophylline and beta-adrenergic agonist bronchodilator therapies should be used in ways that enhance efficacy and reduce toxicity with goals to ensure continuity of sleep at night and top performance during diurnal activity without medication related side effects. A chronotherapeutic approach to the treatment of asthma is particularly warranted for more severely affected patients, those that are likely to be dependent on optimization of bronchodilator medications.

References

1. Dethlefsen U, Repges R. Ein neues therapieprinzip bei nächtlichem asthma. Med Klin 1985; 80:44–47.
2. Turner-Warwick M. Epidemiology of nocturnal asthma. Am J Med 1988; 85(suppl 1B):6–8.
3. Douglas NJ. Asthma at night. Clin Chest Med 1985; 6:663–674.
4. Hetzel MR, Clark TJH, Branthwaite MA. Asthma: analysis of sudden deaths and ventilatory arrest in hospital. Br Med J 1977; 1:808–811.
5. Bateman JRM, Clark SW. Sudden death in asthma. Thorax 1979; 34:40–44.
6. British Thoracic Association. Death from asthma in two regions of England. Br Med J 1982; 285:1251–1255.
7. Joad JP, Ahrens RC, Lindgren SD, Weinberger MM. Relative efficacy of maintenance therapy with theophylline, inhaled albuterol, and the combination for chronic asthma. J Allergy Clin Immunol 1987; 79:78–85.
8. Zwillich CW, Neagley SR, Cicutto L, et al. Nocturnal asthma therapy: inhaled bitolterol versus sustained-release theophylline. Am Rev Respir Dis 1989; 139:470–474.
9. Nassif EG, Weinberger M, Thompson R, Huntley W. The value of maintenance theophylline in steroid-dependent asthma. N Engl J Med 1981; 304:71–75.

10. Neuenkirchen H, Wilkens JH, Oellerich M, Sybrecht GW. Nocturnal asthma: effect of a once per evening dose of sustained-release theophylline. Eur J Respir Dis 1985; 66:196–204.
11. Martin RJ, Cicutto LC, Ballard RD, et al. Circadian variations in theophylline concentrations and the treatment of nocturnal asthma. Am Rev Respir Dis 1989; 139:475–478.
12. D'Alonzo GE, Smolensky MH, Feldman S, et al. Twenty-four hour lung function in adult patients with asthma: chronoptimized theophylline therapy once-daily dosing in evening versus conventional twice-daily dosing. Am Rev Respir Dis 1990; 142:84–90.
13. Smolensky MH, McGovern JP, Scott PH, Reinberg A. Chronobiology and asthma. II. Body-time-dependent differences in the kinetics and effects of bronchodilator medications. J Asthma 1987; 24:90–134.
14. Taylor DR, Duffin D, Kinney CD, et al. Investigation of diurnal changes in the disposition of theophylline. Br J Clin Pharmacol 1983; 16:413–416.
15. Reed RC, Schwartz HJ. Circadian variation in steady-state trough theophylline concentrations. Therap Drug Monitor 1986; 8:155–160.
16. Lesko LJ, Brousseau D, Canada AT, Eastwood G. Temporal variations in trough serum theophylline concentrations at steady state. J Pharm Sci 1980; 69:357–359.
17. Nakano SH, Watanabe K, Nagai K, et al. Time-of-day effect on theophylline kinetics following oral dosing with aminophylline. IRCS Med Sci 1982; 10:798–799.
18. Decourt S, Fodor F, Flouvat B, et al. Pharmacokinetics of theophylline in night-workers. Br J Clin Pharmacol 1982; 13:567–569.
19. Kyle GM, Smolensky MH, Thorne LG, et al. Circadian rhythm in the pharmacokinetics of orally administered theophylline. In: Recent Advances in the Chronobiology of Allergy and Immunology, Smolensky MH, Reinberg A, McGovern JP (eds). Pergamon Press, Oxford, 1980, pp 95–111.
20. Kelly HW, Murphy S. Efficacy of a 12-hour sustained-release preparation in maintaining therapeutic serum theophylline levels in asthmatic children. Pediatrics 1980; 66:97–102.
21. Scott PH, Tabachnik E, MacLeod S, et al. Sustained-release theophylline for childhood asthma: evidence for circadian variation of theophylline pharmaceutics. J Pediatr 1981; 99:476–479.
22. Smolensky MH, Scott PH, Kramer WG. Clinical significance of day-night differences in serum theophylline concentration with special reference to Theo-Dur. J Allergy Clin Immunol 1986; 78:716–722.
23. Barnes PJ, Greening AP, Neville L, et al. Single dose slow-release aminophylline at night prevents nocturnal asthma. Lancet 1982; 1:299–301.
24. Goldenheim PD, Conrad EA, Schein LK. Treatment of asthma by a controlled-release theophylline tablet formulation: a review of the North American experience with nocturnal dosing. Chronobiol Int'l 1987; 4:397–408.
25. Straughn S, Meyer M, Golub A, Gonzalez M. A chronopharmacokinetic model for controlled-release formulations. Ann Rev Chronopharmacol 1984; 1:93–96.

26. Straughn AB, Meyer MC, Golub AL, Gonzalez MA. Administration of Theo-Dur once daily vs twice daily. In: Sustained Release Theophylline and Nocturnal Asthma, Isles AF, vonWichert P (eds). Exerpta Medica, Amsterdam, 1985, pp 116–124.

27. Thuresson S-O, Friberg K, Atze G, Bengtsson B. Unequal divisions of daily doses of theophylline: a pharmacokinetic study. In: Sustained Release Theophylline and Nocturnal Asthma, Isles AF, vonWichert P (eds). Excerpta Medica, Amsterdam, 1985, pp 125–134.

28. Reinberg A, Pauchet F, Ruff F, et al. Comparison of once-daily evening versus morning sustained-release theophylline dosing for nocturnal asthma. Chronobiol Int'l 1987; 4:409–420.

29. Rivington RD, Calcutt L, Child S, et al. Comparison of morning versus evening dosing with a new once-daily oral theophylline formulation. Amer J Med 1985; 79(suppl 6A):67–72.

30. Van den Brande P, Nys J, Tjandramaga TB, et al. Once-daily dosing of a new ultrasustained-release theophylline preparation. Respiration 1987; 52:144–153.

31. Lamont H, Pauwels R, Van der Straeten M. The effect of dosing time on the pharmacokinetics and pharmacodynamics of a "once-a-day" sustained release theophylline preparation. Br J Clin Pharmacol 1987; 24:735–742.

32. Smolensky MH, Scott PH, Harrist RB, et al. Administration-time-dependency of the pharmacokinetic behavior and therapeutic effect of a once-a-day theophylline in asthmatic children. Chronobiol Int'l 1987; 4:435–448.

33. Vaughan L, Milavetz G, Hill M, et al. Food-induced dose-dumping of Theo-24, a "once-daily" slow-release theophylline product. Drug Intell Clin Pharm 1984; 18:510.

34. Karim S. Theophylline with food: Theo-24. Amer Pharmacol 1985; NS25(3):132–133.

35. Hendeles L, Weinberger M, Milavetz G, et al. Food-induced "dose-dumping" from a once-a-day theophylline product as a source of theophylline toxicity. Chest 1985; 87:758–765.

36. Jonkman JHG. Food interactions with once-a-day preparations: a review. Chronobiol Int'l 1987; 4:459–466.

37. Jonkman JHG, Van der Boon WJV, Balant LP, et al. Chronopharmacokinetics of theophylline after sustained release and intravenous administration to adults. Eur J Clin Pharmacol 1984; 26·215–225.

38. Giacona N, Elvin AT, Seligsohn R, et al. Diurnal variation in theophylline elimination. Drug Intell Clin Pharmacol 1983; 17:452.

39. Smolensky MH, Scott PH, McGovern JP, Albright D. Circadian differences in theophylline effect during constant rate infusion with aminophylline. Ann Rev Chronopharmacol 1986; 3:139–142.

40. Goo RH, Moore JG, Greenberg E, Alazraki NP. Circadian variation in gastric emptying of meals in man. Gastroenterol 1987; 93:515–518.

41. Moore JG, Englert Jr E. Circadian rhythm of gastric acid secretion in man. Nature 1970; 226:1261–1262.

42. Vener KJ, Moore JG. Chronobiologic properties of the alimentary canal

affecting xenobiotic absorption. Ann Rev Chronopharmacol 1988; 4:259–283.

43. Thompson PJ, Kemp MW, McAllister WAC, Turner-Warwick M. Slow release theophylline in patients with airways obstruction with particular reference to the effect of food upon serum levels. Br J Dis Chest 1983; 77:293–298.

44. Pedersen S. Delay in the absorption rate of theophylline from a sustained release theophylline preparation caused by food. Br J Clin Pharmacol 1981; 12:904–905.

45. Kalstrand G. The effect of food intake on the absorption of different sustained release formulations. In: Sustained Release Theophylline and Nocturnal Asthma, Isles AF, vonWichert P (eds). Excerpta Medica, Amsterdam, 1985, pp 169–175.

46. Leeds NH, Gal P, Purchet AA, Walter JB. Effect of food on the bioavailability and pattern of release of a sustained release theophylline tablet. J Clin Pharmacol 1982, 22:196–200.

47. Sips AP, Edelbrock PM, Kulstad S, et al. Food does not affect bioavailability of theophylline from Theolin Retard. Eur J Clin Pharmacol 1984; 26:405–470.

48. Osman MA, Patel RB, Irwin DS, Welling PG. Absorption of theophylline from enteric coated and sustained release formulations in fasted and non-fasted subjects. Biopharm Drug Dispos 1983; 4:63–72.

49. Gonzales M, Betlach C, Frost W, et al. Food influence on theophylline pharmacokinetics from three sustained release formulations. Proceedings 39th Amer Pharm Assoc, Acad Pharm Assoc, Acad Pharm Sci, Oct 20, 1985.

50. Welling PG, Lyons TL, Craig WA, Trochta GA. Influence of diet and fluid on bioavailability of theophylline. Clin Pharmacol Ther 1975; 17:475–480.

51. Lagas M, Jonkman JHG. Greatly enhanced bioavailability of theophylline on post-prandial administration of a sustained release tablet. Eur J Clin Pharmacol 1983; 24:761–767.

52. Jonkman JHG, van der Boon WJV, Balant LP, leContonnec JY. Food reduces the rate but not the extent of the absorption of theophylline from an aqueous solution. Eur J Clin Pharmacol 1985; 28:225–227.

53. Jonkman JHG, van der Boon WJV, Balant LP, leContonnec JY. No effect of food on single dose bioavailability of sustained release theophylline (Sabidal): a comparison between Sabidal and choline theophyllinate solution. Internat J Pharmacol 1985; 25:113–117.

54. Pedersen S, Moller-Peterson J. Influence of food on the absorption rate and bioavailability of a sustained release theophylline preparation. Allergy 1982; 37:531–534.

55. Pedersen S, Moller-Peterson J. Erratic absorption of a slow release theophylline sprinkle product. Pediatrics 1984; 74:534–538.

56. Pedersen S, Moller-Peterson J. Influence of food on the absorption of theophylline from a sustained release formulation. Clin Allergy 1985; 15:253–261.

57. Warren JB, Turner C, Dalton A, et al. The effect of posture on the

sympathoadrenal response to theophylline infusion. Br J Clin Pharmacol 1983; 16:405–411.

58. Warren JB, Cuss F, Barnes PJ. Posture and theophylline kinetics. Br J Clin Pharmacol 1985; 19:707–709.
59. Thompson WD, Thompson PK, Dailey ME. The effect of posture on the composition and volume of the blood in man. J Clin Invest 1928; 5:573–604.
60. Smolensky MH, D'Alonzo GE, Kunkel G, Barnes PJ (eds). Circadian rhythm-adapted theophylline schedules for asthma. Chronobiol Int'l 1987; 4:301–466.
61. Darow P, Steinijans VW. Therapeutic advantage of unequal dosing of theophylline in patients with nocturnal asthma. Chronobiol Int'l 1987; 4:349–357.
62. Bruguerolle B, Philip-Joet F, Parrel M, Arnaud A. Unequal twice-daily, sustained-release theophylline dosing in chronic obstructive pulmonary disease. Chronobiol Int'l 1987; 4:381–386.
63. Schulz H-U, Frercks H-J, Hypa F. Vergleichende theophyllin-serumspiegel messungen über 24 stunden nach konventioneller dosierung eines theophyllin-retard-präparation über 4 tage. TherapieWoche 1984; 34:536–543.
64. Rhind GB, Connaughton JJ, McFie J, et al. Sustained release choline theophyllinate in nocturnal asthma. Br Med J 1985; 291:1605–1607.
65. Cole RB, Al-Khadar A. Effect of slow-release oral aminophylline on circadian variation in airflow obstruction in asthmatics. J Int Med Res 1979; 7(suppl 1):40–44.
66. Fairfax AJ, McNabb WR, Davis HJ, Spiro SG. Slow-release oral salbutamol and aminophylline in nocturnal asthma: relation of overnight changes in lung function and plasma drug levels. Thorax 1980; 35:526–530.
67. Pedersen S. Treatment of nocturnal asthma in children with a single dose of sustained-release theophylline taken after supper. Clin Allergy 1985; 15:79–85.
68. Busse WW, Bush RK. Comparison of morning and evening dosing with a 24-hour sustained-release theophylline, Uniphyl, for nocturnal asthma. Am J Med 1985; 79(suppl 6A):62–66.
69. Fairshter RD, Bhola R, Thomas R, et al. Comparison of clinical effects and pharmacokinetics of once-daily Uniphyl and twice-daily Theo-Dur in asthmatic patients. Am J Med 1985; 79(suppl 6A):48–53
70. Johnston IDA, Ayesh R, Alton E, et al. The pharmacokinetics of Uniphyllin in nocturnal asthma. Br J Dis Chest 1986; 80:235–241.
71. Welsh PW, Reed CE, Conrad E. Timing of once-a-day theophylline dose to match peak blood level with diurnal variation in severity of asthma. Am J Med 1986; 80:1098–1102.
72. Bose B, Cater JI, Clark RA. A once daily theophylline preparation in prevention of nocturnal symptoms in childhood asthma. Eur J Pediatr 1987; 146:524–527.
73. Helm SG. Diurnal stabilization of asthma with once-daily evening administration of controlled-release theophylline: a multi-investigator study. Immunol Allergy Practice 1987; 9:414–419.

74. Arkinstall WW, Atkins ME, Harrison D, Stewart JH. Once-daily sustained-release theophylline reduces diurnal variation in spirometry and symptomatology in adult asthmatics. Am Rev Respir Dis 1987; 135:316–321.

75. Grossman J. Multicenter comparison of once-daily Uniphyl tablets administered in the morning or evening with baseline twice-daily theophylline therapy in patients with nocturnal asthma. Am J Med 1988; 85(suppl 1B):11–13.

76. Bierman CW, Pierson WE, Shapiro GG, Furukawa CT. Is a uniform round-the-clock theophylline blood level necessary for optimal asthma therapy in the adolescent patient? Am J Med 1988; 85(suppl 1B):17–20.

77. Helm SG, Meltzer SM. Improved control of asthma in the office setting: a large-scale study of once-daily evening doses of theophylline. Am J Med 1988; 85(suppl 1B):30–33.

78. Rivington RN, Calcutt L, Hodder RV, et al. Safety and efficacy of once-daily Uniphyl tablets compared with twice-daily Theo-Dur tablets in elderly patients with chronic airflow obstruction. Am J Med 1988; 85(suppl 1B):48–53.

79. Fairfax AJ, Clarke R, Chatterjee SS, et al. Controlled-release theophylline in the treatment of nocturnal asthma. J Int'l Med Res 1990; 18:273–281.

80. Arkinstall WW. Review of the North American experience with evening administration of Uniphyl tablets, a once-daily theophylline preparation, in the treatment of nocturnal asthma. Am J Med 1988; 85(suppl 1B):60–63.

81. Fanta CH, McFadden ER. Pharmacokinetics and clinical response to single- and multiple-dose sustained-release theophylline compounds in perennial bronchial asthma. Am J Med 1985; 79(suppl 6A):54–57.

82. Steinijans VW, Schulz H-U, Beier W, Radtke HW. Once daily theophylline: multiple-dose comparison of an encapsulated micro-osmotic system (Euphylong) with a tablet (Uniphyllin). Int'l J Clin Pharmacol 1986; 24:438–447.

83. Townshend MM, Smith AJ. Factors influencing the urinary excretion of free catecholamines in man. Clin Sci 1973; 44:253–265.

84. Soutar CA, Carruthers M, Pickering CAC. Nocturnal asthma and urinary adrenaline and noradrenaline excretion. Thorax 1977; 32:677–683.

85. Barnes P, Fitzgerald G, Brown M, Dollery C. Nocturnal asthma and changes in circulating epinephrine, histamine, and cortisol. N Engl J Med 1980; 303:263–267.

86. Barnes PJ. Autonomic control of the airways and nocturnal asthma. In: Nocturnal Asthma (Symposium 73), Barnes PJ, Levy J (eds). Royal Society of Medicine, London, 1984, pp 69–75.

87. Ungar A, Phillips JH. Regulation of the adrenal medulla. Physiol Rev 1983; 63:787–843.

88. Moore-Ede MC, Czeisler CA, Richardson GS. Circadian time-keeping in health and disease. N Engl J Med 1983; 309:469–476.

89. Barnes PJ, Fitzgerald GA, Dollery CT. Circadian variation in adrenergic responses in asthmatic subjects. Clin Sci 1982; 62:349–354.

90. Busse WW, Bush RK. Comparison of Uniphyl tablets and inhaled albuterol as maintenance therapy in asthmatic adults (abstract). Am J Med 1988; 85(suppl 1B):10.

91. Gaultier C, Reinberg A, Gerbeaux J, Girard F. Circadian changes in lung resistance and dynamic compliance in healthy and asthmatic children. Effects of two bronchodilators. Resp Physiol 1975; 31:169–182.

92. Gaultier C, Reinberg A, Motohashi Y. Circadian rhythm in total pulmonary resistance of asthmatic children. Effects of a β-agonist agent. Chronobiol Int'l 1988; 5:285–290.

93. Brown A, Smolensky M, D'Alonzo G, et al. Circadian chronesthesy of the airways of healthy adults to the β-agonist bronchodilator isoproterenol. Ann Rev Chronopharmacol 1988; 5:163–166.

94. Horn CR, Clark TJH, Cochrane GM. Inhaled therapy reduces morning dips in asthma. Lancet 1984; 1:1143–1145.

95. Sur S, Mohiuddin AA, Vichyanond P, Nelson HS. A random double-blind trial of the combination of nebulized atropine methylnitrate and albuterol in nocturnal asthma. Ann Allergy 1990; 65:384–388.

96. Wolstenholme RJ, Shettan SP. Comparison of a combination of fenoterol with ipratropium bromide (Duovent) and salbutamol in young adults with nocturnal asthma. Respiration 1989; 55:152–157.

97. Maesen FPV, Smeets JJ, Gubbelmans HLL, Zweers PGMA. Formoterol in the treatment of nocturnal asthma. Chest 1990; 98:866–870.

98. Fitzpatrick MF, Mackay T, Driver H, Douglas NJ. Salmeterol in nocturnal asthma: a double-blind, placebo controlled trial of a long acting inhaled β₂-agonist. Br Med J 1990; 301:1365–1368.

99. Maesen FPV, Smeets JJ, Gubbelmans HLL, Zweers PGMA. Bronchodilator effect of inhaled formoterol vs salbutamol over 12 hours. Chest 1990; 97:590–594.

100. Ullman A, Svedmyr N. Salmeterol, a new long-acting inhaled beta₂-adrenoreceptor agonist: comparison with salbutamol in adult asthmatic patients. Thorax 1988; 43:674–678.

101. Williams HL, Lubin A, Goodnow JJ. Impaired performance with acute sleep loss. Psychol Monogr 1959; 73:1–26.

102. Friedman RC, Bigger JT, Kornfield DS. The intern and sleep loss. N Engl J Med 1971; 285:201–203.

103. Fitzpatrick MF, Cheshire K, Whyte KF, et al. Sleep quality and daytime cognitive performance in nocturnal asthma (abstract). Thorax 1990; 45:338P.

104. Eriksson NE, Haglind K, Ljungholm K. A comparison of sustained-release terburtaline and ordinary terbutaline in bronchial asthma. Br J Dis Chest 1982; 76:202–204.

105. Koëter GH, Postma DS, Keyzer JJ, Meurs H. Effect of oral slow-release terbutaline on early morning dyspnea. Eur J Clin Pharmacol 1985; 28:159–162.

106. Postma DS, Koëter GH, v.d. Mark TW, et al. The effects of oral slow-release terbutaline on the circadian variation in spirometry and arterial blood gas levels in patients with chronic airflow obstruction. Chest 1985; 87:653–657.

107. Postma DS, Koëter GH, Keyzer JJ, Meurs H. Influence of slow-release terbutaline on the circadian variation of catecholamines, histamine, and lung function in nonallergic patients with partly reversible airflow obstruction. J Allergy Clin Immunol 1986; 77:471–477.
108. Dahl R, Harving H, Säwedal L, Anchus S. Terbutaline sustained-release tablets in nocturnal asthma—a placebo-controlled comparison between a high and a low evening dose. Br J Dis Chest 1988; 82:237–241.
109. Eriksson L, Jonson B, Eklundh G, Persson G. Nocturnal asthma: effects of slow-release terbutaline on spirometry and arterial blood gases. Eur Respir J 1988; 1:302–305.
110. Stewart IC, Rhind GB, Power JT, et al. Effect of sustained release terbutaline on symptoms and sleep quality in patients with nocturnal asthma. Thorax 1987; 42:797–800.
111. Brander PE, Sovijärvi ARA, Salmi T, et al. Nocturnal oxygen saturation and body movement in asthmatics treated with controlled-release preparations of theophylline or terbutaline. Eur J Clin Pharmacol 1990; 39:117–121.
112. Vikka V, Brander P, Hakulinen A, et al. Once-daily theophylline in the treatment of nocturnal asthma. Eur J Clin Pharmacol 1990; 39:241–243.
113. Heins M, Kurtin L, Oellerich M, et al. Nocturnal asthma: slow-release terbutaline versus slow-release theophylline therapy. Eur Respir J 1988; 1:306–310.
114. Dahl R, Pedersen B, Hägglöf B. Nocturnal asthma: Effect of treatment with oral sustained-release terbutaline, inhaled budesonide, and the two in combination. J Allergy Clin Immunol 1989; 83:811–815.
115. Pedersen BK, Laursen LC, Gnosspelius Y, et al. Bambuterol: Effects of a new anti-asthmatic drug. Eur J Clin Pharmacol 1985; 29:425–427.
116. Persson G, Gnosspelius Y, Anehus S. Comparison between a new once-daily, bronchodilating drug, bambuterol, and terbutaline sustained-release, twice-daily. Eur Respir J 1988; 1:223–226.
117. D'Alonzo GE, Smolensky MH. Once-daily oral bambuterol, a long-acting beta$_2$-agonist, versus placebo for the treatment of asthma (abstract). Am Rev Respir Dis 1992; 145:A66.
118. Powell ML, Weisberger M, Dowdy Y, et al. Comparative steady state bioavailability of conventional and controlled-release formulations of albuterol. Biopharm Drug Distrib 1987; 8:461–468.
119. Storms WW, Nathan RA, Bodman SF, et al. The effect of repeat action albuterol sulfate (Proventil® Repetabs®) in nocturnal symptoms of asthma. J Asthma 1992; 29:205–212.
120. Milledge JS, Morris J. A comparison of slow-release salbutamol with slow-release aminophylline in nocturnal asthma. J Int Med Res 1979; 7(suppl 1):106–110.
121. Simons FER, Luciuk GH, Simons KJ. Sustained-release theophylline for treatment of asthma in preschool children. Am J Dis Child 1982; 136:790–793.
122. Primrose WR. Variation in theophylline levels following morning or evening administration. Ann Rev Chronopharmacol 1984; 1:85–88.

123. Reed RC, Schwartz HJ. Circadian variation in trough theophylline concentrations. Drug Intell Clin Pharmacol 1983; 17:444.
124. Coulthard KP, Birkett DJ, Lines DR, et al. Bioavailability and diurnal variation and absorption of sustained release theophylline in asthmatic children. Eur J Clin Pharmacol 1983; 25:667–672.
125. Samaan SS, Fox RW, Bukantz SC, Lockey RF. Comparison of diurnal and nocturnal theophylline disposition. Ann Rev Chronopharmacol 1984; 1:69–72.
126. Saint-Pierre M, Leeder S, Spino M, et al. Circadian variation in theophylline absorption. Ann Rev Chronopharmacol 1984; 1:81–84.
127. Saint-Pierre MV, Spino M, Isles AF, et al. Temporal variation in the disposition of theophylline and its metabolites. Clin Pharmacol Ther 1985; 38:89–95.
128. Segrestaa JM, Dahan R, Houlbert D, et al. Chronocinique en at stationnaire d'une thephylline a libation prolonge. Thérapie 1984; 39:633–637.
129. Rogers RJ, Kalisker A, Wiener MB, Szefter S. Inconsistent absorption from a sustained-release theophylline preparation during continuous therapy in asthmatic children. J Pediatr 1985; 106:496–501.
130. Kramer WG, Scott PH, Smolensky MH, et al. Circadian rhythms in steady-state theophylline pharmacokinetics in children. Ann Rev Chronopharmacol 1984; 1:77–80.
131. Thompson PJ, Butcher MA, Frazer LA, Marlin GE. Pharmacokinetics of a single evening dose of slow release theophylline in patients with chronic lung disease. Br J Clin Pharmacol 1981; 12:443–445.
132. Kramer WG, Scott PH, Smolensky MH, et al. Effect of time of once-daily dosing on serum theophylline concentrations at steady state. Ann Rev Chronopharmacol 1986; 3:151.

8

Corticosteroids, Oral and Inhaled, in the Treatment of Nocturnal Asthma

William R. Beam, M.D., and Richard J. Martin, M.D.

Introduction

Nocturnal worsening of asthma has been a recognized feature of the asthma syndrome since the first known descriptions of the disease. In 1698, Dr. John Floyer, in describing his own asthma symptoms, reported a 7-year period of predominantly nocturnal episodes of wheezing and breathlessness.[1] More recently, Turner-Warwick and colleagues reported the incidence of nightly asthma symptoms to be almost 40% in an asthmatic population approaching 8,000.[2] The frequency of asthma associated awakenings rose to 64% of subjects when the incidence of awakenings was reportable at a level greater than 3 nights per week. It is noteworthy that no particular medication or combination of medications was identified as being associated with a lower frequency of nocturnal asthma. Table 1 lists the medications and the frequency of persistent nocturnal asthma symptoms. The issue of medication effectiveness

Martin RJ (editor): *Nocturnal Asthma: Mechanisms and Treatment,* © Futura Publishing Co., Inc., Mount Kisco, NY, 1993.

Table 1

Medication and Frequency of Nocturnal Awakenings (% patients)

	Every Night	≥3 Nights/Week
No Treatment	35	14
Beta$_2$ Agonist	35	25
Cromolyn	39	17
Beclomethasone	44	14
Budesonide	47	17
Theophyllines	51	14
Oral Steroids	50	13

Adapted from: Turner-Warwick. Am J Med 1988.

is paramount in the management of nocturnal asthma since the majority of asthma deaths occur at night.[3] Dr. Turner-Warwick makes the point that the frequency of nocturnal worsening in asthma is identical to that found in 1971, prior to the introduction of inhaled corticosteroids. As can be seen in Table 1, even budesonide, the most potent inhaled corticosteroid, does not appear to influence the incidence of nocturnal asthma awakenings. In addition, her data reveals that 63% of asthmatics on oral corticosteroids still complained of awakenings (at least 3 nights per week) due to nocturnal asthma. Do these observations suggest that nocturnal asthma is a corticosteroid insensitive phenomenon?

Traditional homeostatic theories of drug pharmacokinetics would favor the position that nocturnal asthma is a corticosteroid insensitive process, or raise issues of inadequate dosage to maintain therapeutic drug levels and patient compliance as possible explanations. Alternatively, the newer concept of chronopharmacology may allow insight into the relative corticosteroid responsiveness of nocturnal asthma. Chronopharmacology refers to time-dependent differences in a drug effect resulting from the influences of various biological rhythms. Biological rhythms may alter the effectiveness or toxicity of a drug such that a disease process may appear variably responsive, or have an altered dose-response relationship, dependent on drug timing rather than dosage.[4] In the case of nocturnal asthma, circadian variability in plasma hormone levels; inflammatory cell number, distribution, and recruitment; and drug receptor number and function may all be pertinent in establishing the corticosteroid responsiveness of the event. The contribution of circadian proinflam-matory mechanisms to the pathogenesis of nocturnal asthma and consequences of these events regarding corticosteroid therapy are areas of current investigation and interest.

Chronobiology, the study of naturally occurring biological rhythms, encompasses the spectrum of intrinsic molecular and biochemical alterations to integrated whole organ and body func-tions. Chapters 1 and 2 cover this exciting field of investigation. Figure 1 represents some of the characteristic peaks of biological rhythms found in humans over a 24-hour, or circadian, time span. This abbreviated chronobiological clock is synchronized to a typical daylight activity schedule and includes a brief listing of noteworthy circadian events. Most obstetrical physicians, and their spouses, would little argue with the finding that the peak time for the onset of

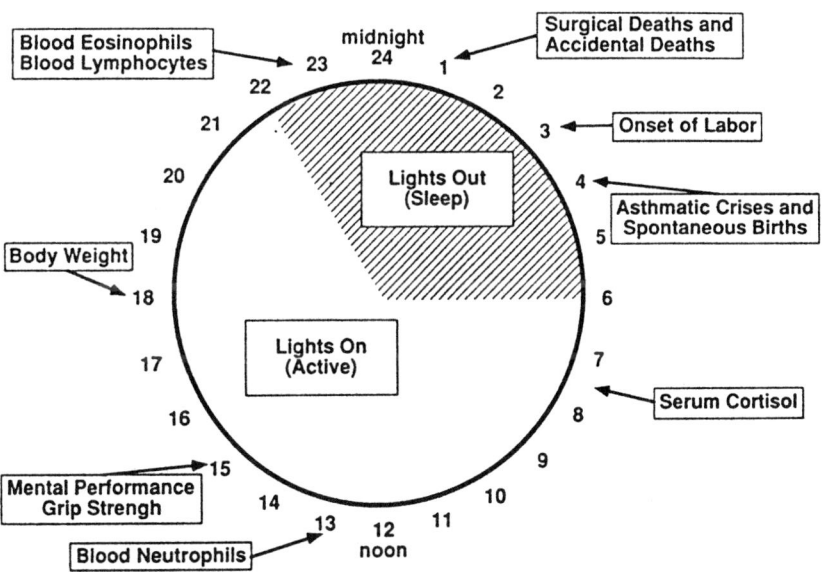

Figure 1. Human circadian rhythms. The peak incidence of asthmatic worsen-ing occurs during the sleep related hours at 4 AM. Serum cortisol is highest upon awakening, while blood eosinophil and lymphocyte numbers are highest during sleep. Adapted from: Morris RW. Chronobiology and Health. Part I. Basic Principles. Skyline Publishers, Inc., Pharm-index, Jan 1989, pp 5–16.

labor is 3 AM, but most first-time mothers would favor the explanation that the peak onset of spontaneous births at 4 AM represents a 25-hour, rather than a 1-hour, time shift from the peak onset of labor. With regard to asthma, the peak incidence of asthmatic worsening, as well as ventilatory arrest and death, occurs at 4 AM. Biological rhythms which may contribute to this phenomenon include a peak serum cortisol level at approximately 8 AM as well as a peak blood lymphocyte and eosinophil count occurring early during sleep at approximately 11 PM. The peak blood neutrophil count occurs much earlier at 1 PM. Circadian variations in inflammatory cell profiles may contribute to both the pathogenesis of nocturnal asthma as well as the time-dependent corticosteroid sensitivity which can be seen.

Historical Development of Corticosteroid Therapy

Corticosteroids have become a major therapy in the treatment of asthma since it has become clear that inflammatory cells and their mediators play a prominent role in the pathophysiology of the syndrome. Compound E, as cortisone was originally labeled, was first isolated in the 1930s and first synthesized by Reichstein and colleagues in 1943.[5] Cortisone was first used clinically in 1949 by Hench and coworkers in the treatment of rheumatoid arthritis.[6] The clinical utility of the compound was quickly recognized, and therapeutic applications were extended to other inflammatory diseases. Cortisone was first used in the treatment of asthma in 1950 by multiple investigators.[7,8] It is noteworthy that within a year following the first published reports of the clinical efficacy of cortisone, Reichstein, Kendall, and Hench received the Nobel Prize in Medicine for their work. The next several years resulted in a search for newer substances with improved antiinflammatory effects and fewer systemic side effects. These included prednisone and prednisolone,[9] methylprednisolone,[10] dexamethasone,[11] and triamcinolone.[12] Although these compounds vary in their relative potencies, durations of action, and salt retaining properties (Table 2), when used in equivalent antiinflammatory doses, they are all effective antiinflammatory drugs.[13] Prednisone and prednisolone remain the most commonly used oral preparations in the treatment of asthma.

As corticosteroids became a prominent component of the medi-

Table 2

Comparison of Common Corticosteroid Preparations

Compound	Equivalent Dose† (mg)	Plasma Half-Life* (min)	Anti-inflammatory Potency**	Salt-retaining Potency**	Relative Receptor Affinity**
Cortisone	25	30	0.8	0.8	0.01
Cortisol	20	90	1.0	1.0	1.0
Prednisone	5	60	4	0.8	0.05
Prednisolone	5	200	4	0.8	2.2
Methyl-prednisolone	4	180	5	0.5	11.9
Triamcinolone	4	300	5	0	1.9
Dexamethasone	0.75	200	30	0	7.1

†Equivalent dose relationships approximate for PO or IV administration and may vary with IM route
*Plasma half-life is not equivalent to biological half-life
**Comparisons made using potency of cortisol as baseline value = 1.0
Adapted from ref. 13.

cal management of various disease states, rapid recognition of their serious side effects and complications was detailed in the medical literature. Because of these hazards, efforts were made by many investigators to develop preparations that could be delivered locally to the airways and thus achieve high end-organ concentrations (i.e., airways) and avoid systemic administration. Betamethasone and beclomethasone were the first topical aerosol preparations which achieved clinical success in the treatment of asthma.[14,15] Their success resulted in a renewed search for more effective compounds, which is still ongoing. Triamcinolone acetonide,[16] flunisolide,[17] and budesonide[18] are the currently available inhaled corticosteroids in the United States or Europe which are popular in the treatment of asthma. The contribution of both oral and aerosol corticosteroids to the management of nocturnal worsening of asthma remains to be defined. Epidemiological data would suggest little impact thus far, although the issue remains open to debate.[2]

Mechanisms of Corticosteroid Action

In spite of the widespread use of corticosteroids as chemotherapy in a number of disease states, the precise mechanisms of action with respect to immunosuppression and antiinflammatory effects remain an enigma. Still, a generally accepted model of the hormone receptor (Fig. 2) allows some insight into the early dynamics of corticosteroid effects. Cells that are targets for a specific hormone contain receptors that bind the hormone and mediate its cellular effects.[19] In the case of corticosteroids, cellular receptors can be found in a diversity of tissues as well as in inflammatory cells. In contrast to catecholamine or thyroid hormone receptors, the corticosteroid hormone receptors are found initially in the soluble intracellular compartment of the cell, rather than on the outer surface of the cell plasma membrane (catecholamine receptors) or in the chromatin of the cell (thyroid hormone receptors). The mechanism by which corticosteroids enter a cell are unknown but thought to involve passive diffusion. The hormone is then rapidly bound by the cytosolic receptors. The interaction between the steroid hormone and receptor results in "activation" or "transformation" such that the complex is conformationally changed to allow further binding to the nuclear chromatin. As a result, and likely involving a DNA directed and RNA mediated mechanism, corticosteroids moderate the expression of a limited number of transcriptionally active genes. Ultimately, phenotypic responses of a cell specific type are mediated by changes in protein synthesis.

In addition to this general model of steroid receptor function, there may be nonDNA directed or nonnuclear directed actions by steroid hormones. Corticosteroids have been shown to inhibit the secretion of some cellular proteins in a manner which is too rapid to be mediated by RNA synthesis. This rapid inhibition of protein secretion may be mediated by a class of receptors that differ from those generally responsible for corticosteroid hormone action. This conclusion is based on the finding that the relationship between steroid structure and activity for the fast actions differs from those for other corticosteroid effects.[19,20]

As previously stated, the ultimate phenotypic response produced by corticosteroids is dependent on tissue or cell specific alterations in protein production or secretion. The link between antiinflammatory effects relevant to asthma and this model of steroid action has been postulated by a number of investigators. Corticosteroids have been shown to induce the synthesis of proteins such as Lipomodulin and Macrocortin which inhibit Phospholipase A_2.[21] Phospholipase A_2 is the protein responsible for enzymatic cleavage of arachidonic acid from membrane phospholipids. By

Steroid Hormone-Responsive Cell

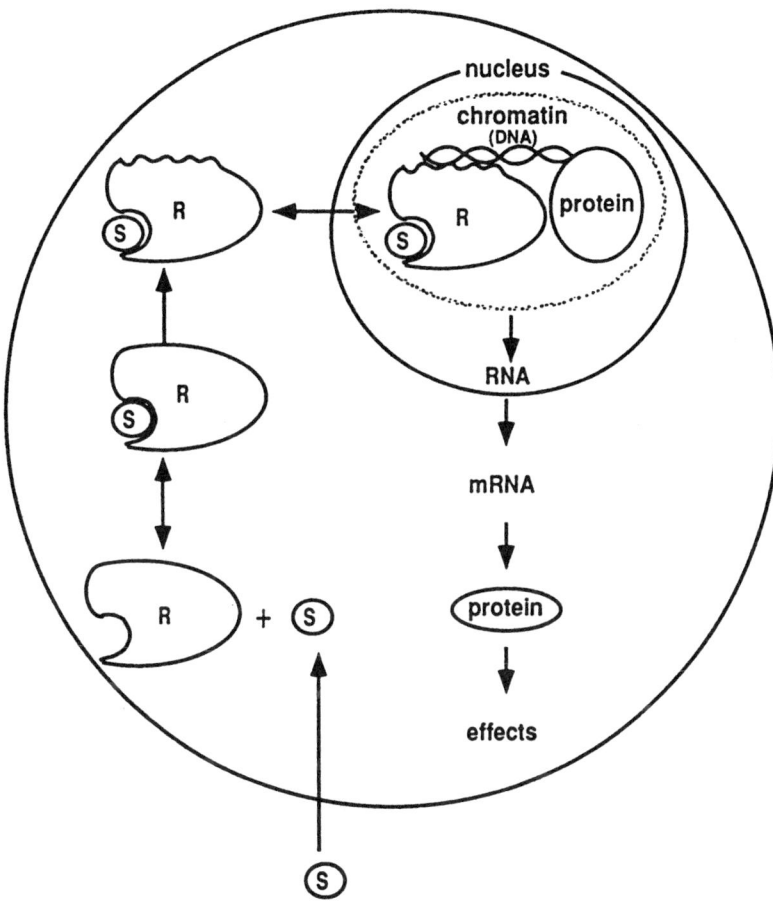

Figure 2. Steps in steroid hormone action. After penetrating the cell membrane, steroid (S) combines with receptor (R), altering its conformation. S-R then binds chromatin and alters mRNA synthesis. The resulting changes in cell specific mRNA synthesis produce alterations in protein synthesis which then mediate the cell's response to the steroid hormone. Adapted from reference 19.

inhibiting the release of arachidonic acid precursors, both the cyclooxygenase and lipoxygenase pathways are depleted of their common precursor compound (Fig. 3). This leads to a reduction in the synthesis of prostaglandins, leukotrienes, and possibly other mediators of inflammation that have been implicated in asthma.[22]

In addition to their effects on the arachidonic acid dependent

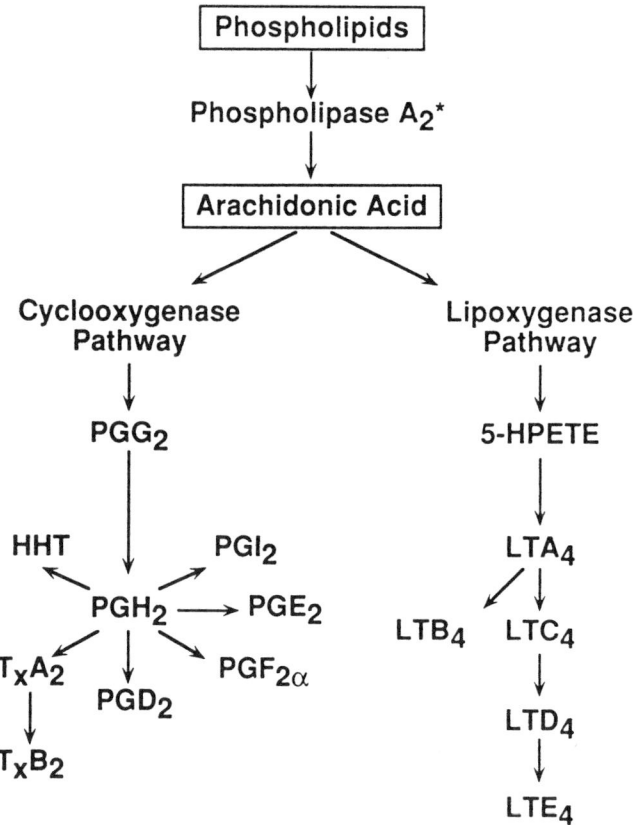

Figure 3. Corticosteroid inhibition of arachidonate pathways. Corticosteroids induce the synthesis of proteins that inhibit phospholipase A_2. Both the cyclooxygenase and lipoxygenase pathways are depleted of their precursor compound, arachidonic acid. * site of activity of corticosteroids due to antiphospholipase A_2 activity. PG = prostaglandins, T_x = thromboxanes, HHT = heptadecatrienoic acid, LT = leukotrienes, 5-HPETE = hydroxyperoxyeicosatetraenoic acid.

pathways, corticosteroids have been shown to have vasoconstrictor activity which may attenuate the microvascular leak phenomenon and subsequent submucosal edema formation found in asthmatic airways.[23,24] Alterations in vascular permeability to proteins and other macromolecules have been shown with both systemic and topical corticosteroid administration and may be related to time-dependent reductions in vascular responsiveness to histamine, bradykinin, and catecholamines.[25] Since airway edema may contribute to bronchial responsiveness,[26] this mode of action may be important in asthma.

Corticosteroids have been shown to increase the β-adrenergic receptor density on cell membranes[27] and improve the degree of agonist-receptor coupling.[28] These observations suggest a potential role for corticosteroids in restoring the responsiveness of down-regulated β-adreneric receptors in asthmatics. Downregulation of β-adrenergic receptors may occur as a result of chronic β-agonist use,[29,30] although the concept is not uniformly accepted.[31,32] With respect to nocturnal asthma worsening, it has been shown that β-adrenergic receptor density and function are reduced at 4 AM with impaired response to isoproterenol.[33] In these situations, corticosteroids may improve asthma by restoring or increasing adrenergic receptor number and affinity. In addition, β-adrenergic receptor mediated smooth muscle relaxation can be potentiated with corticosteroids.[34] The mechanisms involved in corticosteroid induced smooth muscle relaxation are unknown, but may involve direct action on the muscle[35] or enhancement of catecholamine mediated events.[36]

An additional effect of corticosteroids pertinent to the asthma syndrome is their effect on mucous. Severe and fatal asthma, which is most often nocturnal, is associated with diffuse mucous plugging. This situation may be due to both increased secretion and decreased clearance. Corticosteroids have been shown to reduce mucous glycoprotein secretion from airways and to decrease sputum production in stable asthmatics. In addition to altering mucous proteins directly, periciliary fluid may be changed in a way that augments mucociliary clearance. It is likely these events are in part related to corticosteroid inhibition of leukotriene products of inflammation.[37]

Effects of Corticosteroids on Inflammatory Cells

In the past, the focus of asthma therapy was concentrated on mechanisms involved in airway smooth muscle bronchoconstriction with resultant emphasis on bronchodilator drugs. There is mounting evidence to support the contribution of inflammatory phenomena in the pathogenesis of asthma. Animal studies, although without a unifying cellular mechanism, suggest a link between bronchial responsiveness of laboratory "asthma" and the extent of airways inflammation.[38,39] Although this link is less well characterized in humans with clinical asthma, bronchial tissue analysis supports the association. Submucosal infiltration with lymphocytes, eosinophils, and other inflammatory cells in conjunction with epithelial damage and collagen deposition has been described. These histologic findings span the spectrum of asthma severity from fatal to mild or subclinical.[40–42] The inflammatory cell most characteristic of asthma remains the eosinophil,[43] but contributions by macrophages, lymphocytes, and possibly neutrophils remain a focus of current investigation. Activated T-lymphocytes and macrophages can produce a variety of cytokines which may serve to amplify the inflammatory response in patients with asthma. It seems clear that asthma is more than simple bronchoconstriction, and that bronchodilator medications may not satisfactorily influence the inflammatory events associated with increased bronchial responsiveness. Corticosteroids may attenuate the asthmatic inflammatory state by inhibition of individual inflammatory cellular function and interruption of collaboration between cell types.[44]

Mast Cells and Basophils

Corticosteroids do not directly inhibit the release of preformed mediators from human mast cells,[45] but they may inhibit the release of mast cell growth factors from T-lymphocytes, macrophages, and endothelial cells.[24] As previously discussed, corticosteroids can inhibit the vascular permeability and mucous secretion seen in tissue exposed to mast cell derived mediators. Basophils are commonly referred to as the circulating counterpart of the mast cell and share similar functional abilities, although they have different progenitor cell origins. In contrast to mast cells, corticosteroids effectively alter basophil kinetics and consequently produce reduced circulating basophil numbers. Also, antigen-induced release of mediators can be blocked.[46,47] The contribution of mast cell or basophil mediated events in nocturnal worsening of asthma is unknown.

Neutrophils

In contrast to most circulating inflammatory cell types, corticosteroids produce a neutrophilic leukocytosis which reaches a peak 4 to 6 hours after the drug is taken and represents a two- to fourfold increase in neutrophil number.[13] Corticosteroid-induced neutrophilia is likely due to an accelerated release from bone marrow stores and the reduced migration from the intravascular space to sites of inflammation.[48] In addition to altered emigration to inflamed tissue, corticosteroids also alter macrophage and T-lymphocyte "priming" of neutrophils to an activated state.[49] Although many other neutrophil functions have been studied for corticosteroid effects, most of the studies used supraphysiological doses of drugs and may not be applicable to the asthmatic state. Although the role of neutrophils in daytime asthma is uncertain, neutrophil influx into the airways has been demonstrated following antigen challenge,[50] and neutrophils may play a role in ozone-induced airway hyperreactivity in dogs.[51] With respect to nocturnal asthma, Martin and colleagues have demonstrated a circadian cellular recruitment of neutrophils into the airways of asthmatics prone to nocturnal worsening (Fig. 4A).[52] These circadian changes in bronchoalveolar lavage neutrophil counts were not paralleled in the peripheral blood, but did correlate with overnight deterioration in peak expiratory flow rates (Fig. 4B).

Eosinophils

The ability of corticosteroids to produce a significant reduction in blood eosinophil numbers was recognized early in the research involving adrenocortical hormone extracts,[53] and is most likely due to a redistribution of cells out of the circulating pool.[54] It was previously believed that corticosteroids produce eosinopenia due to cell lysis, but the effect is more likely the combination of rapid tissue sequestration and inhibition of release from bone marrow stores. Analogous to the effects on neutrophils, corticosteroids alter eosinophil trafficking to sites of inflammation as well as inhibiting cytokine mediated events important to eosinophil survival and activation.[44] It is likely that the impact of corticosteroids on T-lymphoctyte, macrophage, and endothelial cell derived cytokines as they relate to local control of eosinophil function is relevant to the utility of steroids in asthma. Issues

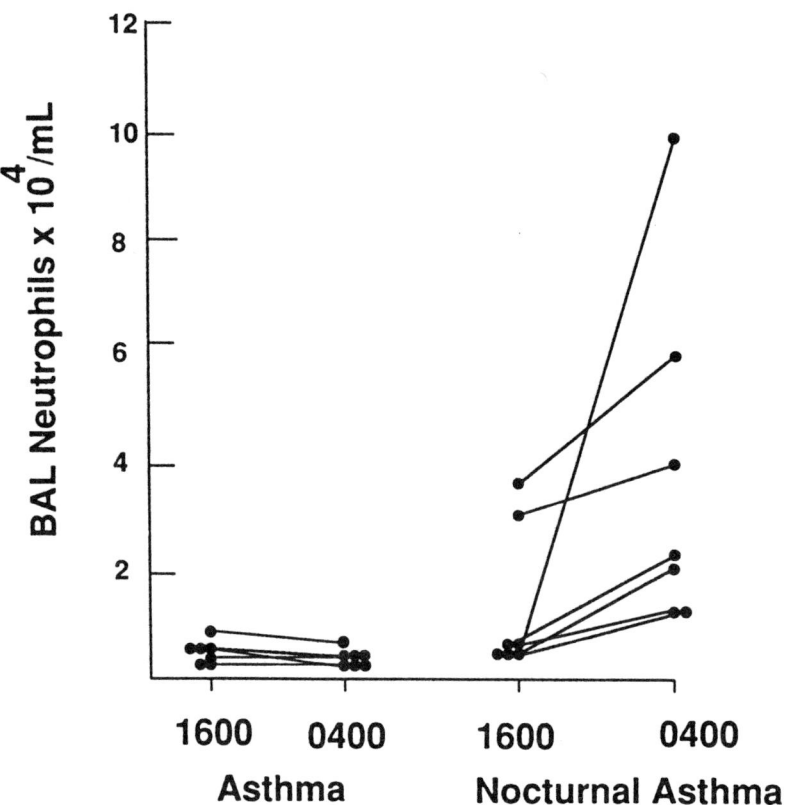

Figure 4A. Bronchoalveolar lavage (BAL) neutrophil recruitment in asthmatics prone to nocturnal worsening. Asthmatics prone to nocturnal worsening demonstrate circadian recruitment of neutrophils into their bronchoalveolar lavage fluid. This phenomenon is not seen in asthmatics without nocturnal worsening. Adapted from reference 52.

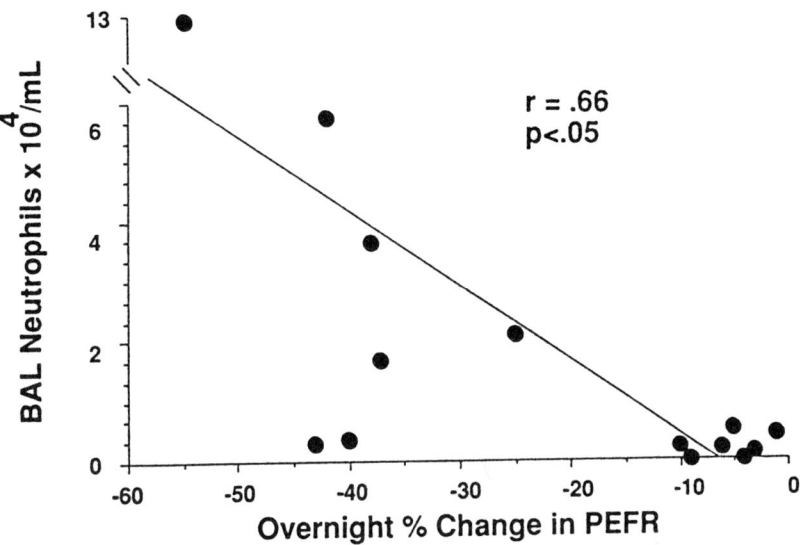

Figure 4B. Correlation of 4 AM brochoalveolar lavage (BAL) neutrophil counts and overnight percent change in peak expiratory flow rates (PEFRs). A significant correlation was demonstrated between the severity of nocturnal asthma and neutrophil recruitment into the asthmatic airways. Adapted from reference 52.

regarding the direct effects of corticosteroids on human eosinophil functions require further investigation. Data supporting a role for eosinophils in asthma continue to grow. Eosinophils are commonly found to be increased in daytime blood, BAL fluid, and bronchial biopsies of asthmatics; and eosinophilic airways inflammation has been shown to correlate with the severity of daytime asthma.[40–43] In a similar manner, Martin and coworkers have implicated eosinophils in the pathogenesis of nocturnal worsening of asthma (Figs. 5A, 5B) and suggested the source of these cells is the blood.[52] Circadian variations in blood eosinophil numbers have been demonstrated in asthmatics with a nadir at 10 AM and a maximum concentration near midnight.[55] Consequently, nocturnal worsening of asthma and circadian eosinophil trafficking into the lung appear to be related events which may be corticosteroid sensitive.

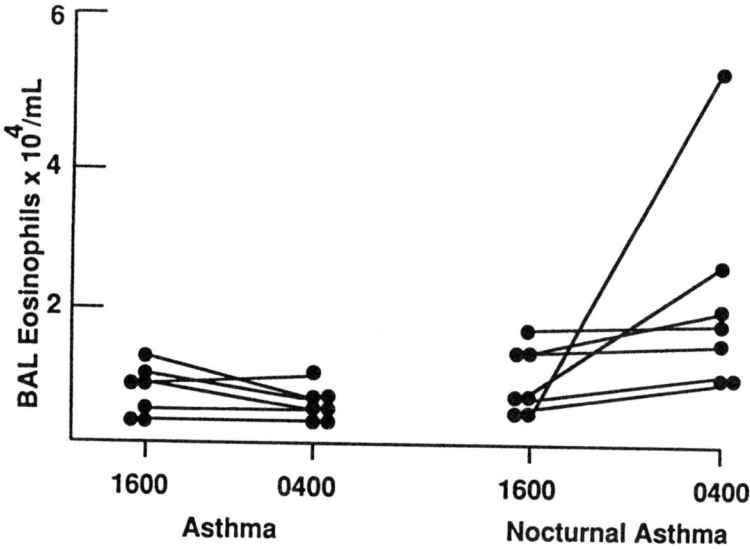

Figure 5A. Bronchoalveolar lavage (BAL) eosinophil recruitment in asthmatics prone to nocturnal worsening. Asthmatics prone to nocturnal worsening demonstrate circadian recruitment of eosinophils into their bronchoalveolar lavage fluid. This phenomenon is not seen in asthmatics without nocturnal worsening. Adapted from reference 52.

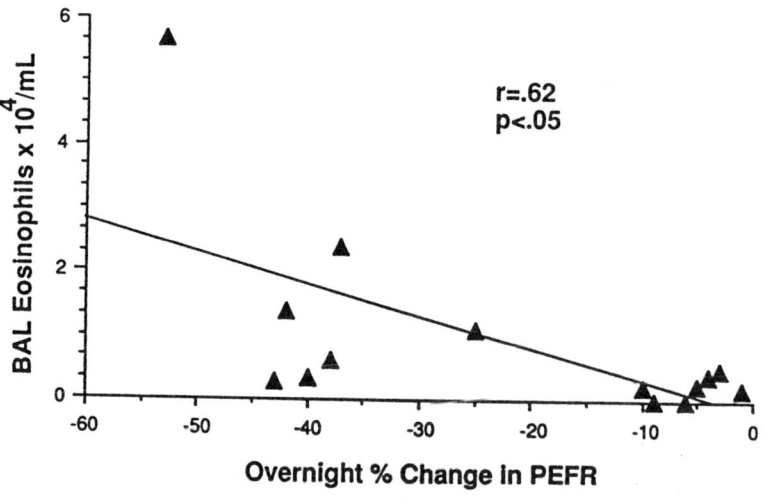

Figure 5B. Correlation of 4 AM bronchoalveolar lavage (BAL) eosinophil counts and overnight percent change in peak expiratory flow rates (PEFRs). A significant correlation was demonstrated between the severity of nocturnal asthma and eosinophil recruitment into the asthmatic airways. Adapted from reference 52.

Lymphocytes

The effects of corticosteroids on lymphocytes appears to be species specific, and a failure to appreciate these species differences has resulted in considerable confusion in the literature. Although corticosteroid-induced lymphocytopenia has been described in both man and several animal models, the mechanisms responsible for the phenomenon are different.[56,13] The mouse, rat, and rabbit are "corticosteroid-sensitive" species and corticosteroid-induced lysis of lymphocyte populations with reductions in splenic, lymph node, and thymic weight, as well as generalized body wasting, can be seen. Man is relatively "corticosteroid-resistant" and corticosteroid-induced lymphocytopenia appears to be the result of redistribution of lymphocytes out of the circulating pool (Fig. 6) rather than actual

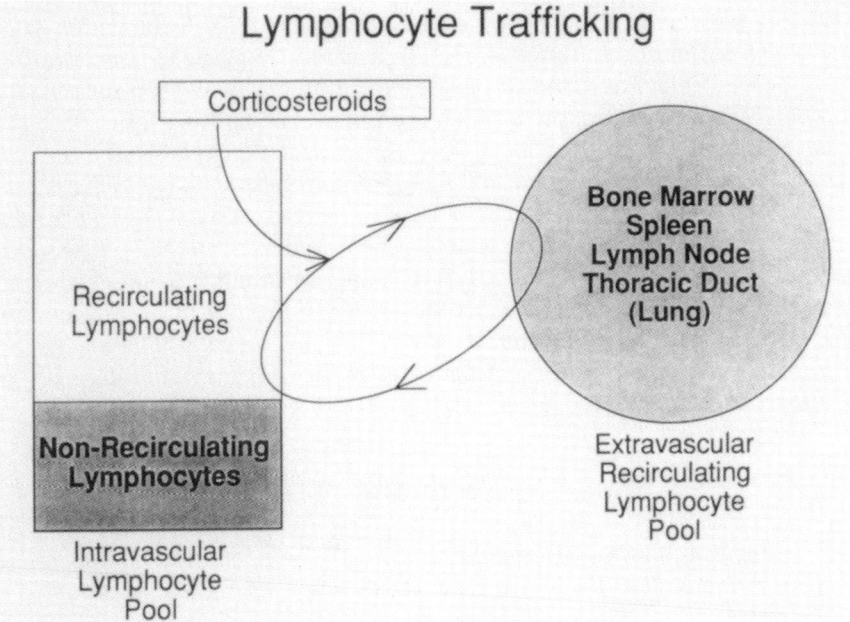

Figure 6. Mechanism of corticosteroid-induced lymphocytopenia. The equilibrium between the intravascular and extravascular portions of the recirculating lymphocyte pool can be altered by corticosteroids. The intravascular recirculating lymphocytes are depleted from the circulation and accumulate in the extravascular compartments of the recirculating lymphocyte pool. Adapted from reference 13.

cell lysis. In man, a single dose of corticosteroid medication results in a transient lymphocytopenia which is maximal in 4 to 6 hours, generally resolves in 24 hours, and involves T-lymphocytes to a greater degree than B-lymphocytes. Thymic-derived or T-lymphocytes are a major source of cytokine mediators which serve to amplify an inflammatory response as well as promote collaborative cellular activity.[44] It is likely that the effects of corticosteroids on lymphocytes in the asthmatic model of inflammation relates to both an inhibition of cytokine formation as well as changes in lymphocyte circulation patterns.

In addition to eosinophils, T-lymphocytes are commonly found in the inflammatory milieu associated with severe asthma.[40–42] The observations that blood T-lymphocytes expressing the interleukin-2 receptor are elevated in patients with acute, severe asthma, that T-lymphocytes appear to emigrate from the circulatory pool into the lung following allergen challenge, and that models of eosinophilic inflammation are T-lymphocyte mediated suggest a role for lymphocytes in asthmatic inflammation.[57] In asthmatics prone to nocturnal worsening, elevated BAL lymphocyte numbers have been demonstrated at 0400 hours, the typical nadir of overnight lung function.[52] In addition, blood lymphocyte populations demonstrate circadian variability in normals such that CD_4-positive T-lymphocytes (Fig. 7) and the CD_4/CD_8 ratio (Fig. 8) of T-lymphocytes increase during the sleep related hours.[58] These observations support a role for lymphocytes in the pathogenesis of nocturnal asthma and suggest a situation in which biological rhythms may influence inflammatory events by altering an available pool of circulating cells. The ability of corticosteroids to alter lymphocyte trafficking may therefore be important in nocturnal asthma.

Monocytes and Macrophages

Macrophages are the predominant cell type found in the bronchoalveolar lavage fluid of both normals and asthmatics. Although their specific contribution to the airways inflammation of asthma is unknown, they must be considered in the discussion based on their numbers alone. Corticosteroids produce an acute monocytopenia in blood, possibly due to altered release from bone marrow stores, and reduced migration of these cells to sites of inflammation.[13] Macrophages are uniquely suited to playing a central role in inflammation

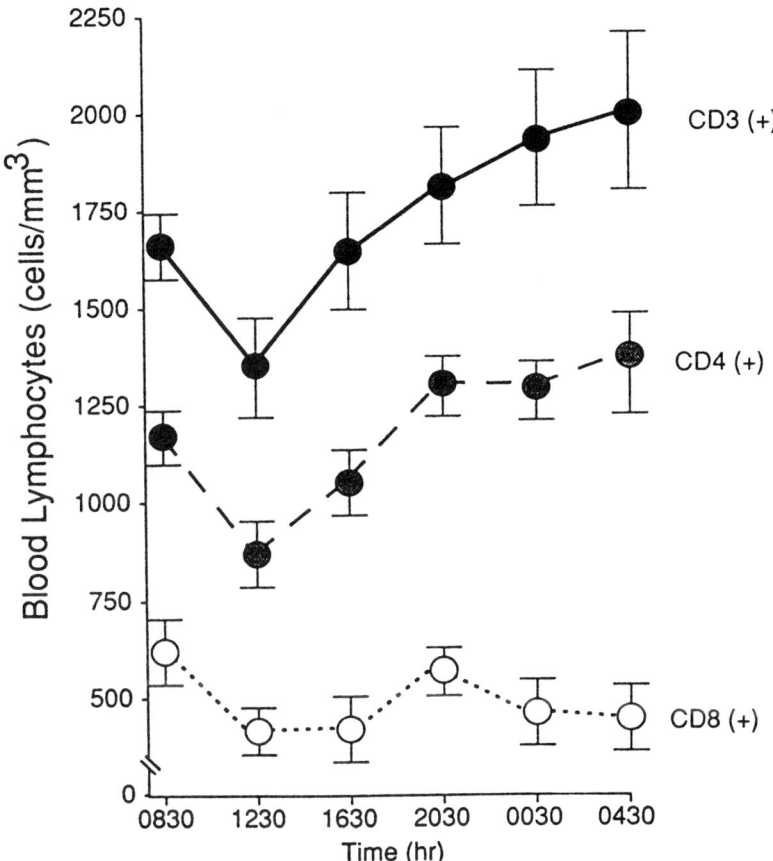

Figure 7. Circadian variations in blood T-lymphocyte populations. Elevations in both CD_3 (+) and CD_4(+) lymphocyte numbers during the sleep related hours can be demonstrated in normals. Adapted from reference 58.

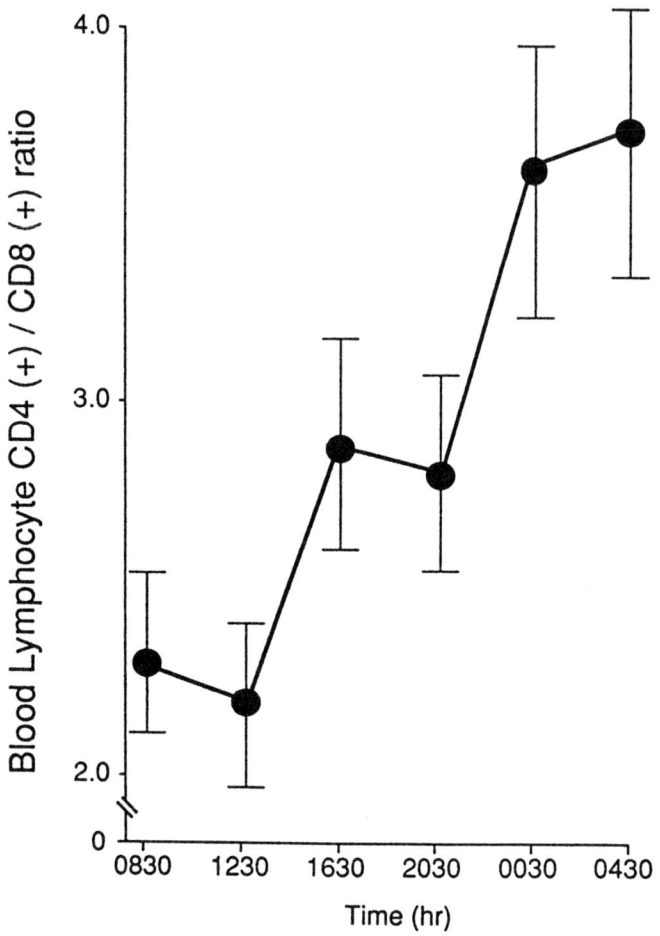

Figure 8. Circadian variations in the CD_4/CD_8 ratio. An increase in the blood T-lymphocyte CD_4/CD_8 ratio during the sleep related hours can be demonstrated in normals. Adapted from reference 58.

since they produce cytokine mediators which are important for localized cellular collaboration as well as cellular recruitment. Monocytes and macrophages are both functionally sensitive to corticosteroids. In addition to blocking the synthesis and secretion of interleukin-1, which is necessary for T-lymphocyte activation and proliferation, corticosteroids also block the formation and secretion of various phospholipid mediators from both the cyclooxygenase and lipoxygenase pathways. Although corticosteroids also inhibit the release of macrophage derived enzymes, it is likely that their predominant antiinflammatory effects are a result of impaired T-lymphocyte and macrophage collaboration and the resultant loss of inflammatory amplification mediated by cytokine production.[44]

Circadian Variation in Plasma Cortisol

The secretion of corticosteroids was one of the first endocrine functions to be recognized as having endogenous biological rhythms.[59] In both healthy and asthmatic subjects active during the daylight hours, the highest plasma cortisol levels occur near the time of awakening from sleep, approximately 8 AM, and then nadir in the middle of the night, approximately at midnight (Fig. 9).[60,61] The natural fluctuation between peak and trough serum cortisol levels may vary by three- to fivefold.[60,61] Multiple investigators have attempted to establish a relationship between circadian variations in airway caliber and plasma cortisol. Reinberg and colleagues

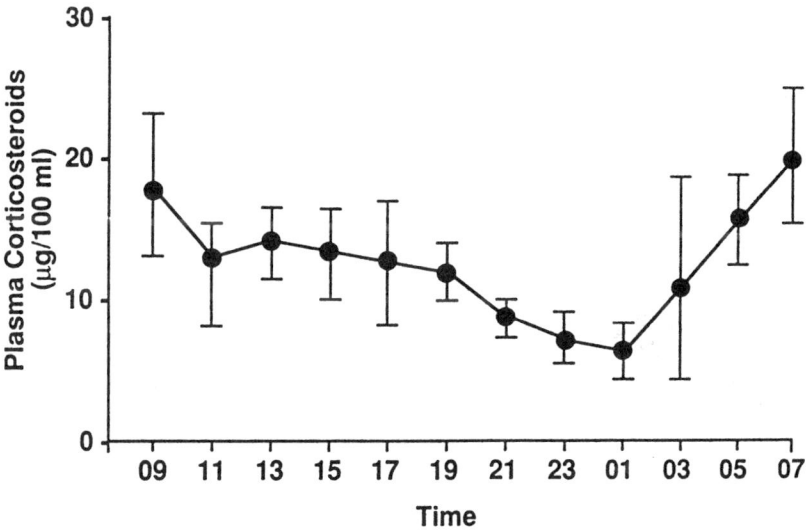

Figure 9. Circadian variations in plasma cortisol levels in asthmatics prone to nocturnal worsening. The highest plasma cortisol levels occur near the time of awakening and then nadir in the middle of the night, approximately at midnight. Adapted from reference 61.

demonstrated synchrony in the timing of nocturnal bronchoconstriction and the lowest urinary excretion of 17-hydroxycorticosteroid over a 24-hour period.[62] Additionally, Barnes and colleagues found that the nadir of plasma cortisol occurred at midnight, while the greatest drop in overnight lung function occurred at 4 AM.[60] No differences in circadian hormone levels were demonstrated between asthmatics and normals to explain the differences in airway patency between the groups. This observation does not obviate the possibility that the overnight decline in plasma cortisol may contribute to the pathogenesis of nocturnal asthma, but it is unlikely to be the sole factor involved.

Spectrum of Corticosteroid Sensitivity in Nocturnal Asthma

Following the observation that nocturnal worsening of asthma coincided with the lowest urinary excretion of corticosteroid metabolites, Souter and colleagues attempted to better define the relationship between plasma cortisol and airway function.[61] In a group of six nocturnal asthmatics, they infused variable physiological doses of intravenous hydrocortisone to eliminate the normal circadian variation and overnight fall in plasma cortisol. Overnight lung function improved an average of 20% with a heterogeneous response pattern (Table 3). In five of the six subjects, physiological doses of corticosteroid did not completely block the nocturnal decrements in lung function. Souter concluded that the circadian variation in cortisol secretion was not the main cause of the circadian variation in asthmatic airway obstruction. However, at physiological dose range, airways inflammation and possible resultant mucosal edema may still play an important role in the development of nocturnal asthma. Day-night alterations in plasma cortisol may attenuate, but not completely abolish, the airways inflammation associated with asthma, and therefore further studies were needed to better define the corticosteroid sensitivity of nocturnal asthma.

To further assess the corticosteroid sensitivity of nocturnal asthma and address the question of whether low overnight plasma cortisol levels contribute to the phenomenon, Beam and coworkers evaluated the effects of supraphysiological doses of intravenous hydrocortisone given overnight to nocturnal asthmatics.[63] Prior to the study all subjects demonstrated clinically stable daytime

Table 3

Nocturnal Fall in Peak Expiratory Flow Rates (PEFRs) Prior to and
During Physiological Hydrocortisone Infusion

| | Nocturnal Fall in PEFR (L/min) | | |
Patient	Prior to HC Infusion	During HC Infusion	Index of Improvement
1	120	240	− 100%
2	60	40	33%
3	120	90	25%
4	100	50	50%
5	125	100	20%
6	130	10	92%

Adapted from Reference 61.

asthma, but persistent nocturnal worsening of spirometry in spite of daytime steroid therapy. Nine of 11 subjects studied were maintained on daytime prednisone therapy with a mean daily dose greater than 25 mg, yet still demonstrated overnight decrements in spirometry between 23% and 66%. As a group, the mean overnight decrement in forced expiratory volume in 1 second (FEV_1) significantly improved from a baseline value of 46% to 12% concurrent with overnight hydrocortisone infusion at 100 mg per hour (Fig. 10). Additionally, fewer subjects awoke from sleep requiring $beta_2$-agonist therapy while receiving the steroid infusion. Overall, the mean improvement between baseline and steroid infusion nights was 67%.

Still, a spectrum of response to overnight corticosteroid infusion at pharmacological dose was evident with two of 11 subjects demonstrating less than 10% change in their nocturnal drop in FEV_1, two of 11 subjects showing complete abolishment of nocturnal bronchoconstriction, and the remaining seven subjects demonstrating between 45% and 87% improvement compared to baseline (Table 4). This heterogeneity among asthmatics prone to nocturnal bronchoconstriction supports the concept of asthma as a syndrome rather than a distinct disease.[64] The contribution of corticosteroid insensitive factors to airway responsiveness in subsets of asthmatics is likely substantive. These factors may include mucociliary clearance, effects of sleep on lung volume and airway resistance, airway cooling, cholinergic and adrenergic tone, and possibly others.[65] Souter et al. demonstrated similar heterogeneity in their population

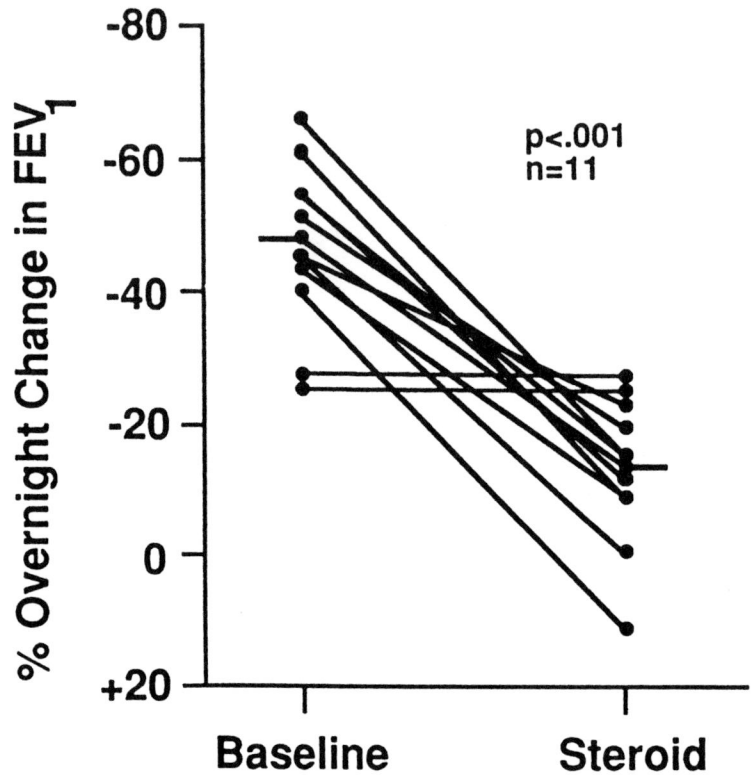

Figure 10. Spectrum of response in nocturnal asthma to supraphysiological dose hydrocortisone infusion. Overnight corticosteroid infusion in asthmatics having nocturnal worsening while receiving daytime corticosteroid therapy yields further improvement in most and no significant change in about 10% of subjects studied. Horizontal bars = mean values. Adapted from reference 63.

of nocturnal asthmatics treated with physiological doses of cortico-steroids.[61] Although Beam et al. demonstrated a greater ability to attentuate the nocturnal fall in FEV_1 with a mean improvement of 67% compared to 20% by Souter et al., the dose of hydrocortisone used was between 10 and 100 times greater. Differences in ability to attenuate nocturnal asthma may represent greater bronchoalveolar concentrations of corticosteroid, directly related to recirculating blood levels.[66] These studies are in agreement that it is unlikely that circadian variability of plasma cortisol is the only factor contributing to the pathogenesis of nocturnal asthma. Additionally, in

Table 4

Nocturnal Change in FEV₁ During Supraphysiological Hydrocortisone Infusion

		Baseline Night		Hydrocortisone Infusion			
Subject	Bedtime $FEV_1(L)$	$\Delta FEV_1(L)$	$\% \Delta FEV_1^*$	Bedtime $FEV_1(L)$	$\Delta FEV_1(L)$	$\% \Delta FEV_1$	Index of Improvement
1	2.13	−0.50	−23	2.35	−0.56	−24	−4
2	2.53	−0.69	−27	2.00	−0.50	−25	+7
3	2.00	−0.80	−40	1.60	−0.35	−22	+45
4	1.70	−0.88	−52	1.45	−0.28	−19	+60
5	1.38	−0.76	−55	1.30	−0.20	−15	+73
6	1.25	−0.82	−66	1.12	−0.17	−15	+77
7	1.23	−0.68	−55	0.90	−0.10	−11	+80
8	2.83	−1.23	−43	2.50	−0.20	−8	+81
9	2.70	−1.65	−61	2.65	−0.20	−8	+87
10	1.00	−0.45	−45	0.80	0	0	+100
11	1.35	−0.54	−40	1.30	+0.15	+11	+128
Mean ± SEM	1.83 ± 0.20	−0.82 ± 0.11	−46 ± 4	1.63 ± 0.20	−0.22 ± 0.06	−12 ± 3	+67 ± 11

*Mean of two baseline nights from bedtime to lowest value throughout the night
$P < 0.05$ comparing the same variable from baseline to steroid infusion night
ΔFEV_1 on each night was significant, $P < 0.05$
Adapted from Reference 63.

agreement with observations made by Turner-Warwick[2] and Souter,[61] Beam and colleagues identified a group of asthmatics with persistent overnight worsening in spite of conventional therapeutic doses of corticosteroids. Recognition that these patients exist is important since they are the asthmatics most at risk for ventilatory arrest and death.[3]

In summary, comparing the responses of nocturnal asthmatics to high doses versus physiological doses of intravenous hydrocortisone during sleep yields similar groupings. A small number of subjects appear highly sensitive and virtually abolish their nocturnal fall in spirometry; a small number of subjects appear relatively resistant, similar to a recognized population of chronic asthmatics[67]; and the majority demonstrate variable degrees of improvement. Supraphysiological doses of corticosteroids given during sleep appear to greatly attenuate nocturnal asthma, which was relatively resistant to conventional steroid therapy, possibly by altering circadian variations in inflammation. The contribution of inflammatory factors to the pathogenesis of nocturnal asthma and the ability of a chronopharmacological approach to therapy to offer optimal effectiveness in moderating this inflammatory response are ongoing areas of contemporary investigation.

Timing of Corticosteroids in Nocturnal Asthma

Although numerous investigators have demonstrated that corticosteroid side effects, such as adrenal suppression, are influenced by the dosing schedule[68,69] as well as the dosage, the alternative approach of evaluating differences in corticosteroid effectiveness by varying dose schedules has received less attention. Normally, corticosteroid secretion is at its circadian peak level at the beginning of the activity span[59,60] and, if dosing of corticosteroids is timed to this natural peak, the risk and magnitude of adrenal suppression is minimized. Conversely, if adrenal suppression is the desired clinical effect, for example in the treatment of congenital adrenal hyperplasia, the optimal dosing time for corticosteroids appears to be at the end of the activity cycle, prior to sleep.[70] Recognition that corticosteroid administration should be adjusted to achieve the most favorable balance between time-dependent variations in both effectiveness and tolerance is central to therapeutic success. With respect to nocturnal worsening of asthma, the epidemiological observation that overnight asthma symptoms were not improved with corticosteroid use[2] may suggest the need for a critical appraisal of the risk/benefit balance of variably timed corticosteroid dose schedules.

Reinberg and colleagues have demonstrated a sustained interest in the time-dependent pharmacology of corticosteroids with respect to effectiveness in managing asthma.[71–74] An initial study in 1974 evaluated the spirometric response of twelve asthmatic boys to a single variably timed dose of methylprednisolone (40 mg) or placebo.[72] The 24-hour mean peak expiratory flow rate (PEFR) improved more over the placebo baseline when the steroid injection was given at 7 AM or optimally at 3 PM. The 7 PM and 3 AM dosing times were less effective (Fig. 11). In a follow-up study, nine adult asthmatics were treated with a combination corticosteroid preparation (Dutimelan® 8–15 Mite) dosed at 8 AM and 3 PM daily for 5 weeks and monitored for improvement in lung function as well as evidence of adrenal suppression.[74] In addition to subjective improvement in dyspnea, the 24-hour mean PEFRs rose significantly over the 5 weeks without the development of significant adrenal suppression as determined by urinary corticosteroid metabolite production. Furthermore, in a similar study involving eight asthmatic adults, Reinberg and coworkers evaluated the response over 8 days to corticosteroids dosed at 8 AM and 3 PM compared to 3 PM and 8 PM.[73] In agreement with their prior studies, the 8 AM and 3 PM dosing schedule was more effective in producing an improvement in the 24-hour mean PEFRs than the alternative 3 PM and 8 PM dosing schedule (Fig. 12). Based on these studies Reinberg concludes that in the treatment of chronic asthma with nocturnal worsening administration of corticosteroids should be restricted to the morning and early afternoon hours for a diurnally active patient.[71] This schedule of dosing appears more effective and better tolerated than divided daily dosing or other schedules that include an evening corticosteroid dose.

To better clarify the contribution of timing of corticosteroids to their ability to block circadian recruitment of inflammatory cells into the lung and attenuate the nocturnal worsening of asthma, Beam and colleagues evaluated the response of blood eosinophil counts, bronchoalveolar lavage cytology, and overnight pulmonary function to a single variably timed dose of prednisone.[75] Seven asthmatic males with stable, well-controlled daytime asthma but persistent nocturnal worsening of spirometry were treated in a double-blind placebo controlled design with a single 50 mg dose of prednisone at dose times

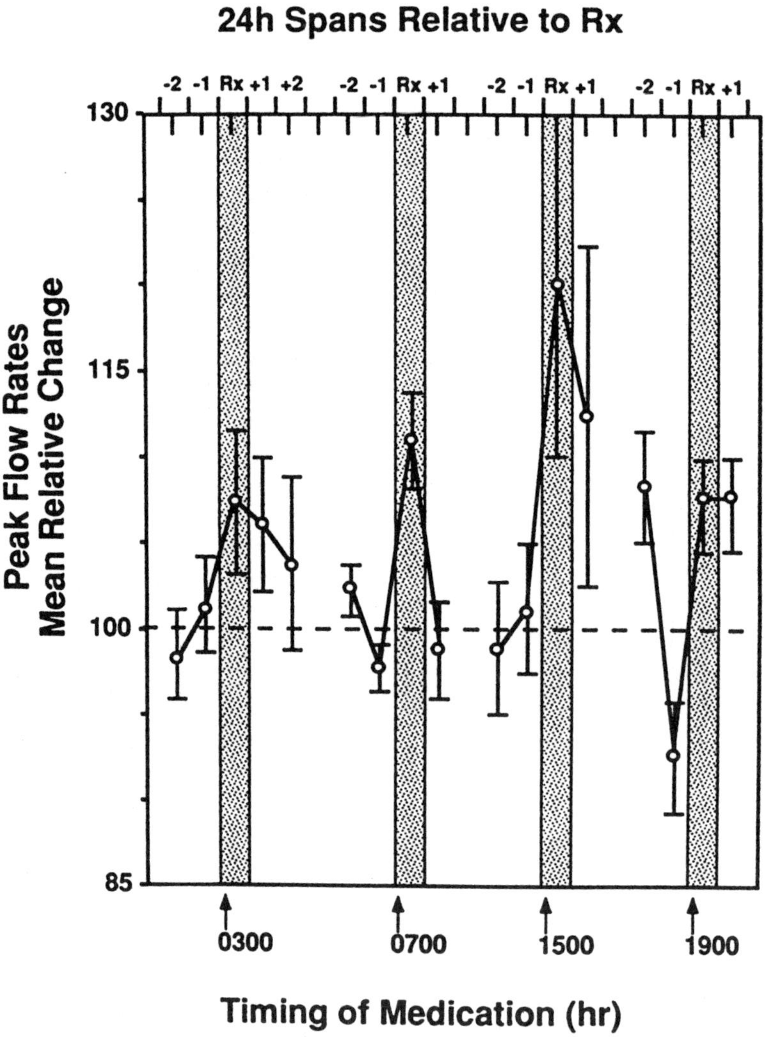

24h Spans Relative to Rx

Figure 11. Chronoeffectiveness of a single corticosteroid dose on peak expiratory flow rates (PEFR) in children. Corticosteroid dosing at 7 AM and 3 PM was associated with significant improvement in the mean PEFR over a 24-hour span; 3 PM dosing was optimal. Adapted from reference 72.

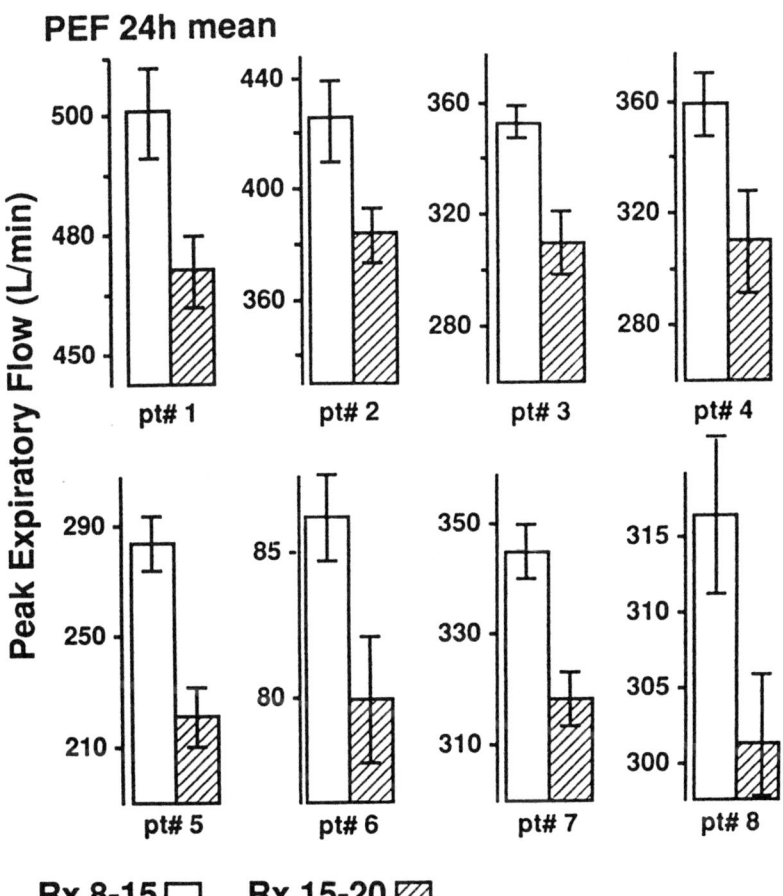

Figure 12. Chronoeffectiveness of 8 AM/3 PM corticosteroid dosing compared to 3 PM/8 PM dosing in adults. Dutimelan® 8-15 resulted in improved mean PEFRs in eight adults compared to Dutimelan® 15-20. PEF = Peak Expiratory Flow rates. Adapted from reference 73.

of 8 AM, 3 PM, and 8 PM. Compared to placebo, a single prednisone dose at 3 PM resulted in a reduction in the overnight percent fall in FEV_1 (Fig. 13) and improvement in the absolute 4 AM FEV_1 of about one liter (Fig. 14). In contrast, neither an 8 AM nor 8 PM prednisone dose when compared to placebo resulted in overnight spirometric improvement. These dose times represent the common components of once daily and twice daily dose schedules. Since a single 50 mg dose of prednisone was used in all three phases, Beam et al.'s results suggest that timing of corticosteroids may be more pertinent than dosage in altering the pathogenesis of nocturnal asthma. Although the study was not a therapeutic trial, it is in agreement with Reinberg and colleagues, as previously discussed, who also found timing of corticosteroid dose relevant in the management of asthma.

Another important observation of Beam et al.'s study was that neither the 8 AM nor 3 PM prednisone doses improved the mean bedtime FEV_1s. This finding suggests that the corticosteroid sensitivity of daytime, prior to bedtime, and nocturnal asthma are

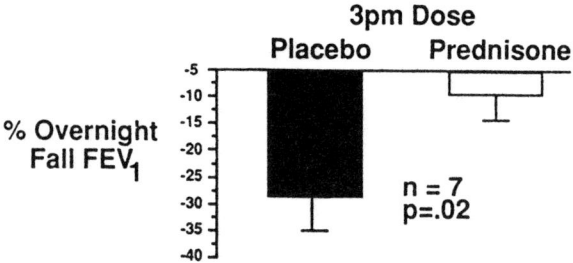

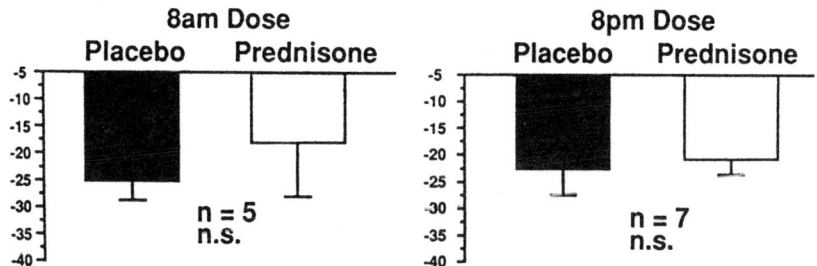

Figure 13. Corticosteroid dose schedule influence on overnight percent fall in FEV_1. Only the 3 PM dose of prednisone was effective in reducing the percent overnight fall in FEV_1. Adapted from reference 75.

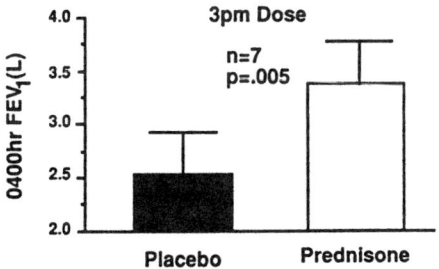

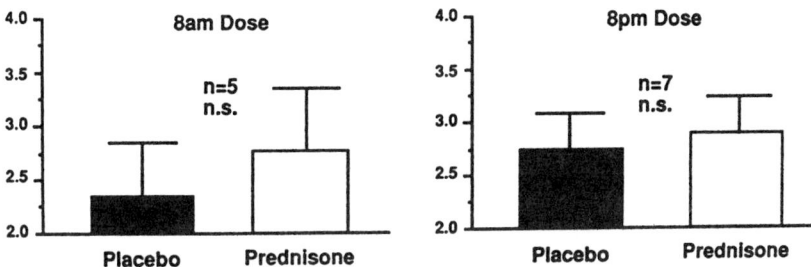

Figure 14. Corticosteroid dose schedule influence on the 4 AM FEV$_1$. Only the 3 PM dose of prednisone significantly increased the absolute 4 AM FEV$_1$ compared to placebo and without significant change in bedtime FEV$_1$. Adapted from reference 75.

dissimilar, and is in agreement with a prior study which demonstrated additional improvement with overnight steroid infusion in spite of therapeutic daytime prednisone doses.[63] If the results represented a simple shift in the timing of a generalized reduction of inflammation, either the 8 AM or 3 PM dose schedules might be expected to produce an improvement in the bedtime FEV$_1$s. The observation that the late asthmatic response (LAR) following nocturnal aerosol challenge is more severe,[76] and that asthmatics prone to nocturnal worsening have a greater airways inflammatory cell load[52] suggests the potential for a heightened inflammatory response at night. Consequently, the influence of corticosteroids on asthmatic lung function may differ in a time-dependent fashion.

Eosinophils are commonly found to be increased in daytime blood, BAL fluid, and bronchial biopsies of asthmatics, and eosinophilic airways inflammation has been shown to correlate with the severity of daytime asthma.[43,77] Martin and colleagues have implicated eosinophils in the nocturnal asthma phenomenon,[52] and suggested the source of these cells is the blood. Circadian variations in blood eosinophil numbers have been demonstrated in asthmatics with a nadir at 10 AM and maximum concentration at midnight.[55] As would be predicted by eosinophil kinetics data, the 8 PM prednisone dose of Beam et al.'s study[75] resulted in a significant reduction of 4 AM blood eosinophil counts to zero. However, this effect was not associated with significant spirometric improvement. The 8 AM dose phase resulted in an early evening reduction of blood eosinophils, zero at 8 PM, but a significant return in number by 4 AM. Again, this effect was not associated with spirometric improvement. In contrast, the 3 PM prednisone dose produced a sustained reduction in blood eosinophil counts at both 8 PM and 4 AM (Fig. 15). These results

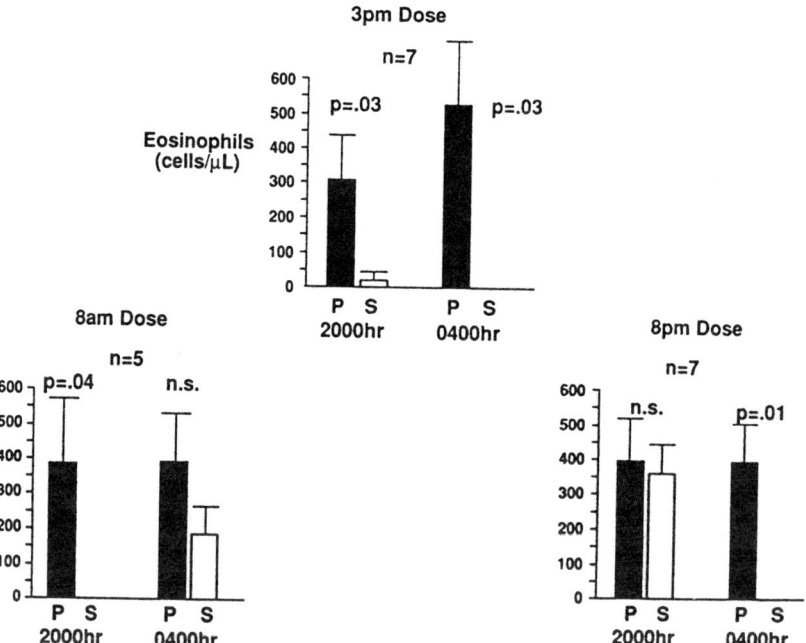

Figure 15. Corticosteroid dose schedule influence on overnight blood eosinophil counts. Only the 3 PM prednisone dose produced a sustained reduction in blood eosinophil counts at both 8 PM and 4 AM. The 8 PM prednisone dose produced a significant reduction in blood eosinophil number at 4 AM, but was not associated with spirometric improvement. Adapted from reference 75.

suggest that eosinophil recruitment must occur at a time much earlier than the expected nadir of spirometry at 4 AM. Beam et al. speculate that influx of blood eosinophils, and possibly other cells or mediators, into the lung at night occurs within a narrow window of time not adequately blocked by either the 8 AM or 8 PM dose schedules. Alternatively, the 3 PM prednisone dose may have altered the degree of eosinophil activation, as well as kinetics, and prevented the release of granule derived proteins which can be cytotoxic to bronchial epithelium.[43]

In addition to effects on overnight spirometry and blood eosinophil counts, the 3 PM dose phase of Beam et al.'s study resulted in a significant pancellular reduction in 4 AM BAL cytology not demonstrated with either alternative dose phase (Fig. 16). Indeed the 8 PM and 8 AM dose phases produced no significant change in any BAL cellular profile (Figs. 17,18). This observation suggests that the 3 PM dose of prednisone interrupted the inflammatory cascade at one or more critical steps in its genesis. Corticosteroids are known to

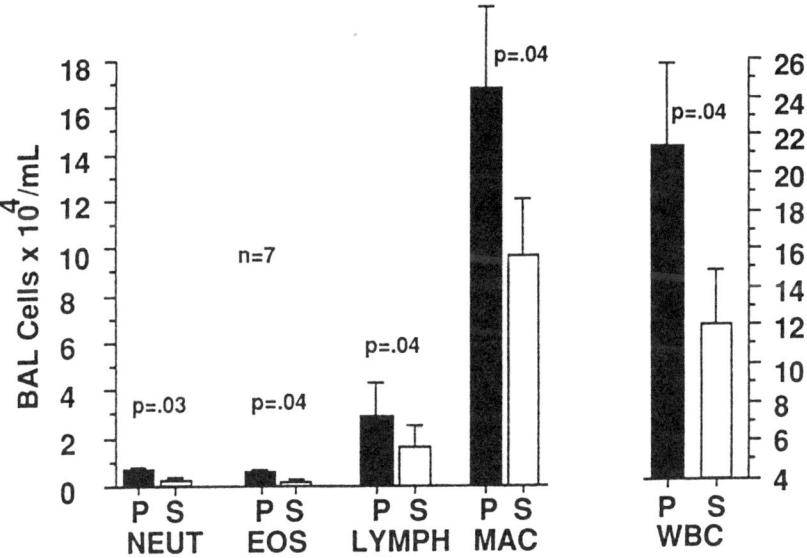

Figure 16. 4 AM bronchoalveolar lavage (BAL) cytology response to a single prednisone dose at 3 PM. The 3 PM prednisone dose resulted in a pancellular reduction in 4 AM BAL cytology. P = placebo, S = steroid, Neut = neutrophils, Eos = eosinophils, Lymph = lymphocytes, Mac = macrophages, WBC = total leukocytes. Adapted from reference 75.

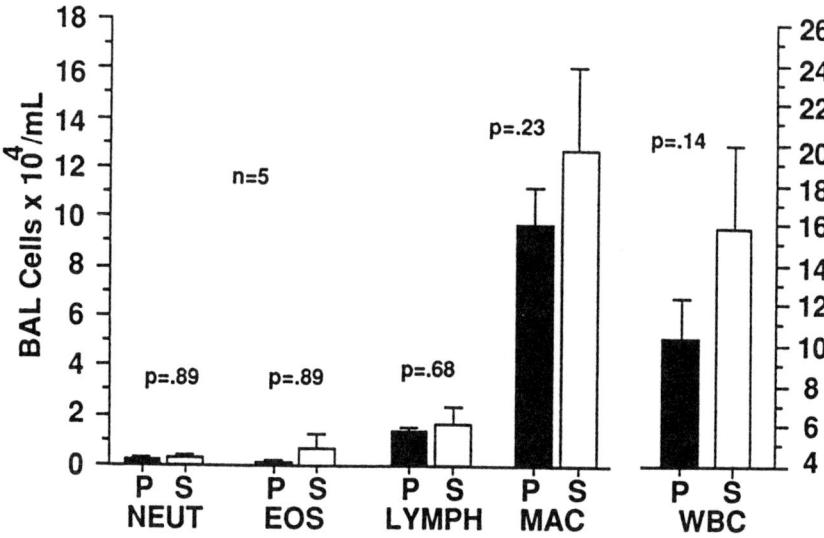

Figure 17. 4 AM bronchoalveolar lavage (BAL) cytology response to a single prednisone dose at 8 AM. No cell line was significantly reduced with 8 AM prednisone dosing. P = placebo, S = steroid, Neut = neutrophils, Eos = eosinophils, Lymph = lymphocytes, Mac = macrophages, WBC = total leukocytes. Adapted from reference 75.

influence both the function and kinetics of all the inflammatory cells represented in the BAL fluid.[44] It is noteworthy that Martin and colleagues demonstrated elevations in total white cell number, neutrophil, eosinophil, and lymphocyte counts in the BAL fluid of the nocturnal asthma cohort when compared to asthmatics without nocturnal worsening.[52] These observations support a collaborative cellular mechanism of inflammation that is corticosteroid sensitive, yet dependent on timing in addition to dosage.

In summary, Beam et al.'s data support the relevance of timing of prednisone dose in altering the inflammatory milieu associated with nocturnal worsening of asthma. In contrast to prednisone dosing at 8 AM and 8 PM, the 3 PM dose produced significant improvement in overnight spirometry as well as a pancellular reduction in the 4 AM BAL cytology. Further studies will be required to clarify the contribution of time-dependent factors as they relate to inflammatory cell recruitment to the lung. Additionally, both short- and long-term studies will be needed to evaluate the physiological

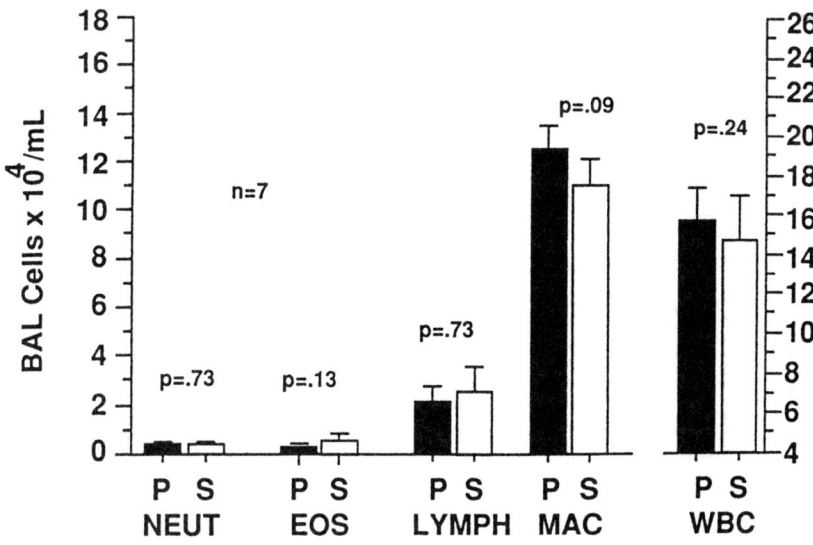

Figure 18. 4 AM bronchoalveolar lavage (BAL) cytology response to a single prednisone dose at 8 PM. No cell line was significantly reduced with 8 PM prednisone dosing. P = placebo, S = steroid, Neut = neutrophils, Eos = eosinophils, Lymph = lymphocytes, Mac = macrophages, WBC = total leukocytes. Adapted from reference 75.

improvement in nocturnal asthma, and asthma overall, with a 3 PM dosing schedule of corticosteroids, as well as to compare the incidence and severity of adrenal suppression and other potential side effects with the classical 8 AM dosing schedule.

Inhaled Corticosteroids in Nocturnal Asthma

In her 1988 epidemiological analysis of nocturnal asthma, Dr. Turner-Warwick concluded that the incidence of nocturnal awakenings due to asthma was unchanged from prior observations done 15 years earlier.[2] Furthermore, she stated that "no drug or drug combination was associated with a significantly lower frequency of nocturnal awakenings," and suggested that the introduction of inhaled corticosteroids in the early 1970s has not impacted on nighttime asthma symptom control. A review of Table 1 supports her contentions. Even budesonide, the most potent inhaled corticosteroid currently used to treat asthma, does not appear to influence the frequency of nocturnal asthma awakenings, with 64% of patients using budesonide still reporting nocturnal awakenings at least 3 nights per week. In a manner analogous to oral corticosteroid use, these observations suggest that nocturnal worsening represents an aspect of the asthma syndrome insensitive to inhaled corticosteroids. Alternatively, issues regarding dosage used and timing of dose may be pertinent to the effectiveness and tolerance profiles of inhaled, as well as oral, corticosteroids. Although numerous investigators have evaluated inhaled corticosteroids in the treatment of asthma, relatively few have specifically addressed the issue to nighttime symptom control. Since the issue of efficacy in this area remains an open one, the observations by Dr. Turner-Warwick call for further investigation to assess the possible contributions of aerosol corticosteroids to the management of nocturnal asthma. The inhaled corticosteroids currently available in the United States or Europe used to treat asthma include beclomethasone, betamethasone, triamcinolone, flunisolide, and budesonide. Although these compounds differ in their relative potencies, both in terms of topical and systemic corticosteroid effects, there is little data to suggest that when used in equivalent antiinflammatory doses, any one medication is clinically superior to another.

In contrast to oral corticosteroids in which the usual dose frequency is once daily, inhaled corticosteroids are routinely prescribed to be used several times a day. In an effort to establish the optimal dose frequency and evaluate relationships regarding compliance, efficacy, and side effects, several investigators have addressed the influence of dosing schedules when using inhaled corticosteroids.[80–82] McGivern and coworkers found that at equivalent total doses, once daily inhaled beclomethasone resulted in lower morning and evening peak expiratory flow rates (PEFRs) and an increase in both daytime and nighttime symptoms compared to a three or four-times-a-day divided dose schedule.[78] Malo and colleagues compared twice-a-day (b.i.d.) to four-times-a-day (q.i.d.) dosing frequency using variable doses of budesonide over a six month period.[79] They concluded that the twice-a-day regimen resulted in almost twice as many days with nocturnal asthma and cough and almost three times as many days with disability due to asthma. Also, overnight variability in PEFRs and relapse frequency were higher during the twice-a-day schedule. These results were in agreement with observations made by Toogood and colleagues who also found the q.i.d. schedule resulted in increased efficacy compared to an equivalent dose b.i.d. schedule.[80] Consequently, although some of the inhaled corticosteroids are known to be effective when used in a b.i.d. schedule (budesonide, flunisolide), the most effective dosing frequency appears to be a q.i.d. schedule, and dosage may be best titrated by altering puffs per dose rather than reductions in dose frequency.[80]

Although patient compliance may lapse with increased dose frequency, this issue alone would not likely limit the use of inhaled corticosteroids in the attempted management of nocturnal asthma. The issue of efficacy remains the key focus. As previously stated, Malo and colleagues found the q.i.d. dosing schedule for budesonide was associated with a lower incidence of nocturnal asthma and cough compared to the b.i.d. dosing schedule.[79] However, the incidence of nocturnal asthma approached 25% of the nights for the asthmatics using q.i.d. budesonide dosing. Also the maximum evening to morning variability in PEFRs remained about 29% on the q.i.d. dosing schedule and 33% during the b.i.d. dosing schedule. Further studies support the limited ability of inhaled corticosteroids, even at high dose and with frequent dosing, to attenuate nocturnal worsening of asthma. Horn and colleagues evaluated the response of asthmatics prone to "morning dipping" to inhaled beclomethasone (400 μg q.i.d.) for two weeks.[81] They identified a spectrum of responses in which about one-half of the subjects significantly reduced their overnight falls in PEFRs while the other half demonstrated no significant improvement in their overnight PEFRs compared to baseline. In a subsequent study, Horn and coworkers prospectively evaluated asthmatics using inhaled beclomethasone at variable doses up to 2000 μg per day. At the end of the 36-week study, one-half the subjects studied continued to have nocturnal awakenings due to asthma.[82] These studies, which suggest inhaled corticosteroids used in high dose and with frequent dosing may be only about 50% effective in controlling nocturnal asthma, are not dissimilar from Turner-Warwick's observations. Also, in the subset of asthmatics who do demonstrate improvement in nocturnal asthma with inhaled corticosteroids, the magnitude of additional improvement beyond that seen with beta-agonist therapy is relatively small.[81,83,84] Thus, although Sly and coworkers have suggested that inhaled triamcinolone may be effective in reducing nocturnal awakening due to wheezing in a pediatric population,[85] adult studies using high dose, frequent dosing regimens have demonstrated limited effectiveness in improving nocturnal symptoms or overnight spirometry. Also, at the dosage range used in Horn's studies, the issue of systemic absorption and subsequent side effects becomes clinically important.[80,86]

As corticosteroids became a common modality in the management of numerous disease states, an appreciation of their side effects and serious complications soon became realized. Because of these hazards, inhaled preparations were developed which many hoped would allow local delivery of the drug to the airway, thus achieving high end-organ steroid concentrations, yet avoiding systemic absorption and resultant side effects. In the clinical setting, this goal has not been fully accomplished. All the inhaled corticosteroids are delivered by metered dose devices which propel particles of drug into the mouth, pharynx, and airways. The majority of the drug is trapped in the mouth and pharynx and only about 3% to 6% of the drug reaches the small peripheral airways and alveoli.[87] Consequently, in addition to possible systemic side effects due to absorption via the lung, gut, and pharyngeal mucosa, inhaled corticosteroid use is uniquely associated with local side effects of candidiasis and dysphonia.[88–91] The incidence of oropharyngeal candidiasis has been reported to range between 9% and 20%,[88] although the risk can be minimized by having patients routinely rinse their mouths after using the medication. A spacer device may also be useful in reducing oral deposition of steroid, and the triamcinolone aerosol unit includes such a chamber. Oral candidiasis as a consequence of inhaled corticosteroids usually responds well to local therapy with an antifungal agent. In contrast, intermittent dysphonia has been reported to be a dose related phenomenon associated with efficient inhaler use resulting in laryngeal deposition of the steroid medication.[89,90] The mechanism of dysphonia may include both laryngeal candidiasis as well as vocal cord dysfunction.[90] Thus, it has a variable response to antifungal therapy. Risk factors for dysphonia include hypothyroidism, habitual throat clearing, and daily laryngeal stress in patients such as singers, public speakers, and teachers.[91] In addition to voice rest and local antifungal therapy, if clinically indicated, dysphonia may also improve with the use of a spacer device. Lastly, unusual but associated complications of inhaled corticosteroid preparations include atrophic glossitis,[91] esophageal candidiasis,[92] and combined herpetic and candida associated esophagitis.[93] These more unusual complications are most often seen with concomitant antibiotic and systemic corticosteroid use.[91–93]

Characteristic skin changes of atrophy and epidermal thinning may result from the topical application of corticosteroids, and the risk of these side effects parallels the potency of the steroid used.[94] These observations raised concern over the effects of long-term inhaled corticosteroid use on bronchial mucosa. Lundgren and colleagues have demonstrated a continued interest in this question, and they have evaluated asthmatics using inhaled corticosteroids at 6 months into therapy[95] and after 10 years of therapy.[96] In Lundgren's initial study, eight asthmatics were evaluated using scanning electron microscopy (SEM) on bronchial mucosal biopsies before and after 6 months of daily inhaled beclomethasone therapy.[95] Analysis of epithelial cell cilia structure by SEM demonstrated improvement in four, no change in three, and slight deterioration in one of the subjects studied. The relative lack of bronchial epithelial damage demonstrated was in agreement with other investigators examining the issue.[97] In their 10-year follow-up study, Lundgren and colleagues evaluated six asthmatics and six nonsmoking controls using SEM analysis.[96] Prior to treatment with inhaled corticosteroids, the asthmatic cohort had increased numbers of inflammatory cells in their bronchial biopsies consisting mainly of lymphocytes and eosinophils. After 10 years of treatment, no differences in the number of inflammatory cells, the thickness of the epithelium, or the thickness of the basement membrane could be demonstrated between the asthmatics and normal controls. Interestingly, the asthmatic cohort still demonstrated increased bronchial responsiveness to methacholine. These studies suggest that concerns over bronchial mucosal damage by inhaled corticosteroids may not manifest clinically, although continued surveillance is warranted as higher doses of these medications are used in clinical practice. Also, these studies are in agreement with Vathenen and colleagues who have suggested that the mechanism of action by inhaled corticosteroids in asthma relates to suppression of inflammation rather than a reversal of the underlying defect causing increased bronchial reactivity.[98]

Inhaled corticosteroid use has been associated with posterior subcapsular cataract formation,[99] increased bone turnover,[100] cushingoid appearance,[101] hypothalamic-pituitary-adrenal (HPA) axis suppression,[86,102,103] and growth retardation.[101,104,105] The population most at risk for these side effects is in the pediatric age groups. A clear dose risk response is lacking, which may suggest differences in susceptibility within individuals or be a consequence of medication specific pharmacokinetics. In either case, more investigation will be required to assess both the short-term and long-term side effect profiles of inhaled corticosteroids, particularly in the pediatric population. Some investigators have even suggested that inhaled corticosteroid use in severely asthmatic children has been associated with an increased risk of asthmatic death.[106]

In summary, the observation by Dr. Turner-Warwick that inhaled corticosteroid use was not associated with a reduced incidence of asthma associated awakenings[2] remains to be refuted. Inhaled corticosteroids, used in relatively high doses of 1600 to 2000 μg per day and employed in the most efficacious dosing schedule of four-times-a-day dosing, demonstrate limited effectiveness in controlling nocturnal asthma.[81,82] Although higher doses or more chronoeffective dosing regimes may expand the role of inhaled corticosteroids in treating nocturnal asthma, current data suggest that about 50% of patients will continue to exhibit overnight worsening of spirometry and asthma symptoms while on the medication.

Corticosteroid Dose Schedule and Adrenal Suppression

In addition to intensive bronchodilator treatment, systemic corticosteroids have become a standard adjunctive therapy in the treatment of acute, severe asthma.[107–109] Although corticosteroids may significantly contribute to accelerated clinical improvement,[110,111] their use chronically is often limited by the well-described cushingoid side effects of truncal obesity, osteoporosis, cataract formation, glucose intolerance, skin fragility, and a myriad of other complications. Concerns over these complications by both patients and their physicians frequently result in an attempt to withdraw systemic corticosteroids after the acute episode of asthma has stabilized. As is the case with other diseases requiring corticosteroid treatment, the initial reductions in dose are most often limited by evidence of deterioration in disease control. In the case of asthma, serial pulmonary function testing provides a more reliable guide to disease control than clinical signs and symptoms.[112,113] Asthmatics warrant close surveillance during corticosteroid withdrawal, since this time is associated with relapse and sudden death.[107,113–115] This association was highlighted in one review of asthma mortality which revealed that the duration of symptoms during the fatal asthmatic exacerbation was less than 3 hours in 90%

of patients who had previously received systemic corticosteroids.[116] Again, the majority of asthma related ventilatory arrests and deaths occur during the sleep related hours.[3] Still, with careful monitoring, the majority of asthmatics are successfully tapered from the pharmacological doses of corticosteroids used during acute exacerbations to doses closer to the physiological range without loss of disease control. At this point in corticosteroid withdrawal, medication-induced suppression of adrenal function becomes important.[117] Similarly, asthmatics requiring chronic corticosteroid therapy are also at risk for suppression of the hypothalamic-pituitary-adrenal (HPA) axis and loss of adrenal responsiveness. In this group, the dosing schedule is one of several factors influencing the risk and severity of iatrogenic adrenal suppression.

The degree of adrenal suppression seen with chronic corticosteroid use and during corticosteroid withdrawal appears dependent upon a number of factors including the total dosage, the duration of therapy, the frequency and timing of the dose schedule, the route of administration, and individual susceptibilities.[117–124] The HPA axis may require up to 9 months to fully recover from corticosteroid-induced suppression, and the usual sequence of recovery involves pituitary production of adrenocorticotropin (ACTH) prior to the adrenal cortex being able to secrete cortisol in response to stress.[125] The clinical signs and symptoms of corticosteroid withdrawal and/or hypoadrenalism include fatigue, lethargy, anorexia, arthralgias, orthostatic hypotension, hypoglycemia, and desquamation of the skin. Since corticosteroid-induced adrenal insufficiency may be prolonged and has been associated with an increased risk of death during periods of stress (surgery, trauma), a protocol for withdrawal that involves periodic evaluation of the HPA axis has been proposed by Byyny.[117] During periods of anticipated stress, supplemental corticosteroids may be required until testing of the HPA axis reveals adequate adrenal production of cortisol at rest and in response to exogenous ACTH or insulin challenge. Byyny recommends 100 mg of hydrocortisone parenterally every 6 to 8 hours for the first 3 to 4 days until the stress is resolved.

Plasma cortisol levels normally demonstrate circadian variability with a peak near the time of awakening and a nadir near midnight.[59–61] As previously mentioned, cortisol production by the adrenal cortex is regulated by the anterior pituitary hormone, ACTH, which in turn is regulated by the hypothalamic hormone, corticotropin-releasing factor (CRF).[117,118] The conventional explanation of adrenal suppression resulting from corticosteroid medications is based on a negative feedback model such that an elevation in plasma corticosteroid levels results in inhibition of ACTH and CRF secretion. Thus, endogenous stimulation of the adrenal cortex is inhibited and secretion of cortisol diminished. Reinberg has suggested that this model is inadequate to explain the time-dependent variables seen in both human and animal experiments. He has proposed that the HPA axis appears to be "programmed-in-time" such that the program can be modified only at certain periods within the 24-hour span.[71,119,120] In Reinberg's model, the system is open to negative feedback at specific hours within the circadian time structure, and only during these hours can a change in the program be made with regard to the secretion of ACTH. It follows from this model that the dose schedule may greatly influence the degree of adrenal suppression evident for a given dosage of corticosteroid medication. A review of currently available literature supports this speculation.

Grant and coworkers evaluated the adrenocortical response of six normal males to 8 mg of triamcinolone given as either a single oral dose at 8 AM or as 2 mg divided oral doses at 8 AM, 1 PM, 6 PM, and midnight for a total of 8 days.[121] They demonstrated a significant reduction in the morning plasma 17-hydroxycorticosteroid (17-OHCS) levels and suppression of the 24-hour urinary 17-OHCS excretion associated with the divided dose schedule. The single morning dose schedule resulted in no significant reduction in either the morning plasma level or urinary excretion of 17-OHCS (Fig. 19). Grant concluded that the divided dose schedule resulted in adrenal suppression, beginning 4 days into treatment and lasting beyond the 8 day trial, while the single morning dose schedule preserved adrenal function both during and immediately after the trial. These results were in agreement with a prior study done by DiRaimondo and Forsham in which a single morning dose of 10 mg of prednisone was compared to a 2.5 mg four-times-a-day dosing schedule.[122] Again, the single morning dose schedule produced negligible suppression of adrenal function, while the divided dose schedule resulted in significant suppression of urinary 17-OHCS levels for the same total daily dose.

Harter and colleagues compared an alternate day morning dose schedule to a daily divided dose schedule in a mixed patient population numbering almost sixty.[123] The daily divided dose schedule resulted in the greatest degree of adrenal suppression. Interest-

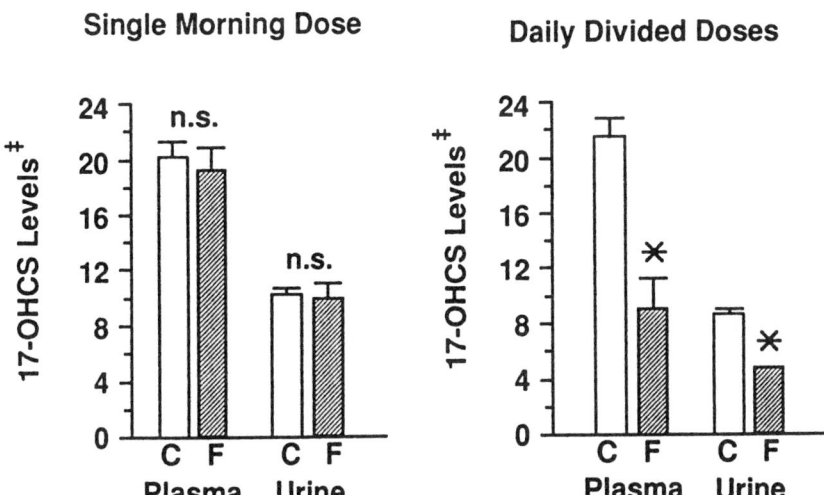

Figure 19. Corticosteroid dose schedule influence on adrenal suppression. The single morning dose of 8 mg triamcinolone resulted in no difference in 17-OHCS levels after 8 days, while the 2 mg four-times-a-day schedule resulted in reduced plasma and urinary 17-OHCS levels for the same duration of treatment. C = control level, F = final level after treatment for 8 days, * = P < 0.01, ‡ = plasma (μg/100 ml), urine (mg/24°). Adapted from reference 121.

ingly, the alternate day morning dose schedule did not produce significant adrenal suppression even though the total corticosteroid dose was between two and six times greater than that used in the divided dose cohort. Harter's data suggest that the risk of adrenal suppression may be more closely linked to dose schedule than to total dosage used. In a similar study, Ackerman and Nolan evaluated corticosteroid-induced adrenal suppression following an alternate day morning dose schedule compared to a daily divided dose schedule.[124] They assessed adrenal function by evaluating resting plasma 17-OHCS levels and levels following insulin-induced hypoglycemic stress. They monitored the rise in plasma 17-OHCS levels in the steroid treated groups and compared them to normal controls. They found the daily divided dose schedule to be associated with a low resting plasma 17-OHCS level and a blunted rise following an insulin tolerance test. In contrast, the alternate day morning dose schedule was associated with a reduced resting plasma 17-OHCS level but a normal rise following hypoglycemic stress (Fig. 20). They

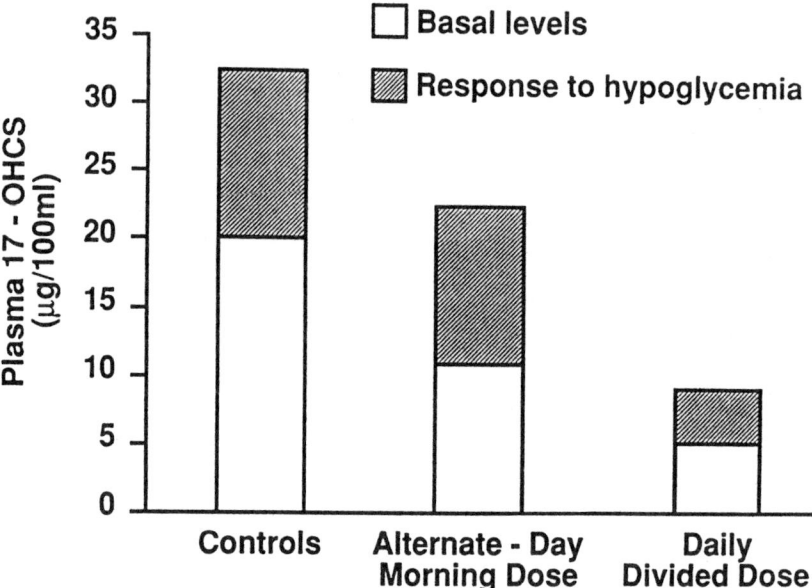

Figure 20. Influence of corticosteroid dosing schedules on basal plasma 17-OHCS levels and response to insulin-induced hypoglycemia. Although the basal levels of 17-OHCS are reduced from control levels by both dose schedules, the alternate day morning dose schedule still allows near normal response to hypoglycemic stress. Adapted from reference 124.

concluded that the alternate day morning dose schedule allowed the maintenance of a normal adrenocortical response to stress, and this adrenal reserve was not seen with the daily divided dose schedule. In summary, these studies suggest that the correlation between adrenal suppression and dose schedule is strong and that the daily divided dose schedule (four-times-a-day) poses the greatest risk for this complication.

Although these studies begin to address the issue of dose schedule influence on corticosteroid-induced hypoadrenalism, they do not allow insight into possible variability of influence within a 24-hour, or circadian based, dose schedule. The underlying assumption in the prior studies cited [121–124] is that an alternate day or daily morning dose schedule would be effective in achieving or maintaining disease control. In the case of nocturnal worsening of asthma, evidence suggests that an early afternoon, 3 PM by Reinberg et al.'s and Beam et al.'s data, [72–75] dosing time is the most effective dose schedule. Although not specifically addressed in Turner-Warwick's analysis, the relative insensitivity of nocturnal asthma to oral corticosteroids may be a result of the routine use of morning dose schedules. Consequently, variations in the risk of adrenal suppression within a 24-hour dose schedule may be important in assessing the overall clinical utility of 3 PM corticosteroid dosing for nocturnal asthma. If the HPA axis is programmed in time, as Reinberg suggests, then the comparative risk of adrenal suppression between an 8 AM and 3 PM dosing schedule may be small. No study currently available for review specifically addresses this issue, although data are available on the influence of corticosteroid dosing within a 24-hour span and the development of adrenal insufficiency.

Nichols, Nugent, and Tyler evaluated the circadian variability in corticosteroid-induced adrenal suppression using dexamethasone dosed orally at 8 AM, 4 PM, or midnight.[126] In a cohort of normal volunteers, they established the baseline circadian pattern of plasma 17-OHCS, then repeated their measurements after a 2-day course of 0.5 mg dexamethasone dosed at one of the three scheduled times. They demonstrated that the 8 AM dose schedule resulted in suppression of plasma 17-OHCS levels during the afternoon and sleep related hours, but insignificant suppression at 8 AM the following day. The 4 PM dose schedule produced suppression of plasma 17-OHCS levels at midnight, with partial suppression at 8 AM, but no suppression evident by 4 PM the following day. The midnight dose schedule resulted in near total adrenal suppression for the subsequent 24-hour period. They concluded that the 8 AM and 4 PM dose schedules resulted in temporary suppression of cortisol secretion, while the midnight dosing resulted in complete suppression of cortisol production for a full 24-hour period. These results are compatible with Reinberg's model in which the HPA axis is open to negative feedback at specific hours within the circadian time structure. Additional data supporting this concept was supplied by Ceresa and colleagues.[120] They analyzed the circadian pattern of urinary 17-OHCS excretion after short duration continuous intravenous infusions of corticosteroids with varying dose schedules. They observed significant suppression of urinary 17-OHCS excretion when the corticosteroids were infused between midnight and 4 AM, but no adrenal suppression when the infusion was scheduled between 8 AM and 4 PM. They were, however, able to produce adrenal suppression, as shown by urinary 17-OHCS excretion, during the 8 AM to 4 PM dosing time by increasing the dose during infusion by two-to fourfold. In summary, these studies suggest that circadian variability exists for risk of adrenal suppression within a 24-hour dosing schedule. Maximum inhibition occurs when corticosteroids are dosed at the end of the activity cycle and during sleep, and minimum inhibition occurs during the morning and early afternoon. However, evidence for adrenal suppression can be found even with morning and early afternoon dosing, if the total dose is high enough. Streck and Lockwood demonstrated transient adrenal insufficiency, lasting about 5 days, following treatment with 50 mg of prednisone dosed as 25 mg at 8 AM and 4 PM daily for 5 days.[127] This compares to the transient 3-day period of suppression shown by Webb and Clark in patients with chronic airflow obstruction given a 3-week course of 40 mg of prednisolone (dosed as 20 mg orally b.i.d.).[128] Unfortunately, neither of these studies address the relative risk of adrenal suppression, comparing pharmacological doses of corticosteroids given at 8 AM versus 3 PM.

In conclusion, the corticosteroid dose schedule is one of several important factors influencing the risk and severity of iatrogenic adrenal insufficiency. Further studies will be needed to assess the risk-benefit balance of afternoon dosing for nocturnal asthma. At this time there is no evidence to support an evening dose schedule since it poses the greatest risk of adrenal impairment without documented efficacy in improving nocturnal asthma.

References

1. Floyer J. A treatise of the asthma. Wilkin & Innis, London, 1698.
2. Turner-Warwick M. Epidemiology of nocturnal asthma. Am J Med 1988; 85(Suppl 1B):6–8.
3. Hetzel MR, Clark TJH, Branthwaite MA. Asthma: analysis of sudden deaths and ventilatory arrests in hospital. Br Med J 1977; 1:808–811.
4. Smolensky MH, D'Alonzo GE. Biologic rhythms and medicine. Am J Med 1988; 85(Suppl 1B):34–46.
5. Reichstein T, Shoppee CW. The hormones of the adrenal cortex. Vitam Horm 1943; 1:346–413.
6. Hench PS, Kendall EC, Slocumb CH, et al. The effect of a hormone of the adrenal cortex (compound E) and of pituitary adrenocorticotropic hormone on rheumatoid arthritis. Proc Staff Meet Mayo Clinic 1949; 24:181–197.
7. Carryer HM, Prickman LE, Mayturn CK, et al. Effects of cortisone on bronchial asthma and hay fever occurring in subjects sensitive to ragweed pollen. Proc Mayo Clinic 1950; 25:282–295.

8. Randolph TG, Rollins JP. Effect of cortisone on bronchial asthma. J Allergy 1950; 21:288–293.
9. Herzog HL, Nobile A, Tolksdorf S, et al. New anti-arthritic steroids. Science 1955; 121:176.
10. Spero GB, Thompson JL, Mayerlein BJ, et al. Adrenal hormones and related compounds. J Am Chem Soc 1956; 78:6213–6214.
11. Bernstein S, Lenhard RH, Allen WS, et al. 16-Hydroxylated steroids. J Am Chem Soc 1956; 78:5693–5697.
12. Peak MD, Cayton RM, Howard P. Triamcinolone in corticosteroid resistant asthma. Br J Dis Chest 1979; 73:39–44.
13. Fauci AS. Glucocorticosteroid therapy: Mechanisms of action and clinical considerations. Ann Int Med 1976; 84:304–315.
14. Clark THJ. Effect of beclomethasone diproprionate delivered by aerosol in patients with asthma. Lancet 1972; 1:1361–1364.
15. Frears JF, Wilson LC, Friedman M. Betamethasone 17-valerate by aerosol in childhood asthma. Arch Dis Child 1973; 48:856–863.
16. Williams MH. Treatment of asthma with triamcinolone acetonide aerosol. Chest 1975; 68:765–768.
17. Lowell FC, Ohman JL, Williams M. Double-blind trial of flunisolide aerosol in bronchial asthma (abstract). J Allergy Clin Immunol 1976; 57:257.
18. Thalen A, Brattsand R. Synthesis and anti-inflammatory properties of budesonide. Arzn-Forsch/Drug Res 1979; 29:1687–1690.
19. Baxter JD, Funder JW. Hormone receptors. N Eng J Med 1979; 301:1149–1161.
20. Jones MT, Gillham B, Mahmoud S. Hypothalamus and ACTH secretion, the endocrine function of the human adrenal cortex. James VHT, Siero M, Giusti G, et al. (eds). Academic Press, New York, 1978, pp 55–85.
21. Flower RJ. Lipocortin and the mechanism of action of the glucocorticoids. Br J Pharmacol 1988; 94:987–1015.
22. Barnes PJ, Chung KF, Page CP. Inflammatory mediators and asthma. Pharmacol Rev 1988; 40:49–84.
23. Tsurufuji S, Sugio K. Molecular mechanism in the manifestation of anti-inflammatory activity of glucocorticosteroids. Eur J Rheum Inflam 1978; 1:226–231.
24. Persson CGA. Role of plasma exudation in asthmatic airways. Lancet 1986; 2:1126–1129.
25. Svensjo E, Roempke K. Time-dependent inhibition of bradykinin and histamine-induced increase in microvascular permeability by local glucocorticosteroid treatment. Glucocorticosteroids, Inflammation, and brochial hyperreactivity. Hogg JC, Ellul-Micallef R, Brattsand R (eds). Excerpta Medica, Amsterdam, pp 136–144.
26. Hogg JC, Pare PD, Moreno R. The effect of submucosal edema in airways resistance. Am Rev Respir Dis 1987; 135:S54–S56.
27. Mano K, Akbarzadeh A, Townley RG. Effect of hydrocortisone on beta-adrenergic receptors in lung membrances. Life Sci 1979; 25:1925–1930.

28. Davies AO, Lefkowitz RJ. Agonist promoted high affinity state of the β-adrenergic receptor in human neutrophils: Modulation by corticosteroids. J Clin Endocrinol Metab 1981; 53:703–708.
29. Conolly ME, Tashkin DP, Hui KKP, et al. Selective subsensitization of beta-adrenergic receptors in central airways of asthmatics and normal subjects during long-term therapy with inhaled salbutamol. J Allergy Clin Immunol 1982; 70:423–431.
30. Weber RW, Smith JA, Nelson HS. Aerosolized terbutaline in asthmatics: Development of subsensitivity with long-term administration. J Allergy Clin Immunol 1982; 70:417–422.
31. Harvey JE, Baldwin CJ, Wood PJ, et al. Airways and metabolic responsiveness to intravenous salbutamol in asthma. Effect of regular inhaled salbutamol. Clin Sci Mol Med 1981; 60:579–585.
32. Repsher LH, Anderson JA, Bush RK, et al. Assessment of tachyphylaxis following prolonged therapy of asthma with inhaled albuterol aerosal. Chest 1984; 85:34–38.
33. Szefler SJ, Ando R, et al. Plasma histamine, epinephrine, cortisol, and leukocyte β-adrenergic receptors in nocturnal asthma. Clin Pharmacol Ther 1991; 49:59–68.
34. Townley RG, Reeb R, Fitzgibbon T. The effect of corticosteroids on the beta-adrenergic receptors in bronchial smooth muscle. J Allergy 1970; 45:A118.
35. Aviado DM, Carrillo LR. Anti-asthmatic action of corticosteroids. J Clin Pharmacol 1970; 10:3–11.
36. Shenfield GM, Hodson ME, Clark SW, et al. Interaction of corticosteroids and catecholamines in the treatment of asthma. Thorax 1975; 30:435–439.
37. Ellul-Micallef R. Pharmacokinetics and pharmacodynamics of glucocorticoids. In: Drug Therapy for Asthma, Jenne JW, Murphy S, (eds). Marcel-Dekker, New York, 1987, pp 463–516.
38. Chung KF. Role of inflammation in the hyperreactivity of the airways of asthma. Thorax 1986; 41:657–662.
39. O'Byrne PM, Hargreave FE, Kirby JG. Airways inflammation and hyperresponsiveness. Am Rev Respir Dis 1987; 136:S35–S37.
40. Dunnill MS. The pathology of asthma with special reference to changes in the bronchial mucosa. J Clin Pathol 1960; 13:27–33.
41. Beasley R, Roche WR, Roberts JA, Holgate ST. Cellular events in the bronchi in mild asthma and after bronchial provocation. Am Rev Respir Dis 1989; 139:806–817.
42. Laitinen LA, Heino M, Laitinen A, et al. Damage of the airway epithelium and bronchial reactivity in patients with asthma. Am Rev Respir Dis 1985; 121:599–606.
43. Bousguet J, Chavez P, Lacoste JY, et al. Eosinophilic inflammation in asthma. N Eng J Med 1990; 323:1033–1039.
44. Schleimer RP. Effects of glucocorticosteroids on inflammatory cells relevant to their therapeutic applications in asthma. Am Rev Respir Dis 1990; 141:S59–S69.
45. Schleimer RP, Schulman ES, MacGashan DW, et al. Effects of dexamethasone on mediator release from human lung fragments and purified lung mast cells. J Clin Invest 1983; 71:1830–1835.

46. Leonard EJ. Two populations of human blood basophils; effect of prednisone on circulating numbers. J Allergy Clin Immunol 1987; 79:775–780.
47. Schleimer RP, Davidson DA, Peters SP, et al. Inhibition of basophil leukotriene release by anti-inflammatory steroids. Int Arch Allergy Appl Immunol 1985; 77:241–243.
48. Bishop CR, Athens JW, Boggs DR, et al. Leukokinetic studies. J Clin Invest 1968; 47:249–260.
49. Petroni KC, Shen L, Guyre PM. Modulation of human polymorphonuclear leukocyte lgG Fc receptors and Fc receptor mediated functions by IFN-gamma and glucocorticoids. J Immunol 1988; 140:3467–3472.
50. Metzger WJ, Zavala D, Richerson HB, et al. Bronchial allergen challenge and bronchoalveolar lavage of allergic asthmatic lungs. Am Rev Respir Dis 1987; 135:433–440.
51. O'Byrne PM, Walters AEH, Gold BD, et al. Neutrophil depletion inhibits airway hyperresponsiveness induced by ozone exposure. Am Rev Respir Dis 1984; 130:214–219.
52. Martin RJ, Cicutto LC, Smith HR, Ballard RD, Szefler SJ. Airways inflammation in nocturnal asthma. Am Rev Respir Dis 1991; 143:351–357.
53. Sanders RH, Adams E. Changes in circulating leukocytes following administration of adrenocortex extract (ACE) and adrenocorticotropic hormone (ACTH) in infectious mononucleosis and chronic lymphatic leukemia. Blood 1950; 5:732–741.
54. Anderson V, Bro Rasmussen F, Hougaard K. Autoradiographic studies of eosinophil kinetics, effects of cortisol. Cell Tissue Kinet 1969; 2:139–146.
55. Dahl R. Diurnal variation in the number of circulatory eosinophil leukocytes in normal controls and asthma. Acta Allergologics 1977; 37:301–303.
56. Claman HN. Corticosteroids and lymphoid cells. N Engl J Med 1972; 287:388–397.
57. Corrigan CJ, Kay AB. Activated T-lymphocytes in acute and severe asthma: A primary target for both new and conventional asthma therapy. Immunol Allergy 1990; 12:209–215.
58. Levi FA, Canon C, Touton Y, et al. Circadian rhythms in circulating T-lymphocyte subtypes and plasma testosterone, total and free cortisol in five healthy men. Clin Exp Immunol 1988; 71:329–335.
59. Weitzman ED, Fukushima D, Nogeire C, et al. Twenty-four hour pattern of the episodic secretion of cortisol in normal subjects. J Clin Endocrinol & Metab 1971; 33:14–22.
60. Barnes P, Fitzgerald G, Brown M, et al. Nocturnal asthma and changes in circulating epinephrine, histamine, and cortisol. N Eng J Med 1980; 303:263–267.
61. Souter CA, Costello J, Ijaduola O, et al. Nocturnal and morning asthma: relationship to plasma corticosteroids and response to cortisol infusion. Thorax 1975; 30:436–440.
62. Reinberg A, Ghata J, Sidi E. Nocturnal asthma attacks: their relationship to the circadian cycle. J Allergy 1963; 34:323–330.

63. Beam WR, Ballard RD, Martin RJ. Spectrum of corticosteroid sensitivity in nocturnal asthma. Am Rev Respir Dis 1992;145:1082–1086.
64. Snapper JR. Inflammation and airway function; the asthma syndrome. Am Rev Respir Dis 1990; 141:531–533.
65. Douglas NJ, Flenley DC. Breathing during sleep in patients with obstructive lung disease. Am Rev Respir Dis 1990; 141:1055–1070.
66. Braude AC, Rebuck AS. Pulmonary disposition of cortisol. Ann Int Med 1982; 97:59–60.
67. Carmichael J, Paterson IC, Crompton GK, et al. Corticosteroid resistance in chronic asthma. Br Med J 1981; 282:1419–1422.
68. Grant SD, Forsham PH, DiRaimondo VC. Suppression of 17-hydroxycorticosteroids in plasma and urine after single and divided doses of triamcinolone. N Engl J Med 1965; 273:1115–1118.
69. Harter JG, Reddy WJ, Thorn GW. Studies on an intermittent corticosteroid dosage regimen. N Engl J Med 1963; 269:591–596.
70. Moeller H. Chronopharmacology of hydrocortisone in the treatment of congenital adrenal hyperplasia. Eur J Pediatr 1985; 144:370–373.
71. Reinberg AE. Chronopharmacology of corticosteroids and ACTH. In: Chronopharmacology. Cellular and Biochemical Interactions, Lemmer B, (ed). Marcel-Dekker, New York, 1989, pp 137–167.
72. Reinberg AE, Halberg F, Falliers CJ. Circadian timing of methylprednisolone effects in asthmatic boys. Chronobiologia 1974;1:333–347.
73. Reinberg AE, Gervais P, Ghaussade M, et al. Circadian changes in effectiveness of corticosteroids in eight patients with allergic asthma. J Allergy Clin Immunol 1983; 71:425–433.
74. Reinberg AE, Guillet P, Gervais P, et al. One month chronocorticotherapy. Control of the asthmatic condition without adrenal suppression and circadian rhythm alterations. Chronobiologia 1977; 4:295–312.
75. Beam WR, Weiner DE, Martin RJ. Timing of prednisone and alterations of airways inflammation in nocturnal asthma. Am Rev Respir Dis 1992;146:1524–1530.
76. Mohiuddin AA, Martin RJ. Circadian basis of the late asthmatic response. Am Rev Respir Dis 1990; 142:1153–1157.
77. Diaz P, Galleguillos FR, Gonzalez MC, et al. Bronchoalveolar lavage in asthma. J Allergy Clin Immunol 1984; 74:42–48.
78. McGivern DV, Ward M, MacFarlane JT, et al. Failure of once daily inhaled corticosteroid treatment to control chronic asthma. Thorax 1984; 39:933–934.
79. Malo JL, Cortier A, Merland N, et al. Four-times-a-day dosing frequency is better than twice-a-day regimen in subjects requiring a high dose inhaled steroid, budesonide, to control moderate to severe asthma. Am Rev Respir Dis 1989; 140:624–628.
80. Toogood JH, Baskerville JC, Jennings B, et al. Influence of dosing frequency and schedule on the response of chronic asthmatics to aerosol steroid, budesonide. J Allergy Clin Immunol 1982; 70:288–298.
81. Horn CR, Clark TJH, Cochrane GM. Inhaled therapy reduces morning dips in asthma. Lancet 1984; 1:1143–1145.

82. Horn CR, Clark TJH, Cochrane GM. Can the morbidity of asthma be reduced by high dose inhaled therapy? A prospective study. Respir Med 1990; 84:61–66.
83. Dahl R, Pedersen B, Hagglof B. Nocturnal asthma: Effect of treatment with oral sustained-release terbutaline, inhaled budesonide, and the two in combination. J Allergy Clin Immunol 1989; 83:811–815.
84. Haahtela T, et al. Comparison of a β_2-agonist, terbutaline, with an inhaled corticosteroid, budesonide, in newly detected asthma. N Engl J Med 1991; 325:388–392.
85. Sly MR, Imseis M, Frazer M, et al. Treatment of asthma in children with triamcinolone acetonide aerosol. J Allergy Clin Immunol 1978; 62:76–82.
86. Prahl P, Jensen P, Bjerregaard-Andersen H. Adrenocortical function in children on high dose aerosol therapy. Allergy 1987; 42:541–544.
87. Newman SP, Pavia D, Moren F, et al. Deposition of pressurized aerosols in the human respiratory tract. Thorax 1981; 36:52–55.
88. Smith MJ, Hodson ME. High-dose beclomethasone inhaler in the treatment of asthma. Lancet 1983; 1:265–269.
89. Toogood JH, Lefcoe NM, Haines DSM, et al. A graded dose assessment of the efficacy of beclomethasone dipropionate aerosol for severe chronic asthma. J Allergy Clin Immunol 1979; 59:298–308.
90. Williams AJ, Baghut MS, Stableforth DE, et al. Dysphonia caused by inhaled steroids: recognition of a characteristic laryngeal abnormality. Thorax 1983; 38:813–821.
91. Toogood JH, Jennings B, Greenway RW, et al. Candidiasis and dysphonia complicating beclomethasone treatment of asthma. J Allergy Clin Immunol 1980; 65:145–153.
92. Stein MR, Shay SS, Jacobson K. Monilial esophagitis in asthmatic patients treated with beclomethasone. J Allergy Clin Immunol 1979; 63:172A.
93. Hemstreet MP, Reynolds DW, Meadows J. Oesophagitis—a complication of inhaled steroid therapy. Clin Allergy 1980; 10:733–738.
94. Freinkel RK, Freinkel N. Cutaneous manifestations of endocrine disorders. In: Dermatology in General Medicine. 3rd Ed, Fitzpatrick TB, Eisen AZ, Wolff K, Freedberg IM, Austen KF (eds). McGraw-Hill, New York, pp 267–270.
95. Lundgren R. Scanning electron microscopic studies of bronchial mucosa before and during treatment with beclomethasone dipropionate inhalations. Scand J Respir Dis 1977; Suppl 101:179–187.
96. Lundgren R, Soderberg M, Horstedt P, et al. Morphological studies of bronchial mucosal biopsies from asthmatics before and after ten years of treatment with inhaled steroids. Eur Respir J 1988; 1:883–889.
97. Thiringer G, Eriksson N, Malmberg R, et al. Bronchoscopic biopsies of bronchial mucosa before and after beclomethasone dipropionate therapy. Postgrad Med J 1975; Suppl4:30–31.
98. Vathenen AS, Knox AJ, Wisniewski A, et al. Time course of change in bronchial reactivity with inhaled corticosteroids in asthma. Am Rev Respir Dis 1991; 143:1317–1321.

99. Rooklin AR, Lampert SI, Jaeger EA, et al. Posterior subcapsular cataracts in steroid-requiring asthmatic children. J Allergy Clin Immunol 1979; 63:383–386.

100. Ali NJ, Capewell S, Ward MJ. Bone turnover during high dose inhaled corticosteroid treatment. Thorax 1991; 46(3):160–164.

101. Priftis K, Everard ML, Milner AD. Unexpected side effects of inhaled steroids. Eur J Pediatr 1991; 150:448–449.

102. Tabachnik E, Zadik Z. Diurnal cortisol secretion during therapy with inhaled beclomethasone dipropionate in children with asthma. J Pediatr 1991; 118:294–297.

103. Prahl P. Adrenocortical suppression following treatment with beclomethasone and budesonide. Clin Exp Allergy 1991; 21:145–146.

104. Godfrey S, Balfour-Lynn L, Tooley M. A three to five year follow-up of the use of the aerosol steroid beclomethasone dipropionate in childhood asthma. J Allergy Clin Immunol 1978; 62:335–339.

105. Graff-Lonnevig V, Kraepelien S. Long-term treatment with beclomethasone in asthmatic children with special reference to growth. Allergy 1979; 34:57.

106. Mellis CM, Phelan PP. Asthma deaths in children—a continuing problem. Thorax 1977; 32:29–34.

107. Haskell RJ, Wong BM, Hansen JE. A double-blind, randomized clinical trial of methylprednisolone in status asthmaticus. Arch Int Med 1983; 143:1324–1327.

108. Fiel SB, Swartz MA, Glanz K, et al. Efficacy of short-term corticosteroid therapy in outpatient treatment of acute bronchial asthma. Am J Med 1983; 75:259–262.

109. Chapman KR, Verbeek PR, White JG, et al. Effect of a short course of prednisone in the prevention of early relapse after the emergency room treatment of asthma. N Engl J Med 1991; 324:788–794.

110. Ellul-Micallef R, Borthwick RC, McHardy GJR. The effect of oral prednisone on gas exchange in chronic bronchial asthma. Br J Clin Pharmacol 1980; 9:479–482.

111. Fanta CH, Tossing TH, McFadden ER. Glucocorticoids in acute asthma. A critical controlled trial. Am J Med 1983; 74:845–851.

112. Reubuck AS, Read R. Assessment and management of severe asthma. Am J Med 1971; 51:788–798.

113. Kelson SG, Kelson DP, Fleegler BF, et al. Emergency room assessment and treatment of patients with acute asthma. Am J Med 1978; 64:622–628.

114. MacDonald JB, Seaton A, Williams DA. Asthma deaths in Cardiff 1963–1974: 90 deaths outside hospital. Br Med J 1976; 1:1493–1495.

115. MacDonald JB, MacDonald ET, Seaton A, et al. Asthma deaths in Cardiff 1963–1974: 53 deaths in hospital. Br Med J 1976; 2:721–723.

116. Speizer FE, Doll R, Heaf P, et al. Investigation into use of drugs preceding death from asthma. Br Med J 1968; 1:339–343.

117. Byyny RL. Withdrawal from glucocorticoid therapy. N Engl J Med 1976; 295:30–32.

118. Editorial. Steroid therapy and the adrenals. Lancet 1975; 2:537–538.

119. Reinberg A, Smolensky M. Biologic Rhythms and Medicine. Springer-Verlag, Berlin, 1983.

120. Ceresa F, Angeli A, Boccuzzi G, et al. Once-a-day neurally stimulated and basal ACTH secretion phases in man and their responses to corticoid inhibition. J Clin Endocrinol and Metab 1969; 29:1074–1082.

121. Grant SD, Forsham PH, DiRaimondo VC. Suppression of 17-hydroxycorticosteroids in plasma and urine by single and divided doses of triamcinolone. N Engl J Med 1965; 273:1115–1118.

122. DiRaimondo VC, Forsham PH. Some clinical implications of spontaneous diurnal variation in adrenal cortical secretory activity. Am J Med 1956; 21:321–323.

123. Harter JG, Reddy WJ, Thorn GW. Studies on an intermittent corticosteroid dosage regimen. N Engl J Med 1963; 269:591–596.

124. Ackerman GL, Nolan CM. Adrenocortical responsiveness after alternative-day corticosteroid therapy. N Engl J Med 1968; 278:405–409.

125. Graber AL, Ney RL, Nicholson WE, et al. Natural history of pituitary-adrenal recovery following long-term suppression with corticosteroids. J Clin Endocrinol Metab 1965; 25:11–16.

126. Nichols T, Nugent CA, Tyler FH. Diurnal variation in suppression of adrenal function by glucocorticoids. J Clin Endocr 1965; 25:343–349.

127. Streck WF, Lockwood DH. Pituitary adrenal recovery following short-term suppression with corticosteroids. Am J Med 1979; 66:910–914.

128. Webb J, Clark TJH. Recovery of plasma corticotrophin and cortisol levels after a three-week course of prednisolone. Thorax 1981; 36:22–24.

9

Other Pharmacological Interventions in Nocturnal Asthma

Malcolm Hill, Pharm, D., and Stanley J. Szefler, M.D.

Introduction

The purpose of this chapter is to discuss the role of anticholinergics, cromolyn/nedocromil, and antihistamines, as well as miscellaneous drugs, in the treatment of nocturnal asthma. Additionally, the role of allergic factors and sinusitis on nocturnal asthma are discussed. Although the categories of drugs are distinct, understanding the effects of these medications and their mechanisms of action are useful to help gain insight into the pathogenesis of nocturnal asthma. Anticholinergics and cromolyn/nedocromil have been studied in asthma. The information conveyed in this chapter is limited to that which directly relates to the overnight fall in pulmonary function that clinically typifies nocturnal asthma. Other important parameters that have been evaluated include nocturnal or morning asthma symptom scores and nighttime inhaled beta-agonists use. This has been discussed in other chapters in detail.

The circadian patterns in pulmonary function and airways hyperresponsiveness (Fig. 1) have been documented in the literature

Martin RJ (editor): *Nocturnal Asthma: Mechanisms and Treatment,* © Futura Publishing Co., Inc., Mount Kisco, NY, 1993.

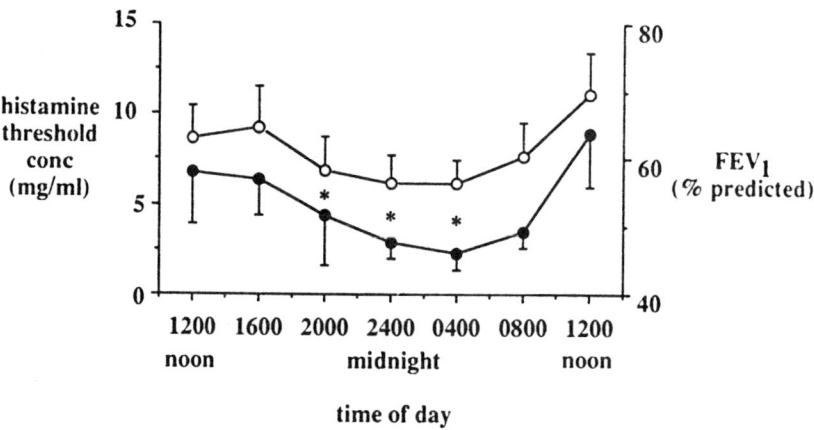

Figure 1. Histamine threshold concentration (an index of airways hyperrespon-siveness, lefthand vertical axis) and pulmonary function (FEV$_1$, righthand vertical axis) versus time in patients with asthma. The values for histamine threshold concentration and FEV$_1$ at 2000, 2400, and 0400 are significantly (P<0.005) different from values at other time points. Adapted from De Vries.[1]

for many years.[1] These patterns are generally unrecognized by physician and patient alike, until dramatic bronchoconstriction occurs overnight and symptoms are quite pronounced. Recently, data gathered from a variety of fields have been constructed to give us a more accurate view of the pathophysiology of nocturnal asthma. Relevant to this chapter is the understanding of the nervous system in asthma, the effect of certain leukocytes on nocturnal changes in pulmonary function, and the effects of collateral diseases such as sinusitis on the development of asthma.

Inhaled Anticholinergics

The Nervous System and Asthma

Two important categories of nervous system control of airway tone in man are the parasympathetic nervous system and the sympathetic nervous system.[2] Other neural mechanisms such as the nonadrenergic, noncholinergic nervous system are present, but their roles in asthma remain unclear. The nonadrenergic, noncholinergic nervous system may be most important in asthma pathogenesis.[3]

Local release of neuropeptides and reflex mechanisms are involved in regulation of microvascular leakage and mucus secretion as well as bronchoconstriction.[3] Pharmacological manipulation of neurotransmitters and mediators involved in the nonadrenergic, noncholinergic nervous system is not complete, and the role in nocturnal asthma is currently undefined.

The parasympathetic nervous system is highly involved in regulation of airway tone in man, and is also thought to be most important in maintenance of bronchial obstruction in asthma as well.[4] The therapeutic role of anticholinergic drugs has been widely studied in nocturnal asthma.

When compared to normal man, all asthmatics have profound bronchoconstriction in response to cholinergic agonists such as methacholine or carbachol, suggesting that the parasympathetic nervous system is involved in the asthmatic response.[5] Sensory nerve endings along the airways travel via the afferent vagus nerve to the central nervous system. Nerve impulses return to the lung via the efferent vagus and arrive at ganglia located throughout the lung. Postganglionic nerves release acetylcholine, which binds to cholinergic receptors located in airways smooth muscle resulting in bronchoconstriction. It is at this postganglionic cholinergic receptor that the anticholinergic bronchodilator atropine and its derivatives such as ipratropium bromide have their site of action. Experimental models have shown that airways inflammation produces an alteration of the sensory nerve endings which may lead to bronchial hyperresponsiveness.[6] Inflammatory processes probably enhance sensitivity to cholinergic agonists seen in asthmatics.

In contrast to the parasympathetic innervation of the airway, the sympathetic nervous system is not as well-defined, and there appears to be little or no direct innervation of airway smooth muscle.[7] There is, however, a rich supply of beta-adrenergic receptors in most airway smooth muscle.[8] This suggests that the importance of the sympathetic nervous system is not in maintaining airway tone, but in reversing bronchoconstriction in response to adverse stimuli.

Imbalance between the sympathetic nervous system and parasympathetic nervous system could partially mediate circadian patterns in airway caliber. Decreased sympathetic activity, measured by lower plasma concentrations of epinephrine,[9] coupled with decreased β-adrenergic receptors and impaired response to β-adrenergic stimulation,[10] reflect decreased tendency for bronchodilation.

Increased parasympathetic nervous system activity mediated through vagal efferents is also present.[11] It has recently been demonstrated that there is a tremendous increase in nonspecific airways hyperresponsiveness that occurs in the early morning hours as compared to afternoon, as well as a circadian pattern in pulmonary function.[1] Additionally, the asthmatic response to allergen challenge is also dramatically increased when the allergen exposure occurs in the evening as compared to allergen exposure in the morning.[12] This increased vagal activity increases bronchoconstrictor tone as well, as it takes higher doses of intravenous atropine to produce a maximum change in airways conductance at 4 AM as compared to 4 PM in patients with asthma.[13]

There is evidence of circadian patterns on the neural control of other systems, such as the cardiovasculature. The increased vagal tone is manifested in the cardiovascular system with the lowest heart rate and increased sinus arrhythmia gap (the difference between fast and slow components of sinus arrhythmia during quiet breathing, expressed in beats/minute) occurring at 4 AM. This finding is more marked in patients with airways obstruction than in normals.[13] Taken together, these data suggest that the parasympathetic nervous system plays an important role in terms of regulating pulmonary physiology as well as in other systems. There is a sound rationale for the application of anticholinergic drugs to obviate the overnight fall in pulmonary function that occurs.

The effects of inhaled anticholinergics have been summarized in Table 1. Coe and Barnes[14] studied the effects of the quarternary anticholinergic, oxitropium bromide, in a low-dose regimen of 0.2 mg administered at bedtime and a high-dose regimen of 0.4 mg administered at bedtime, both for two weeks' duration in a double-blind randomized placebo-controlled fashion. The patients were monitored overnight following a peak expiratory flow rate, and maintained an asthma symptom diary. The overall mean (± S.E.) fall in peak expiratory flow rate, expressed as percentage of evening peak expiratory flow rate was 17.3% ± 2.0% on placebo, which was significantly reduced to 10.3% ± 3.3% after high-dose oxitropium bromide. Nine of the 18 patients studied responded to both low- and high-dose regimens more favorably and were identified as a subgroup of "responders." In this responder subgroup, the percent fall in peak expiratory flow rate was 19.1% ± 3.2% on placebo, 11.5% ± 4.4% on oxitropium 0.2 mg., and only 5.0% ± 4.5% on oxitropium 0.4 mg. This subgroup of responders was not able to be identified in

Table 1

Effect of Anticholinergics on Nocturnal Asthma

Type of patient (n) Reference	Drug/Route	Dose (mg)/Duration	Parameters Assessed	Comments
Adults NA (18) Coe, 1986[14]	Oxitropium bromine MDI	0.2 mg/single doses 0.4 mg/single dose	overnight fall in PEF	Only high dose offered significant protection. Responders and non-responders identified, suggesting increased vagal tone more important in certain individuals.
Asthmatic children (31) Sly, 1987[15]	Ipratropium bromide MDI	0.04 mg three times daily for 4 weeks duration	overnight fall in PEF/symptom diary	Ipratropium failed to modify overnight fall in PEF, symptom control, or use of maintenance medications.
Adults NA (10) Catterall, 1988[16]	Ipratropium bromide nebulized	1.0 mg at 10 PM and 2 AM, single dose	overnight fall in PEF	Treatment with ipratropium increased baseline PEF, but did not alter overnight fall in PEF.
Adults NA (17) Wolstenholme, 1989[17]	Ipratropium bromide MDI	0.08 mg four times daily × 4 weeks	overnight fall in PEF, symptom diary	Addition of ipratropium to regular β_2-agonist therapy did not reduce overnight fall in PEF or alter symptom scores.
Adults (11), moderate to severe asthma with NA Sur, 1990[18]	Atropine methylnitrate nebulized	0.09 mg/kg up to 2 mg max/single dose	overnight fall in PEF, FEV_1	Addition of nebulized atropine methyl-nitrate did not significantly alter the overnight fall in PEF or FEV_1 when added to regular β_2-agonist therapy.

NA = documented nocturnal asthma; MDI = metered dose inhaler; PEF = peak expiratory flow rate (L/min); FEV_1 = forced expiratory volume in 1 second.

terms of asthma characteristics, atopic status, or age. These authors suggest that addition of inhaled steroids does not effectively control asthmatic symptoms. It may be worthwhile attempting to classify individual patients according to whether they are responders or nonresponders to inhaled anticholinergics as well.

Sly and coworkers[15] studied asthmatic children (age 8 to 18 years, mean 12 years), who received ipratropium bromide 0.04 mg (two puffs from metered dose inhaler) t.i.d. for four weeks duration in a double-blind placebo-controlled fashion. The patients were assessed for overnight fall in peak expiratory flow rate; asthma symptom diary scoring including use of as needed beta-agonists; and had histamine bronchoprovocation prior to and at the end of each study period performed as well. There was a small, but significant, improvement in histamine PD-20 from a geometric mean of 0.49 mg/mL to 0.78 mg/mL. However, treatment with the ipratropium bromide did not affect diurnal variation in peak expiratory flow rate, bronchodilator responsiveness to inhaled albuterol, asthma symptoms, or additional doses of albuterol that were required during treatment periods.

Catterall[16] studied the effects of nebulized ipratropium bromide 1 mg administered at 10 PM and 2 AM in a single dose, double-blind, randomized, placebo-controlled study. These investigators determined that peak flow rates were higher throughout the night following treatment with ipratropium bromide as compared to placebo. The fall in the peak expiratory flow rate overnight, however, was similar between treatment groups. These investigators concluded that although ipratropium bromide was able to provide a small, but significant, bronchodilator effect, the inability to decrease the overnight fall in pulmonary function suggests that nocturnal bronchoconstriction is not solely due to an increase in cholinergic activity during the nighttime hours.

Wolstenholme studied ipratropium bromide 0.08 mg q.i.d. via metered dose inhaler for four weeks duration in a double-blind randomized crossover study.[17] In this investigation, ipratropium was administered in combination with fenoterol, also in the metered dose inhaler. The control treatment was albuterol 200 μg via metered dose administered four times daily. Seventeen patients between ages 19 to 35 years with documented nocturnal asthma were evaluated. Peak expiratory flow rate and nocturnal asthma symptom diary were gathered. Prebronchodilator expiratory flow rate measured in the evening while patients were receiving albuterol was (mean ± S.D.) 349 ± 116, which decreased to 286 ± 96 L/min. A similar pattern was observed while patients were receiving combination ipratropium bromide and fenoterol. Prebronchodilator peak expiratory flow rate measured in the evening was 344 ± 112 L/min, which decreased to 278 ± 99 L/min. These differences were not statistically significant. Addition of the anticholinergic was unable to provide a clinically relevant degree of overnight protection.

Sur and coworkers[18] evaluated the effect of atropine methylnitrate 0.05 mg/kg up to 2.0 mg maximum dose was administered via jet nebulizer at 10 PM over four different study nights in a double-blind randomized placebo-controlled crossover study. The eight patients who completed the trial received regular treatment with oral theophylline, an inhaled β-agonist. Seven out of eight patients were receiving inhaled glucocorticoid, and three out of eight were receiving oral glucocorticoids or cromolyn. The percent fall in FEV_1 overnight while the patients were receiving albuterol alone ranged from a mean of 6.4 ± 2.9 to 22.9 ± 14.5 (mean ± S.E.M.). While receiving atropine methylnitrate the percent fall in FEV_1 that occurred overnight ranged from .86 ± 9.8 to 28.9 ± 12.0%. There was no wash-in phase where patients received no albuterol at 10 PM in order to observe percent fall in FEV_1 that occurred overnight without treatment. Therefore, a relatively high dose of nebulized atropine methylnitrate was unable to provide a substantial bronchoprotective effect in adults with moderately severe asthma.

Overall, inhaled anticholinergics have little effect on the overnight fall in pulmonary function that occurs in nocturnal asthma (Table 1). The reasons for this are unclear, but most likely due to the fact that the doses used, or topical rather than systemic administration, were unable to alter vagal tone. Since there is increased parasympathetic tone in asthmatics as compared to normals, application of higher doses of anticholinergics should have some effect. Although in one study,[18] individualized doses of up to 2.0 mg of nebulized atropine methylnitrate was no more effective than albuterol alone.

Perhaps the answer lies in the lack of specificity of these agents for muscarinic receptor subtypes. At least three muscarinic subtypes have been identified,[19] (Fig. 2). Located on autonomic ganglia, M_1 receptors are thought to mediate vagal tone, and antagonism of these receptors results in bronchodilation. M_2 receptors are thought to have a negative feedback effect on the cholinergic presynaptic junction, and antagonism of these receptors may increase acetylcholine release. M_3 receptors are thought to be located on

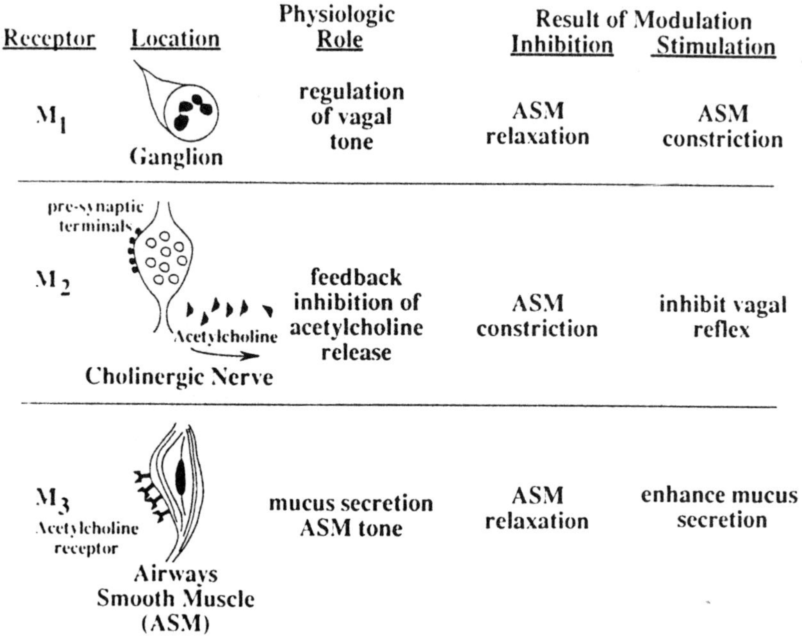

Figure 2. Muscarinic receptor subtype, anatomical locations, and physiological result of pharmacological modulation are shown.

postsynaptic receptors. Antagonism of M_3 receptors probably results in bronchodilation.[19] Therefore, a pure M_3 antagonist, or a drug that could selectively antagonize M_1 and M_3 may have more desirable effects on the vagally mediated overnight fall in pulmonary function than any of the currently available nonspecific antagonists such as ipratropium or atropine derivatives.

Mast Cell Stabilizers/Antihistamines

The mast cell has been linked to asthma for many years.[20] Mast cells are found throughout the walls of the respiratory tract, and increased numbers of these cells are found in the airways of patients with allergic asthma.[21] Decreased numbers of granulated mast cells are found in the airways of patients who died from asthma, suggesting that release of mast cell granule was related to the severity of the acute asthma episode.[21] Histamine is one of the most prodigious mediators found in the mast cell, and it is a potent bronchoconstricting mediator of importance in the pathogenesis of asthma.[20]

There is a circadian pattern in plasma histamine which correlates with the temporal changes in pulmonary function, and these changes are much more dramatic in patients with nocturnal asthma.[9] It has been postulated that the circadian changes in plasma histamine are due to an inverse pattern in circulating concentrations of epinephrine.[9] As plasma epinephrine concentrations decrease, there is increased release of histamine, and presumably other preformed mediators, from mast cells and basophils resulting in bronchoconstriction and the subsequent circadian pattern in pulmonary function. It has been demonstrated in asthmatic children that there is greater recovery of urinary histamine metabolites in those with an overnight fall in pulmonary function as compared to asthmatic children without an overnight fall in pulmonary function.[22] A logical extension of these data is that prevention of mast cell degranulation should attenuate the overnight fall in pulmonary function. The effects of the mast cell stabilizers cromolyn and the related compound nedocromil on nocturnal asthma are summarized in Table 2.

The Drug Committee of the American Academy of Allergy in 1972 published the results of a clinical trial evaluating cromolyn 20 mg administered q.i.d. via spinhaler over 4 weeks in a double-blind, randomized, placebo-controlled, crossover study.[23] In this trial, 252 patients (ages 5 to 72 years) received study medications in one of two groups (active followed by placebo, placebo followed by active). Only patients receiving active drug first reported a substantial and significant decrease in symptoms of nighttime wheezing, which began within 2 weeks after initiation of drug therapy and continued to improve until 4 weeks after treatment. This effect was maintained even though patients switched to placebo, suggesting a carry-over effect. Delayed onset of effect and persistence of effect after discontinuance of drug is not easily explained, but may be related to the antiinflammatory properties of cromolyn and the related drug nedocromil.

Hetzel[24] evaluated the effects of 2 weeks of treatment with cromolyn in adult patients (age range 22 to 66 years) with documented nocturnal asthma (overnight fall in peak expiratory flow rate of 25%). Patients received placebo or a low-dose regimen consisting of nebulized cromolyn 20 mg on waking at 6 AM and at bedtime, and 20 mg by spinhaler at noon (total dose = 60 mg/day). A high-dose regimen consisted of 80 mg by nebulizer t.i.d. and 40 mg by

Table 2

Effects of Cromolyn and Nedocromil on Nocturnal Asthma

Type of patient (n) Reference	Drug/Route	Dose (mg)/Duration of Therapy	Parameters Assessed	Comments
Adult NA (23) Hetzel, 1985[24]	Cromolyn/nebulizer and spinhaler	60 mg/ + 280 mg/ dose/2 weeks	Overnight fall in PEFR symptom scores	No improvement in overnight fall in PEFR. Some improvement in symptom scores in high dose regimen only.
Adult NA (8) Morgan 1986[25]	Cromolyn nebulizer	160 mg/single dose at bedtime	Overnight fall in FEV_1	Cromolyn had no effect.
Adult asthma (48) Ruffin 1986[26]	Nedocromil/MDI	4 mg four times daily/ 4 weeks	Overnight fall in PEFR, symptom scores.	No effect on overnight fall in PEFR. Nedocromil treated patients had lower use of β_2-agonists at night.
Adult asthma (34) Williams 1986[27]	Nedocromil/MDI	4 mg four times daily/ 12 weeks	Diurnal variation in PEFR/symptom scores	Reduction in PEFR diurnal variation by week 4; maintained through week 12. There was no improvement in symptom score.
Adult asthma (29) Harper 1990[28]	Nedocromil/MDI vs. beclomethasone dipropionate/MDI	4 mg four times daily/ 8 weeks. Beclomethasone dipropionate 400 µg/day	Overnight fall in PEFR, symptom score	Both nedocromil and beclomethasone decreased overnight fall in PEFR, however only beclomethasone decreased nocturnal symptom scores and β_2-agonist usage.

dry powder in spinhaler at noon (280 mg/day). Patients received these active treatments, or placebo, in a double-blind randomized, crossover design. During the study there was no attenuation of the overnight fall in peak expiratory flow rate. Prebronchodilator peak expiratory flow while receiving placebo fell from 338 L/min to 296 L/min from bedtime to awakening. With low-dose cromolyn, mean expiratory flow rate at bedtime was 358 L/min, and on waking was 313 L/min. On the high-dose regimen, bedtime peak expiratory flow rate was 349 L/min and decreased to 305 L/min. None of these differences were significant. Despite the lack of improvement in overnight fall in peak expiratory flow, there was significant decrease in the number of nights the patients were awakened per week and the number of β-agonist aerosols recorded during high-dose and low-dose treatment periods, respectively.

Eight adult patients (ages 18 to 60 years) with documented nocturnal asthma were evaluated by Morgan and coworkers.[25] These patients received a single dose of cromolyn, 160 mg, via nebulizer and face mask at bedtime. Spirometry was measured at that time and upon awakening in a double-blind, randomized, placebo-controlled, crossover fashion. In every patient, FEV_1 fell overnight after treatment with placebo, from a (mean $\pm$ SD) of 2.6 ± 1.6 liters before falling asleep to 1.8 ± 1.7 upon awakening. While on the cromolyn treatment, the FEV_1 at bedtime was 2.4 ± 1.5, falling to 1.9 ± 1.2 liters. This study demonstrated that cromolyn in very high single dose application did not influence overnight fall in pulmonary function in patients with chronic asthma.

In a longer term study, Ruffin and coworkers[26] studied the effects of nedocromil 4 mg or placebo via metered dose inhaler q.i.d. for 4 weeks duration. This investigation studied 48 asthmatic patients with mild to moderate asthma, aged 14 to 70 years. Although no significant effect on overnight fall in peak expiratory flow rate was detected in the study, there was a general increase in nighttime β-agonist use in patients on placebo, while patients treated with nedocromil had no change in overnight bronchodilator use. These authors were unable to demonstrate that 4 weeks of nedocromil had any statistically significant or clinically relevant effect on nocturnal asthma.

Williams and coworkers investigated 34 patients (ages 15 to 76 years) with mild to moderate, yet stable, asthma who were treated with nedocromil 4 mg administered via metered dose inhaler q.i.d. or matching placebo for a study period of 12 weeks in a randomized double-blind parallel, placebo-controlled study.[27] Diary card records of peak expiratory flow rate generally improved in both groups, although much more dramatically in the group receiving nedocromil (Fig. 3). During the course of treatment with nedocromil, circadian variation in peak expiratory flow rate tended to decrease by week 4, and became significant by week 12. There was no significant change in symptom scores with regard to nighttime asthma, morning tightness, daytime asthma, or cough. The 12-week study period allowed for the maximum effect of nedocromil to develop when improvement in the circadian variation in peak expiratory flow rate was observed. This study provides the most convincing data demonstrating the beneficial effect of nedocromil on nocturnal asthma.

Most recently, Harper[28] compared the effects of moderate dose inhaled beclomethasone dipropionate (400 μg/day) with inhaled nedocromil sodium 16 mg/day, with both treatments being administered four times daily. Thirteen subjects (ages 25 to 68 years) completed the trial which consisted of two 8-week study periods. The study was a double-blind randomized placebo-controlled trial with the final 2 weeks of each trial arm compared for the change in overnight fall between the evening and morning peak expiratory flow rate and symptom scores. During the baseline wash-in period, mean peak expiratory flow rate in the evening was 333 L/min, which fell to 318 L/min while receiving beclomethasone dipropionate; evening peak expiratory flow rate was 363 L/min which fell to 350 L/min during nedocromil therapy. The evening peak expiratory flow rate was 357 L/min which fell to 341 L/min. The small effect and

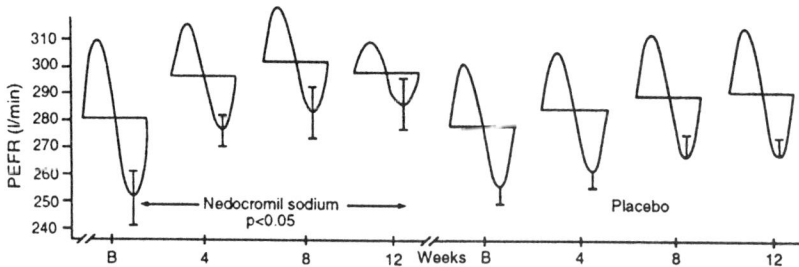

Figure 3. Effect of nedocromil 4 mg four times daily via metered dose inhaler administered to adult asthmatics. Note the progressive decrease in the diurnal variation in peak expiratory flow rate (PEFR). Adapted with permission from Williams et al.[27]

perhaps clinically unimportant change in peak expiratory flow rate was significant for both drugs. These patients had minimal nocturnal asthma during the baseline period, with no documentation of overnight fall in pulmonary function documented prior to enrollment. Both beclomethasone and nedocromil gave significant improvement in the recorded symptom scores when compared to baseline. Beclomethasone appeared to be significantly more effective in reducing β-agonist inhaler use, nocturnal asthma score, and symptoms of morning tightness when compared to nedocromil. In this study, nedocromil seemed to be a minimally effective drug for the treatment of nocturnal asthma, but was not as effective as low to moderate doses of inhaled glucocorticoids.

Although more effective than anticholinergics in preventing the overnight fall in pulmonary function, the effects of cromolyn and nedocromil are limited. Clearly the high-dose single treatment studies[24] had minimal effects. These findings suggest a limited role for mast cell degranulation *per se* as a cause of nocturnal asthma, as cromolyn is thought to be a potent inhibitor of histamine release from these cells.[29] With the longer term studies,[23,27] there was a significant beneficial effect after 4 to 8 weeks of therapy. This time course of response most likely suggests that as the underlying asthma improves with prolonged nedocromil therapy, the overnight fall in pulmonary function is diminished. These data suggest that inhibition of mast cell degranulation plays a limited role in nocturnal asthma, and that the beneficial response that is observed over time is most likely due to other useful antiasthma properties of cromolyn and nedocromil, most likely the result of diminished airways hyperresponsiveness.

Since overnight bronchoconstriction is associated with a rise in plasma histamine concentration,[9] it would follow that antihistamines should be able to attenuate the overnight fall in pulmonary function in nocturnal asthma. The effects of a single 1 mg dose of ketotifen on nocturnal asthma was investigated in a double-blind, randomized crossover trial.[30] Ten adult patients (ages 18 to 55 years) with previously documented nocturnal asthma were studied. While receiving placebo, four patients were awakened by asthma symptoms and requested inhaled beta-adrenergic, while no patients receiving ketotifen awakened. In the remaining six patients on placebo, FEV_1 fell from a baseline of 2.9 ± 0.4 L (mean $\pm$ SEM) by an average of 0.6L on awakening. While receiving ketotifen, FEV_1 fell from a baseline of 2.9 ± 0.4 by an average of 0.9 L. Therefore, ketotifen was unable to prevent the overnight fall in FEV_1 from occurring, while increasing slightly the duration of sleep as compared to placebo. These latter effects are due to the sedative properties of ketotifen. In many other studies with ketotifen, days or weeks of therapy are required before beneficial effects are noted. In a single-dose design as employed in this study, the antihistaminic and mast cell stabilizing properties of ketotifen may not have been optimized.

Miscellaneous Therapy

Mucolytics

Desquamation of bronchial epithelium, and infiltration of mast cells, eosinophils, and lymphocytes into the airway lining are characteristic of asthma.[31] Histologic studies performed on patients with mild intermittent to moderate chronic disease have shown marked inflammatory changes within the airway along with extensive epithelial damage.[32] Similar, but more severe, changes have also been seen in patients who have died from acute asthma attacks.[33] These epithelial changes, along with the findings of mucosal edema, airway inflammation, and mucosal hypersecretion, are relatively consistent features of asthma. The primary role of the mucociliary process is to protect lung tissues from the environment. With the pathology of asthma, it would not be surprising to expect the mucociliary process to be adversely affected.

Studies of mucociliary function in patients with asthma during nighttime sleep suggest that there is a marked circadian pattern, but this parameter is not different in asthma as compared to healthy volunteers.[34] The use of expectorants such as guafenesin[35] has not been demonstrated to be beneficial in terms of increasing mucociliary clearance, although mucolytic therapy to assist removal of impacted mucus plugs in a large bronchus has been occasionally useful. The β_2-agonists directly stimulate ciliary beat frequency, resulting in increased mucociliary clearance.[36] Both theophylline and β_2-agonists increase fluid secretion, decreasing the viscosity of mucus and glucocorticoids and decrease mucus production. In addition, the β_2-agonists directly stimulate ciliary beating. Therefore, these standard antiasthma medications may be playing a role in increasing mucociliary clearance in asthma where there is substantial epithelial damage and mucus hypersecretion.

Methotrexate

Similar to the success of low-dose methotrexate in the treatment of psoriatic and rheumatic disorders, the use of this agent in the treatment of severe asthma has met with some success in controlled studies.[37,38] In these trials, methotrexate was not used alone, but rather as a "glucocorticoid-sparing" agent. Whether methotrexate would have efficacy when used without systemic glucocorticoids in patients with milder asthma is not known. Although its primary mechanism of action is not understood, methotrexate may be acting as an antiinflammatory agent.[39] Methotrexate may also have immunomodulatory effects at very low and relevant concentrations *in vitro*.[40] The potential beneficial effects of methotrexate have raised questions whether other chemotherapeutic agents might be effective in the treatment of asthma.[41] There has not been any assessment of methotrexate in nocturnal asthma, but it is likely that patients who are considered good responders will have improvement in nighttime asthma symptoms, allowing for glucocorticoid dose reduction.

Gold Therapy

Similar to the findings with methotrexate, early studies with gold suggest its usefulness in the treatment of steroid-requiring asthma. Early trials with injectable gold demonstrated improvement in asthmatic symptoms, reduction in glucocorticoid requirements,[42] and diminished airways reactivity.[43] A long-term open trial with oral gold (Auranofin®, Smith, Kline and French) conducted in a similar population of severe asthmatics demonstrated a significant reduction in glucocorticoid use and decreased airways reactivity in about half of the patients evaluated.[44] Gold appears to require months to show efficacy, suggesting that its action might be via attenuation of the inflammatory response, as in arthritis. The relatively slow and incremental response observed, however, may be related to the asthma severity in the patients studied. At this time, large multicenter controlled studies are being conducted to determine efficacy.

Troleandomycin

Troleandomycin (TAO), a macrolide antibiotic, came into clinical use in the 1950s. In the 1960s, TAO was used for "infectious" asthma when it was felt that bacterial infection played an important role in the pathogenesis of asthma.[43] In controlled blind studies, Itkin,[46] followed by Spector,[47] demonstrated substantial antiasthma efficacy of TAO, most notably in steroid-sparing effects in patients with severe asthma when used in combination with the glucocorticoid methylprednisolone. Revised guidelines employing a lower dose of TAO have been associated with a lower incidence of hepatotoxicity.[48] The exact mechanism of action of TAO remains unknown. A portion of the steroid-sparing effect comes from the pharmacokinetic interaction between methylprednisolone and TAO, resulting in delayed clearance of this steroid.[49] More recently, substantial data have been gathered suggesting that troleandomycin and related macrolides such as erythromycin have antiasthma[50] and antiinflammatory effects.[51] Troleandomycin needs to be applied in carefully selected patients with severe asthma, managed by physicians familiar with its use. Exacerbation of glucocorticoid associated adverse effects is a concern with combination TAO/methylprednisolone.[52] Application of this drug has been primarily directed toward glucocorticoid dose reduction rather than improvement in asthma symptomatology, potentially masking direct effects on nocturnal asthma.

Intravenous Gamma Globulin

Availability and relative safety of intravenous gamma globulin has led to its application as a replacement for hypogammaglobulinemia, idiopathic thrombocytopenic purpura, and Kawasaki's disease, as well as several other autoimmune disorders.[53] There has been some evidence presented that patients with asthma and selective hypogammaglobulinemia improve with intravenous gamma globulin infusions.[54,55] Mazer administered high-dose intravenous gamma globulin over a period of 6 months to patients with very severe, glucocorticoid-requiring asthma in open label fashion.[56] Patients underwent substantial systemic glucocorticoid dose reduction, some improvement in pulmonary function, but without improvement in airways hyperresponsiveness. The relative safety of this therapy is offset by high costs and lack of data from larger blind, controlled studies. The data gathered thus far have examined steroid-sparing effects rather than improvement in pulmonary function and decreased nighttime asthma symptoms that may occur on a fixed-dose glucocorticoid regimen.

Treatment of Collateral Diseases on Development of Nocturnal Asthma

Sinusitis

A link between asthma and sinusitis has existed for many years. In 1925, Gottlieb published his findings in which he documented 31 cases of substantial sinus disease in 117 patients with asthma.[57] Since this time, there has been little work to substantiate these findings, or to help delineate the mechanisms involved. A partial list of factors that relate upper respiratory tract disease and asthma includes[58]: postnasal drip of mucus, mediators, or chemotactic factors into the lower airways and their subsequent absorption which may then produce airways inflammation and hyperresponsiveness; mouth breathing of cold or dry air due to nasal obstruction which elicits asthma via heat and water loss from the lower airways; activation of the nasopharyngeal-bronchial reflex due to stimulation of the nose, sinuses, or pharynx. Despite this information, there is little objective data which proves that sinusitis causes or worsens asthma.[59] Nonetheless, there is interesting clinical data to suggest that the potential for sinus disease exists in patients with nocturnal asthma.

Rachelefsky[60] published observations on 48 children with a history of chronic asthma and Waters sinus radiographs that demonstrated the presence of chronic sinusitis. These children were treated empirically with amoxicillin three times daily or trimethoprim/sulfamethoxazole twice daily in the case of penicillin allergy. If no clinical improvement in asthma was noted after 2 weeks of therapy, antibiotics (erythromycin/sulfisoxaziole or erythromycin and trimethoprim/sulfamethoxazole) were continued for an additional 3 weeks. After 2 to 5 weeks of antimicrobial therapy, 38 children (79%) had reverted to normal sinus radiographs, and 38 children (79%) were considered clinically well, with improvements in nighttime and daytime cough and wheeze, and nasal symptoms. Although these data were gathered in an uncontrolled and unblinded fashion, they are certainly suggestive of a link between upper and lower airways disease.

Somewhat conflicting data have been presented by Zimmerman and coworkers.[61] These investigators corroborated earlier findings that asthmatics have a markedly higher incidence of sinus x-rays indicative of maxillary, ethmoid, or frontal sinus disease. Forty-three of 138 asthmatics did have x-ray evidence of sinusitis, with the nonasthma control group (dental patients) having 0 of 50 x-rays with findings suggestive of sinusitis. Interestingly, there was not a higher incidence of positive sinus x-rays as asthma severity increased. These data suggest that sinus disease *per se* is not related to increased antiasthma medication use and increased asthma severity. Clearly, a blinded, randomized placebo-controlled trial of antibiotics in the treatment of asthmatics with sinusitis is required to answer these questions. In the meantime, however, it is important to search for the presence of sinus disease in patients with a nocturnal component to their asthma. Cough (especially nocturnal), mucopurulent nasal discharge, and postnasal drip are common clinical features of chronic sinusitis in adults and children, and may be worse during sleep. In patients with x-ray and clinical evidence of sinusitis, consideration should be given to a 2-3 week course of empiric antibiotic therapy in patients who are not responding optimally to a conventional antiasthma approach.

Allergic considerations

Can nocturnal asthma be a manifestation of the late asthmatic response following allergen exposure? The biphasic physiological response following allergen exposure has been the focus of intensive research over the past several years. Although the precise pathogenesis of this response is not fully understood, the time course of response is fairly clear. When sensitized individuals are exposed to adequate amounts of inhaled allergen, two general patterns of airway obstruction may be observed. The immediate asthmatic response occurs shortly after exposure, and may last from 30 minutes to 2 hours.[62] The late asthmatic response occurs in many patients within several hours after allergen exposure, with airway obstruction lasting up to 12 hours, and increased bronchial hyperresponsiveness lasting for days or weeks following exposure.[63] Furthermore, when exposure to relevant allergen occurs in the evening as compared to the morning, the bronchoconstriction that occurs as a result of the late asthmatic response is much more profound.[12] The immediate asthmatic response to allergen exposure can occur alone or in conjunction with late asthmatic response, and the late asthmatic response can occur as an isolated event. What is important with regard to the time course is the fact that allergen exposure in the late afternoon or evening hours can lead to a late asthmatic response, which will be displayed as nocturnal asthma. The first step in addressing this important issue is identification of allergic factors in the occupational or home environment of which a late asthmatic response could occur. Good examples of exposure to relevant allergen is the presence of the house dust mite *D. pteronyssinus* in the bedding or bedroom,[64] or Western red cedar in the work place.[65] Of equal importance to the development of bronchoconstriction hours after allergen exposure is the persistence of airway hyperresponsiveness that is thought to be sustained for days or weeks following allergen exposure.[66] Once identified, removal of the allergen source is the next step in minimizing the effects on pulmonary function. Pharmacotherapy of the late asthmatic response is the next approach. The medications that can prevent the late asthmatic response from occurring are primarily the glucocorticoids and cromolyn/nedocromil. Immediate pretreatment with single application of cromolyn/nedocromil, or several weeks of therapy with inhaled glucocorticoids, are considered adequate to prevent the late asthmatic response from occurring.[67] When avoidance is not possible, and optimal clinical responses are not achieved pharmacologically, allergy immunotherapy should be considered.[68]

Summary

It appears the most effective medications for decreasing the overnight fall in pulmonary function that occurs in nocturnal asthma are theophylline and oral beta-agonists by virtue of the prolonged onset and duration of action in sustained-release dosage forms. In this situation, the timing of the maximum plasma concentration and pharmacodynamic effect are aimed at offsetting the lowest pulmonary function of the 24-hour period. Inhaled glucocorticoids and nedocromil are useful in the treatment of nocturnal asthma on a long-term basis because of their ability to improve airways hyperresponsiveness and therefore the tendency for overnight fall in pulmonary function. The role of antihistamines as therapeutic adjutants in the treatment of nocturnal asthma, especially modern agents such as astemizole, terfenadine, and loratidine, remain relatively unexplored. Although never evaluated, other antiasthma agents such as gold, methotrexate, troleandomycin, and intravenous gamma globulin may potentially have an effect on nocturnal asthma by the same mechanism, but this issue has been clouded as these agents have been used to reduce systemic glucocorticoids as a primary index of efficacy rather than the ability to reduce airways hyperresponsiveness, or improve nocturnal symptoms. Allergic factors and sinusitis are important entities to identify and manage in patients with nocturnal asthma. This should be done relatively early in patients who are not easily controlled on minimal conventional antiasthma therapy.

References

1. De Vries K, Goei JT, Booy-Noord H, Orie NGM. Changes during 24 hours in the lung function and histamine hyperreactivity of the bronchial tree in asthmatic and bronchitic patients. Int Arch Allergy 1962; 20:93–101.
2. Barnes PJ. Airway Receptors. In: Jenne JW, and Murphy S (eds). Drug Therapy for Asthma: Research and Clinical Practice. Marcel Dekker, New York, 1987, pp 67–95.
3. Barnes PJ. Asthma as an axon reflex. Lancet 1986; 1:242–245.
4. Boushey HA, Holtzman MJ. Autonomic regulation of airways: parasympathetic nervous system. In: Weiss EB, Segal MS, Stein M (eds). Bronchial Asthma: Mechanisms and Therapeutics. Little, Brown and Co., Boston, 1985, pp 111–122.
5. Cockcroft DW. Airway hyperresponsiveness: therapeutic implications. Ann Allergy 1987; 59:405–414.
6. Holtzman MJ, Fabbri LM, O'Byrne PM, Gold BD, Aizawa H, Walters EH, Alpert SE, et al. Importance of airway inflammation for hyperresponsiveness induced by ozone. Am Rev Resp Dis 1983; 127:686–690.
7. Richardson JB. Nerve supply to the lungs. Am Rev Resp Dis 1979; 119:758–803.
8. Zaagsma J, van der Heijden PJCM, van der Schaar MWG, Bank CMC. Comparison of functional beta-adrenoceptor heterogeneity in central and peripheral airways smooth muscle of guinea-pig and man. J Receptor Res 1983; 3:89–106.
9. Barnes P, Fitzgerald G, Brown M, Dollery C. Nocturnal asthma and changes in circulating epinephrine, histamine and cortisol. New Engl J Med 1980; 303:263–267.
10. Szefler SJ, Ando R, Cicutto LC, Surs W, Hill MR, Martin RJ. Plasma histamine, epinephrine, cortisol, and leukocyte β-adrenergic receptors in nocturnal asthma. Clin Pharmacol Ther 1990; 49:59–68.
11. Postma DS, Keyser JJ, Koeter GH, Sluiter HJ, De Vries K. Influence of the parasympathetic and sympathetic nervous system on nocturnal bronchial obstruction. Clin Sci 1985; 69:251–258.
12. Mohiuddin AA, Martin RJ. Circadian basis of the late asthmatic response. Am Rev Respir Dis 1990; 142:1153–1157.
13. Morrison JFJ, Pearson SB. The effect of the circadian rhythm of vagal

activity on bronchomotor tone in asthma. Br J Clin Pharmacol 1989; 28:545–549.

14. Coe CI, Barnes PJ. Reduction of nocturnal asthma by an inhaled anticholinergic drug. Chest 1986; 90:485–488.

15. Sly PD, Landau LI, Olinsky A. Failure of ipratropium bromide to modify the diurnal variation of asthma in asthmatic children. Thorax 1987; 42:357–360.

16. Catterall JL, Rhind GB, Whyte KF, Shapiro CM, Douglas NJ. Is nocturnal asthma caused by changes in airway cholinergic activity? Thorax 1988; 43:720–724.

17. Wolstenholme RJ, Shettar SP. Comparison of fenoterol with ipratropium bromide (Duovent) and salbutamol in young adults with nocturnal asthma. Respir 1989; 55:152–157.

18. Sur S, Mohuiddin AA, Vichyanond P, Nelson HS. A random double-blind trial of the combination of nebulized atropine methylnitrate and albuterol in nocturnal asthma. Ann Allergy 1990; 65:384–388.

19. Barnes PJ. Muscarinic subtypes: implications for lung disease. Thorax 1989; 44:161–167.

20. Kaliner M. Asthma and mast cell activation. J Allergy Clin Immunol 1989; 83:510–520.

21. Holgate ST, Hardy C, Robinson C, Agius R, Howarth PH. The mast cell as a primary effector cell in the pathogenesis of asthma. J Allergy Clin Immunol 1986; 77:274–280.

22. van Aalderen WMC, Postma DS, Koeter GH, Knol K. Nocturnal airflow obstruction, histamine, and the autonomic central nervous system in children with allergic asthma. Thorax 1991;46:366–371.

23. Bernstein IL, Siegel SC, Brandon ML, Brown EB, Evans RR, Feinberg AR, Friedlander S, et al. A controlled study of cromolyn sodium sponsored by the Drug Committee of the American Academy of Allergy. J Allergy Clin Immunol 1972; 50:235–245.

24. Hetzel MR, Clarke JH, Gilliam SJ, Isaac P, Perkins M. Is sodium cromoglycate effective in nocturnal asthma? Thorax 1985; 40:793–794.

25. Morgan AD, Connaughton JJ, Caterall JR, Shapiro CM, Douglas NJ, Flenley DC. Sodium cromoglycate in nocturnal asthma. Thorax 1986; 41:39–41.

26. Ruffin R, Alpers JH, Kromer DK, Rubinfeld AR, Pain MCF, Czarny D, Bowes G. A 4-week Australian multicentre study of nedocromil sodium in asthmatic patients. Eur J Respir Dis 1986; 69(Suppl 147):336–339.

27. Williams AJ, Stableforth D. The addition of nedocromil sodium to maintenance therapy in the management of patients with bronchial asthma. Eur J Respir Dis 1986; 69(Suppl 147):340–343.

28. Harper GD, Neill P, Vathenen AS, Cookson JB, Ebden P. A comparison of inhaled beclomethasone dipropionate and nedocromil sodium as additional therapy in asthma. Respir Med 1990; 84:463–469.

29. Wells E, Mann J. Phosphorylation of a mast cell protein in response to treatment with antiallergic compounds: implications for the mode of action of sodium cromoglycate. Biochem Pharmacol 1983; 32:837–842.

30. Caterall JR, Calverly PMA, Power JT, Shapiro CM, Douglas NJ, Flenley DC. Ketotifen and nocturnal asthma. Thorax 1983; 38:845–848.

31. Frigas E, Gleich GJ. The eosinophil and the pathology of asthma. J Allergy Clin Immunol 1986; 77:527–537.
32. Laitinen LA, Heino M, Laitinen A, Kava T, Haahtela T. Damage of airway epithelium and bronchial reactivity in patients with asthma. Am Rev Respir Dis 1985; 131:599–606.
33. Dunhill MS, Massarella GR, Anderson J. A comparison of the quantitative anatomy of the bronchi in normal subjects, in status asthmaticus, in chronic bronchitis and emphysema. Thorax 1969; 24:176–179.
34. Pavia D. Mucociliary clearance at night: effect of physical activity, posture and circadian rhythm. In: Nocturnal Asthma (Symposium 73), Royal Society of Medicine, London, 1984, pp 29–38.
35. Pavia D, Sutton PP, Lopez-Vidriero MT, Agnew JE, Clarke SW. Drug effects on mucociliary clearance. Eur J Respir Dis 1983; 66(suppl 126):304–317.
36. Pavia D, Agnew JE, Sutton PP, Lopez-Vidriero MT, Clay MM, Killip M, Clarke SW. Effect of terbutaline administered from metered dose inhaler and subcutaneously on tracheobronchial clearance in mild asthma. Br J Dis Chest 1987; 81:361–370.
37. Mullarkey MF, Blumenstein BA, Andrade WP, Bailey GA, Olason I, Wetzel CE. Methotrexate in the treatment of corticosteroid dependent asthma. New Engl J Med 1988; 318:603–607.
38. Shiner RJ, Nunn AJ, Chung KF, Geddes DM. Randomised double blind, placebo controlled trial of methotrexate in steroid dependent asthma. Lancet 1990; 2:137–140.
39. Anderson PA, West ST, O'Dell JR, Via CS, Claypool RG, Kotzin BL. Weekly pulse methotrexate in rheumatoid arthritis: Clinical and immunologic effects in a randomized, double-blind study. Ann Intern Med 1985; 103:489–496.
40. Glynn-Barnhart AM, Chan MA, Gelfand EJ. Effect of methotrexate on in-vitro lymphocyte proliferation [abstract]. Pharmacother 989; 9:194.
41. Lange EB, Kamen BA, Sullivan TJ. Remissions of asthma during cancer chemotherapy [abstract]. J Allergy Clin Immunol 1989; 83:218.
42. Muranaka M, Miyamoto T, Shida T, Kabe J, Makino S, Okumara H, Takeda K, et al. Gold salt in the treatment of bronchial asthma—a double-blind study. Ann Allergy 1978; 40:132–137.
43. Muranaka M, Nakajima K, Suzuki S. Bronchial responsiveness to acetylcholine in patients with bronchial asthma after long-term treatment with gold salt. J Allergy Clin Immunol 1981; 67:350–356.
44. Bernstein DI, Bernstein DL, Bodenheimer SS, Pietrusko RG. An open study of Auranofin in the treatment of steroid dependent asthma. J Allergy Clin Immunol 1988; 81:6–16.
45. Fox JL. Infectious asthma treated with triacetyloleandomycin. Penn Med J 1961; 64:634–635.
46. Itkin IH, Menzel M. The use of macrolide antibiotic substances in the treatment of asthma. J Allergy 1970;45:146–162.
47. Spector SL, Katz FH, Farr RS. Troleandomycin: effectiveness in steriod dependent asthma and bronchitis. J Allergy Clin Immunol 1974; 54:367–379.
48. Wald JA, Friedman BF, Farr RS. An improved protocol for the use of

troleandomycin (TAO) in the treatment of steroid-requiring asthma. J Allergy Clin Immunol 1986; 78:36–43.

49. Szefler SJ, Rose JQ, Ellis EF, Spector SL, Green A, Jusko WJ. The effect of troleandomycin on methylprednisolone elimination. J Allergy Clin Immunol 1980; 66:447–455.

50. Miyatake H, Take F, Taniguchi H, Suzuki R, Tagaki K, Satake T. Erythromycin reduces the severity of bronchial hyperresponsiveness in asthma. Chest 1991; 99:670–674.

51. Greos LS, Vichyanond P, Bloedow DC, Irvin CG, Larsen GL, Szefler SJ, Hill MR. Methylprednisolone achieves greater concentrations in the lung than prednisolone: A pharmacokinetic analysis. Am Rev Respir Dis 1991; 144:586–592.

52. Harris R, German D. The incidence of corticosteroid side effects in chronic steroid-dependent asthmatics on TAO (troleandomycin) and methylprednisolone. Ann Allergy 1989; 63:110–111.

53. Stiehm ER. The use of intravenous immune globulin in immunoregulatory disorders and in the newborn period. Therapeutic advantages in clinical immunology. Immunol Allergy Clin North Am 1988; 8:39–50.

54. Smith TF. Hypogammaglobulinemia and asthma: Do any patients with asthma have deficiency of antibody? J Asthma 1989; 26:5–13.

55. Page R, Friday G, Stillwagon P, Skoner D, Caliguiri L, Fireman P. Asthma and selective immunoglobulin subclass deficiency: Improvement of asthma after immunoglobulin replacement therapy. J Pediatr 1988; 112:127–131.

56. Mazer BD, Gelfand EW. An open-label study of high-dose intravenous immunoglobulin in severe childhood asthma. J Allergy Clin Immunol 1991; 87:983–986.

57. Gottlieb MJ. Relation of intranasal disease in the production of bronchial asthma. JAMA 1925; 85:105–107.

58. Adinoff AD, Irvin CG. Upper respiratory tract disease and asthma. Seminars Resp Med 1987; 8:308–314.

59. McFadden ER. Nasal-sinus-pulmonary reflexes and bronchial asthma. J Allergy Clin Immunol 1986; 78:1–3.

60. Rachelefsky GS, Katz RM, Siegel SC. Chronic sinus disease with associated reactive airway disease in children. Pediatr 1984;73:526–529.

61. Zimmerman B, Stringer D, Feanny S, Reisman J, Hak H, Rashed R, deBenedictis F, et al. Prevalence of abnormalities found by sinus x-rays in childhood asthma: Lack of relation to severity of asthma. J Allergy Clin Immunol 1987; 80:268–273.

62. Pelikan Z, Pelikan M, Kruis M, Berger MPF. The immediate asthmatic response to allergen challenge. Ann Allergy 1986; 56:252–260.

63. Cartier AC, Thompson NC, Frith PA, Roberts R, Hargreave FE. Allergen-induced increase in bronchial responsiveness to histamine: Relationship to the late asthmatic response and change in airways caliber. J Allergy Clin Immunol 1982; 70:170–177.

64. Platts-Mills TAE, DeWeck AL, et al. Dust-mite allergens and asthma: a world-wide problem. J Allergy Clin Immunol 1989; 83:416–426.

65. Cockcroft DW, Hoeppner VH, Werner GD. Recurrent nocturnal asthma

after bronchoprovocation with Western Red Cedar sawdust: association with acute increases in non-allergic bronchial responsiveness. Clin Allergy 1984; 14:61–68.

66. Cartier A, Thompson NC, Frith PA, Roberts R, Hargreave FE. Allergen induced increase in bronchial responsiveness: Relationship to the late asthmatic response and change in airway caliber. J Allergy Clin Immunol 1983; 70:170–177.

67. Peliken Z, Pelikan-Filipek M, Remejer L. Effects of disodium cromoglycate and beclomethasone dipropionate on the asthmatic response to allergen challenge II. Late response. Ann Allergy 1988; 60:217–225.

68. National Heart, Lung, and Blood Institute National Asthma Education Program Expert Panel Report. Guidelines for the diagnosis and management of asthma. J Allergy Clin Immunol 1991; 88(2):425–534.

10

Nonpharmacological Interventions for Nocturnal Asthma

Richard J. Martin, M.D.

Introduction

When asthma treatment is considered, medications being given via the inhaled, oral, or parenteral routes are implied. Obviously, this is the appropriate approach to the therapeutic control of asthma. However, consideration of other nonpharmacological interventions must be undertaken and, as has been stressed throughout chapters in this text, nighttime asthma is markedly different from daytime asthma. This difference is due not only to the naturally occurring circadian rhythms of many different biological functions (see Chapters 1 to 3), physiological responses (see Chapter 4), cellular mechanisms (see Chapters 5 to 6), and medication responses (see Chapters 7 to 9), but also to the patient being prone, loss of upper airway muscle tone, changes in nasal airway diameter, decreased clearance and protective mechanisms, and other factors.

This chapter will present the nonpharmacological processes

Martin RJ (editor): *Nocturnal Asthma: Mechanisms and Treatment,* © Futura Publishing Co., Inc., Mount Kisco, NY, 1993.

that need to be considered in the treatment of nocturnal asthma. The focus is on mechanical or other approaches to the treatment of sleep related asthma (Table 1). Although these interventions will only affect a small proportion of the asthmatic population, for any given patient they may be of significant importance in controlling the nocturnal component and can have a beneficial carryover to the daytime lung function and symptom control.

Nasal Continuous Positive Airway Pressure

Continuous positive airway pressure (CPAP) via a nasal mask applied during sleep is the treatment of choice for patients with obstructive sleep apnea. Since sleep apnea occurs in approximately 2% to 5% of the general population, and asthma in approximately 10%, one would expect to find a subset of patients who have both obstructive apnea and asthma. In these individuals, one may have concern that using nasal CPAP at night to treat sleep apnea would make the patients' asthma worse during the night. This concern would be in regard to several factors: temperature changes in the airways due to high flow rates of air being delivered to the posterior oropharynx, producing the pneumatic splint that keeps the upper airway patent; direct irritation to the airways; reflex mechanisms from the upper airway (nasal, oropharyngeal) to the lower airway producing bronchoconstriction; and bringing more airborne allergens to the lower airways.

Due to these concerns about worsening of asthma in sleep apnea patients, Chan et al. evaluated the use of nasal CPAP in nine asthmatics with concomitant asthma.[1] These patients all suffered from frequent nocturnal asthma attacks, which in three patients had resulted in respiratory arrests. Despite maximal bronchodilator therapy, including oral corticosteroids, the frequency and severity of the nocturnal asthma attacks remained unchanged. These patients, additionally, had symptoms of sleep apnea with heavy snoring during sleep. It was noted that this sonorous snoring started prior to the onset of the unstable asthma in most patients. These individuals also noted sleep fragmentation, i.e., frequent nocturnal awakenings at times associated with choking and acute shortness of breath. These later symptoms could be due to asthma and/or sleep apnea. After the diagnosis of obstructive sleep apnea was made (apnea/

Table 1

Nonmedication Approaches to Nocturnal Asthma

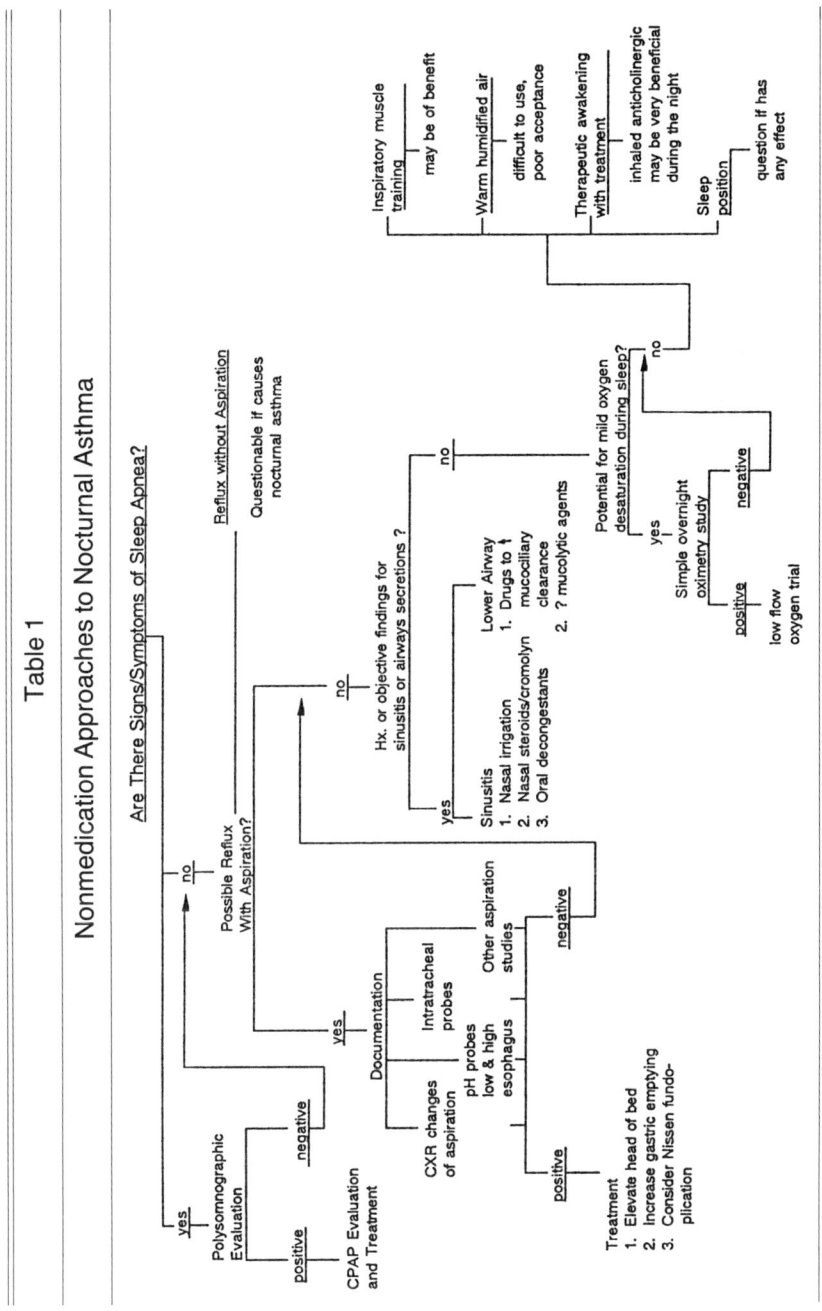

hypopnea index between 5 and 67 per hour), these patients were enrolled in a 6-week study, a 2-week baseline evaluation of morning and evening peak expiratory flow rates (PEFR), a nasal CPAP period with PEFR measurements, and then another 2-week period with no nasal CPAP. Of great interest was the finding that nasal CPAP significantly improved both the morning and evening PEFR pre- and postinhaled bronchodilator (Fig. 1). Additionally, although the postnasal CPAP 2-week evaluation showed a decline in PEFRs, there continued to be a beneficial effect on the asthma compared to the initial baseline period. Clinically, all patients reported a marked improvement in nocturnal and daytime asthma symptoms on nasal CPAP. All patients also used their bronchodilators less frequently during both the day and night with the nasal CPAP. The average nocturnal inhaled bronchodilator usage prior to nasal CPAP was twice nightly. During the period of nasal CPAP, only one patient needed any bronchodilator inhalation during the night, and this was markedly reduced from the nonnasal CPAP period.

The time it took to see improvement in lung function once nasal CPAP was initiated was rapid and usually seen within the first week. Once nasal CPAP was discontinued, the decline in PEFR occurred rapidly. Figure 2 shows the marked circadian lung function changes off and on nasal CPAP and the even more marked improvements that occur during the night at 0300 hour, in a single patient.

This study documented that patients with obstructive sleep apnea and coexisting asthma can be safely treated with nasal CPAP therapy. Perhaps of greater interest and importance is that the relief of the sleep apnea in some manner improves the asthma not only during the night, but also has a carryover daytime effect. The mechanism(s) by which this occurs is not known. Chan et al.[1] postulated that the recurrent episodes of upper airway obstruction and snoring act as a chronic irritant which, when eliminated by nasal CPAP therapy, improved the asthma. Neural receptors at the glottic inlet and in the laryngeal region have been shown to have potent bronchoconstrictive reflex activity.[2] With the repeated stimulation from heavy snoring and apnea of the oropharynx and glottic inlet or larynx receptors during the night, a neural reflex arc could be initiated producing bronchoconstriction. These investigators[1] suggested that the use of nasal CPAP may stabilize the upper airway and remove the chronic nightly irritation to the oropharyngeal area with subsequent elimination of the reflex bronchoconstriction.

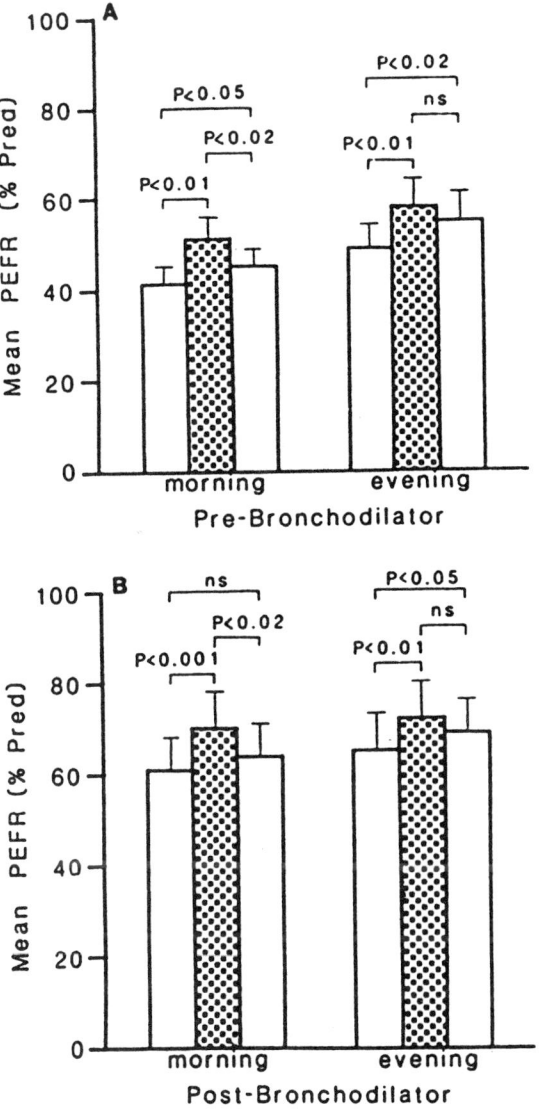

Figure 1. Mean peak expiratory flow rate (PEFR) of 9 patients with both nocturnal asthma and sleep apnea during control periods (unshaded areas) and nasal CPAP period (shaded areas). Figures 1A and 1B show the mean PEFR before and after bronchodilator treatment, respectively; ns = not significant. From reference 1, with permission.

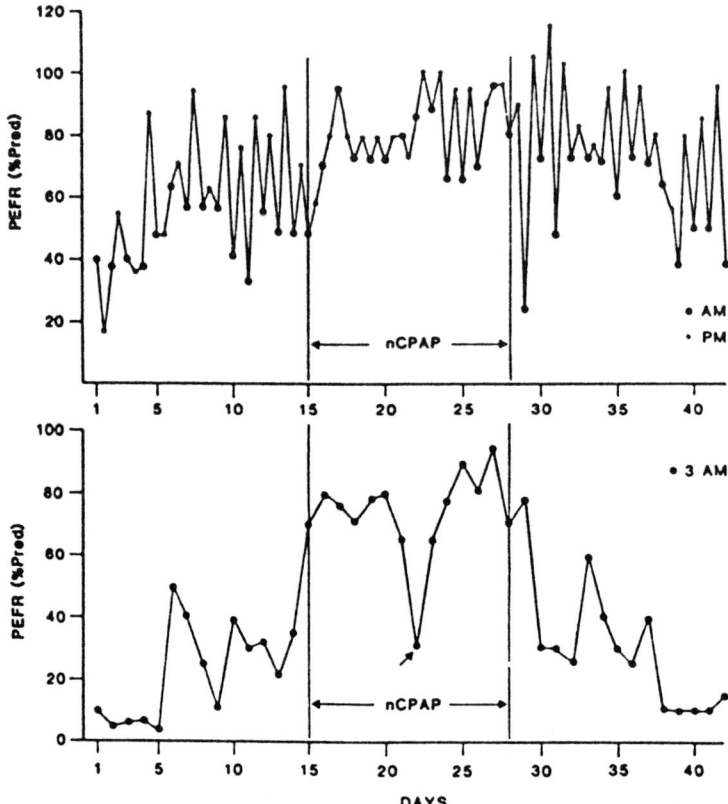

Figure 2. PEFR recordings in a single nocturnal asthma/sleep apnea patient showing improvement during the period of nasal CPAP therapy (upper panel). However, the greatest improvement was recorded at 3 A.M. (lower panel). The arrow indicates the night that the patient did not use nasal CPAP. From reference 1, with permission.

There are other potential mechanisms which could explain why sleep apnea and heavy snoring produce worsening of asthma. Hypoxia through carotid body stimulation could induce reflex bronchoconstriction.[3] Mild daytime hypoxemia in asthmatics has also been shown to increase bronchial reactivity in nonsleep apnea mild asthmatic patients.[4] Patients with sleep apnea can have mild to profound oxygen desaturation which is eliminated with nasal CPAP.

There is a possibility that the sleep fragmentation seen with sleep apnea induces humoral mediator release with resultant bronchoconstriction. If this actually occurs, then normalization of the sleep pattern that occurs with nasal CPAP[5] would control the process, thereby stabilizing the airways. Lastly, vagal tone is markedly increased during obstructive apneas.[6] This is seen by the associated bradycardia during the apnea and postapneic tachycardia as the vagal tone is released. In asthma, increase in vagal tone can produce significant bradycardia, particularly at night.[7] Thus, by eliminating the marked fluctuation in vagal tone by interrupting the upper airway obstruction, it is possible that bronchoconstriction would not be as prominent during sleep.

Guilleminault, et al. felt that the reduction in vagal tone was the cause for the improvement in nocturnal asthma in patients with snoring, small oropharynx, and the use of nasal CPAP.[8] They studied two groups of asthmatic patients, adults with typical sleep apnea and adolescents with loud snoring and craniomandibular abnormalities. It is this latter group of patients that is of interest. These investigators make an assumption that the inspiratory increases in esophageal pressure during the night represented a Müeller maneuver from the narrowed upper airway, and were not the result of worsening of asthma during sleep.[9] Assuming the former is correct, then a consequence of a partial or complete upper airway obstruction, i.e., a Müeller maneuver, is increased intrathoracic pressure and reflex bradycardia via vagal innervation.[10]

The Müeller maneuver with resulting vagal stimulation is apparently involved in the sleep related hemodynamic changes noted with obstructive apnea or hypopnea[11-13]. Anticholinergic medication such as atropine, or autonomic nervous system lesions eliminate the cardiovascular alterations seen with these obstructive processes during sleep.[12-13] Thus, the vagal cholinergic system is important in the sleep apnea syndrome and may play a key role in the nocturnal worsening of asthma associated with upper airway obstruction during sleep.

If elimination of partial or complete upper airway obstruction by nasal CPAP improves associated nocturnal asthma, can nasal CPAP also improve nocturnal asthma in those individuals without sleep apnea? Martin and Pak evaluated this question in nonsnoring, nonapneic asthmatic patients with reproducible nocturnal asthma.[14] There were several interesting observations made in this study. First, each individual had markedly worse sleep when using nasal CPAP as compared to baseline. How well the group slept (sleep efficiency) was $83.1\% \pm 4.9\%$ on the baseline night and only

66.4% ± 4.3% on the nasal CPAP night (P = 0.007). Thus, on the baseline night there was significantly less awake time (P = 0.007). Also shown in regard to sleep staging was greater REM sleep (P = 0.003) on the baseline night (14.1% ± 1.5%) versus the nasal CPAP night (3.4% ± 1.3%). Sleep staging variables are shown in Figure 3. To determine if better adaptation to nasal CPAP would improve the sleep architecture, two patients used nasal CPAP for 1 week (chronometer verification) and were restudied. Again, their sleep was poor.

For the above study group, the overnight change in lung function with nasal CPAP was variable among patients. However, the overnight decrement in the forced expiratory volume in one second (FEV_1) for the group was not improved between the baseline and nasal CPAP nights. Using heart rate as an indicator of changes in vagal tone, it appeared in this particular patient population that nasal CPAP did not decrease vagal input, as the mean heart rates

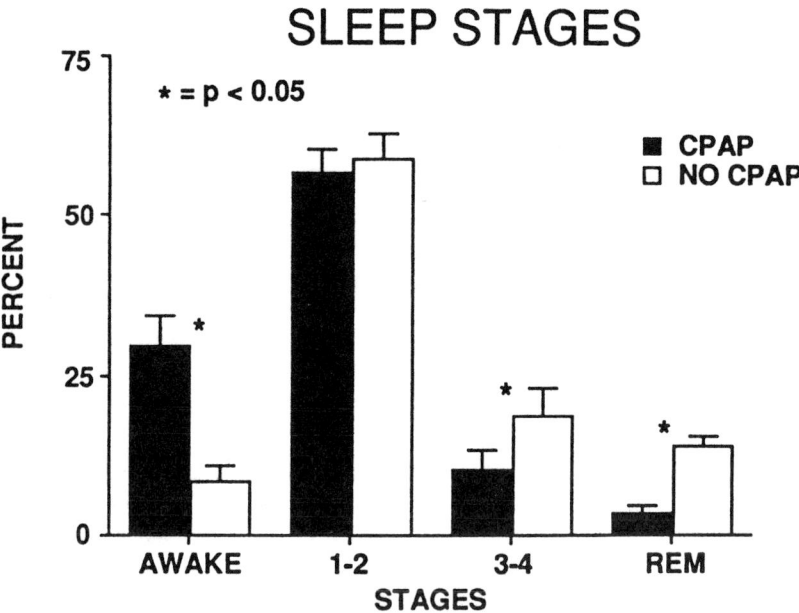

Figure 3. Demonstrates that in patients with nocturnal asthma, but no associated sleep apnea, there is a significant increase in awake time and less stage 3–4 and rapid eye movement (REM) sleep when using CPAP (CPAP—closed bars, baseline night—open bars). From reference 14, with permission.

were essentially the same on both nights. This study further suggested that hypoxia may be playing a role in the development of nocturnal asthma. This is discussed under the section on Oxygen Therapy.

The above studies focus our attention on an important subset of asthmatics with worsening of lung function overnight, i.e., those with corresponding sleep apnea. The exact percentage of patients with the two processes occurring simultaneously is not known, but is probably 1% to 3% of the asthmatic population. Consideration of this possibility should be made with each asthmatic patient so as not to miss a relatively easily treatable cause of nocturnal asthma. Thus, if the asthmatic patient has a history of loud snoring, observation by the bed partner of pauses in respiration during sleep, daytime somnolence, restless sleep, morning headaches, or several of the other numerous signs and symptoms of sleep apnea, then a full polysomnographic evaluation should be undertaken in a sleep laboratory. Mandatory in this evaluation are the spirometric measurements at bedtime, during the night if the patient awakens, and in the morning at the end of the study.

Warm Humidified Air

Daytime studies have shown that a fall in body temperature of 0.7°C secondary to a short duration cold exposure produces acute asthma attacks in the majority of asthmatic patients.[15,16] These investigations postulated that body cooling leads to vasoconstriction and cooling of the respiratory mucosa. This decrease in temperature of the mucosa consequently initiates bronchoconstriction in predisposed patients. This sequence of events was blocked when airway mucosa temperature was maintained by breathing warm humidified air during body cooling.[17] Since body temperature can normally decrease by approximately 1°C during sleep,[18] this potential mechanism of nocturnal asthma was recently tested to see if the nocturnal fall in pulmonary function could be reduced by breathing warm humidified air.[19]

This study showed that the mean fall in temperature from baseline to awakening in seven asthmatics was significant. This decrement was unaffected by taking or withholding bedtime medications. The temperature drop was blocked by breathing warm humidified air (36.6°C before sleep, 36.6°C awakening). This air was delivered via a face mask at 37°C and 100% relative humidity. When breathing room air, FEV_1 decreased slightly but significantly, from 2.9L to 2.7L (P = < 0.05), and the forced expiratory flow between 25% and 75% fell from 2.4 L/sec to 2.0 L/sec (P = < 0.05). When breathing warm humidified air, there was no significant decrement in morning pulmonary function. Using the heated air without regular evening medication did not prevent bronchoconstriction, but substantially reduced it compared to room air breathing. Thus, these results suggest that nocturnal airway cooling may play a role in triggering sleep-induced asthma.

Although breathing warm humidified air may produce some degree of improvement in the overnight lung function of many asthmatic patients, caution is needed, as some asthmatics may have exacerbation of symptoms by either heat or humidification. Thus, this modality needs to be initially tested under supervision before the patient uses it at home. Additionally, in over 30 patients tested at our center, none would use this form of therapy (Fig. 4) at home, even though some improvement was seen in lung function.

Reflux

The relationship between gastroesophageal reflux and nocturnal asthma is an intriguing topic. Obviously, in the recumbent position reflux would be potentiated. If reflux is an important factor in nocturnal asthma, then the worsening would actually be due to reflex mechanisms. That is, the gastric contents irritate the esophageal mucosa, and a reflex bronchoconstriction occurs via the vagal system. Aspiration of gastric contents is another possible mechanism.

Several surgical reports relate that gastroesophageal reflux with possible tracheobronchial aspiration is a trigger factor in asthma.[20-23] The predominant mechanism initiating the condition identified in these reported cases was an incompetent lower esophageal sphincter with or without an associated hiatal hernia. The medications that asthmatics are treated with, e.g., bronchodilators and steroids, tend to cause or potentiate the decreased tone in the gastroesophageal sphincter. The cause and effect of reflux asthma was stated to occur in several of these reports[20-22] when asthmatic symptoms were abolished following the surgical restoration of effective lower esophageal sphincter function.

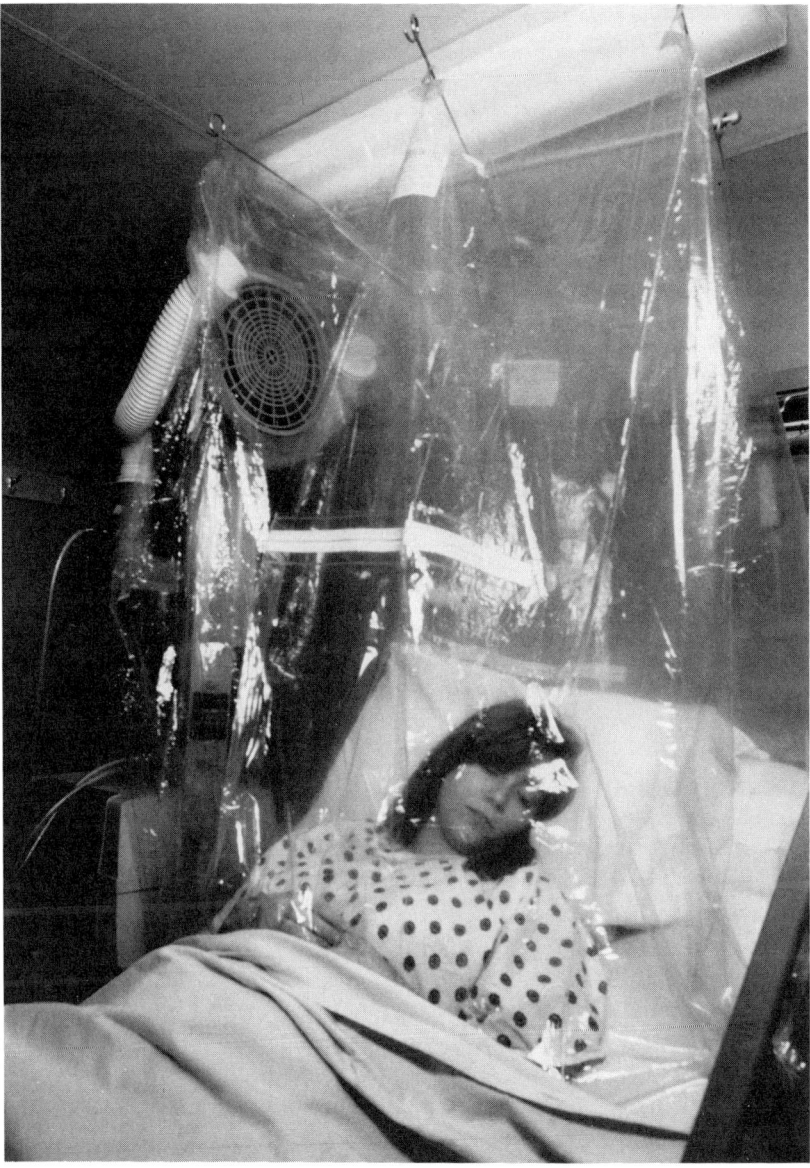

Figure 4. This demonstrates a patient in the tent that produces warm humidified air. Most patients are uncomfortable in such an environment, and thus it is not overall a viable treatment option.

In seeking the relationship between gastroesophageal reflux and asthma, Mays[24] compared 29 patients with severe asthma with 468 hospitalized nonasthmatic patients. Three of the asthmatic patients had lower lung field infiltrates of unknown origin, possibly related to aspiration. Using upper gastrointestinal radiography, 46% of the asthmatic patients had demonstrable reflux and 64% had hiatal hernia. Among the control patients only 5% had reflux and 19% hiatal hernias. The differences between the groups were significant. The high prevalence of hiatal hernia and reflux extended across the entire age spectrum of the asthmatic patients (20 to 68 years). In contrast, the peak incidence among control patients was the sixth decade. Since the asthmatic group was relatively small, these age incidences must be viewed cautiously.

Five asthmatics from the above study[24] had clinical symptoms compatible with the occurrence of reflux, but had no demonstrable reflux on upper gastrointestinal series. Three of these five did have hiatal hernias. This brings up the point that gastrointestinal x-rays are very limited in truly detecting the incidence of reflux. Probably the best way to establish this process would be with esophageal probes that detect changes in pH values. Also, tracheal probes are one of the most sensitive methods to diagnose aspiration. Six more of Mays' patients had demonstrable reflux without historical evidence of clinical reflux, and three had hiatal hernias. Only five of the 29 asthmatics had no clinical or roentgenographic evidence compatible with reflux or hiatal hernia.

The relationship between asthma, reflux, and nocturnal symptoms is further supported by the study of Goodall et al.[25] Eighteen of 20 patients with nocturnal asthma completed a double-blind crossover study using cimetidine. The severity of reflux was graded using a symptom score of heartburn and regurgitation, in addition to objective evaluation of upper gastrointestinal radiograph, fiberoptic endoscopy, and biopsy, manometry, pH monitoring of the distal esophagus, and acid infusion testing. Significant improvement was seen in reflux and nighttime symptoms with cimetidine. Fourteen of the 18 patients felt that their chest symptoms had improved markedly during the period of cimetidine use.

In all probability, direct aspiration of gastric contents plays only an occasional role in nocturnal asthma, as this problem appears to be relatively uncommon. More likely, a vagal reflex mediated by esophageal receptors that are stimulated by reflux contents is the primary factor. In 15 patients with documented asthma and reflux,

Mansfield and Stein[26] infused 0.1 N hydrochloric acid into the esophagus during the day. There was both subjective and objective evidence of bronchoconstriction associated with this daytime acid load. The bronchoconstriction improved after antacids relieved the symptoms.

In a similar study during sleep, Martin et al.[27] infused saline (control) and 0.1 N hydrochloric acid into the distal third of the esophagus. The asthmatic patients were divided into two groups based on the presence or absence of esophagitis as determined by the Bernstein test. The respirator pattern was monitored continuously using an inductance vest. Saline infusion had no effect on the respiratory pattern. In patients with esophagitis, the acid infusion produced changes in the respiratory pattern. These changes included: decreased mean expiratory flow, decreased ratio of inspiratory to total breath duration, and decreased inspiratory to expiratory timing. It was felt that acid in the esophagus alone triggered this altered respiratory pattern and indirectly was indicative of bronchoconstriction. This apparently occurs only in those patients who have preexisting esophagitis. However, other studies have shown only minimal increase in total respiratory resistance[28] or no association between low esophageal pH and worsening of asthma.[29–30]

The above studies supporting a cause and effect for gastroesophageal reflux producing nocturnal asthma have been inadequate in one or more ways. They have lacked reliable, objective, and direct indicators of bronchoconstriction such as continuous lower airway resistance measurements, reliable documentation of esophageal reflux such as continuous esophageal pH, and/or have not included sleeping nocturnal asthmatic subjects.

Tan et al. studied nocturnal asthma patients with and without clinical esophagitis using simultaneous and continuous measurements of lower airway resistance and esophageal pH.[31] The goals were to establish if acid in the esophagus triggered bronchoconstriction, to see if the presence of esophagitis was necessary for such an effect, and to determine if there was a difference between the airway responses to spontaneous reflux and intraesophageal acid infusion.

In all subjects acid infusion to lower the esophageal pH to less than 2 was performed during sleep. In Bernstein positive subjects the mean lower airway resistance during the 30 minutes before infusion was 2.41 ± 0.78 cm $H_2O/L/sec$. This was not significantly different from that obtained during the 30 minutes of acid infusion, 2.24 ± 0.48 cm $H_2O/L/sec$. Additionally, the increase in lower airway resistance over the night was not affected by the presence or absence of acid in the esophagus (Fig. 5). This noneffect was also seen in the overnight decrement in FEV_1 staying constant in the presence or absence of esophageal acid in Bernstein positive or negative nocturnal asthma patients. (Fig. 6).

Tan et al.'s study suggests that gastroesophageal reflux does not have a significant role as a direct trigger of asthma, either in general or in the nocturnal exacerbation of the disease. This study appears to be the first in which airway patency was measured directly with simultaneous monitoring of the esophageal pH during episodes of spontaneous and simulated reflux. Prior weaknesses of previous investigations have thus been taken into account. One aspect was not studied by Tan et al. This would be aspiration of gastric contents which, although uncommon, could play a role in a subset of asthmatics.

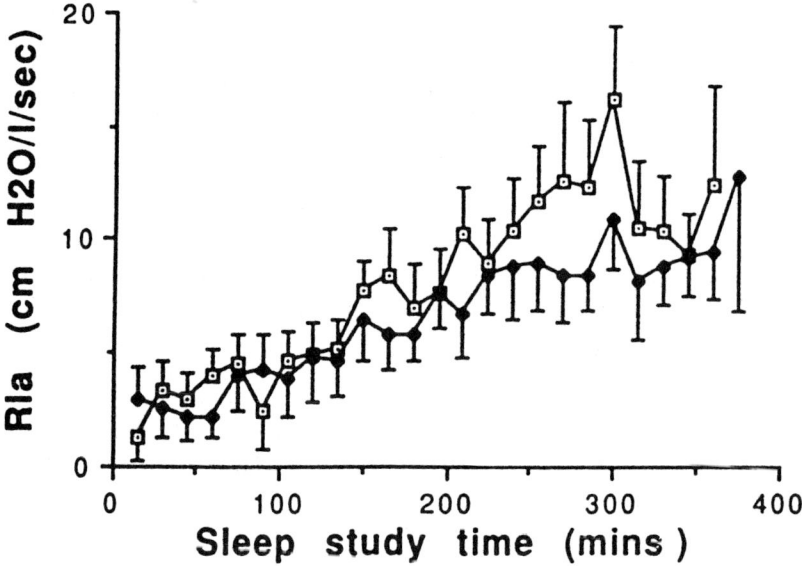

Figure 5. Effect of acid on overnight rise in lower airway resistance (R_{la}). R_{la} rose significantly and similarly in the presence of intraesophageal acid (closed diamonds, includes spontaneous reflux and acid infusion) and absence of intraesophageal acid (open squares, includes ranitidine inhibition of acid or no spontaneous reflux). P-value for the slope of R_{la} over time for each curve was 0.0001. Difference in slopes not significant (P>0.2). From reference 31, with permission.

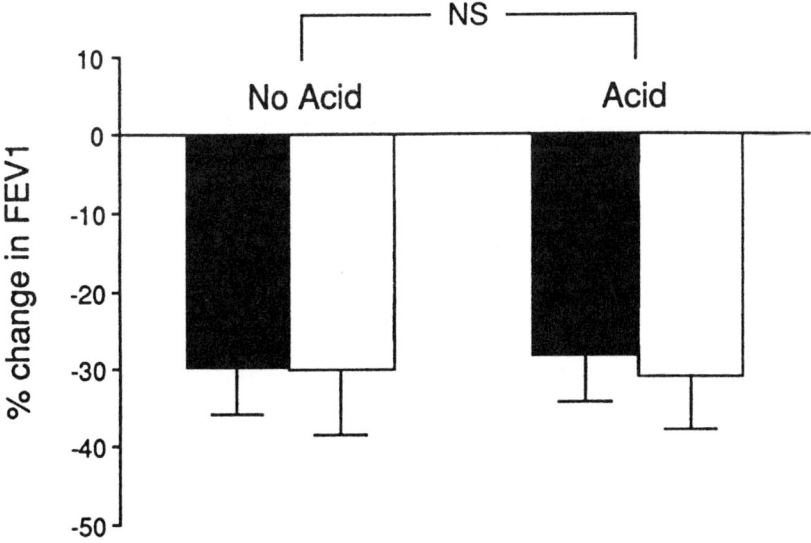

Figure 6. Effect of acid on overnight falls in FEV_1 in 10 patients who had positive Bernstein tests (closed bars) and five patients with negative Bernstein tests (open bars). There was no difference in percent fall FEV_1 between the two groups on the two nights ($P>0.7$). From reference 31, with permission.

Presently, treatment of gastroesophageal reflux in nocturnal asthma probably should be based on symptoms of reflux and not worsening of asthma. However, if the patient complains of a sour taste in their mouth upon arising or has unexplained infiltrates on the chest radiograph, then the possibility of reflux with aspiration should be considered. In this case, elevation of the head of the bed on 4 to 6 inch blocks and a medication such as metoclopramide to speed gastric emptying should be used. Antacids and H-2 blockers will not be of particular benefit in this specialized situation. If significant aspiration is documented, then strong consideration of a Nissen fundoplication would be appropriate.

Planned Nocturnal Awakenings in Conjunction with Treatment

Airways resistance progressively increases throughout sleep in asthmatic patients with nocturnal worsening of their asthma (Fig.

7).[9] A possibility exists that by awakening the patients during the night and using an inhaled bronchodilator, this progressive nature of asthma could be beneficially altered. The home use of a peak flow meter at bedtime, in the morning, and with any awakening during the night would document the nadir peak expiratory flow rate. If the patient slept through the night, artificial awakenings at 0200 and 0400 hours (of different nights) would give the physician an idea of when the worst lung function was occurring. Then waking the individual up approximately 1 hour prior to this nadir in lung function and using an inhaled beta$_2$-adrenergic agonist and/or an inhaled anticholinergic agent could interrupt the cycle of nocturnal asthma.

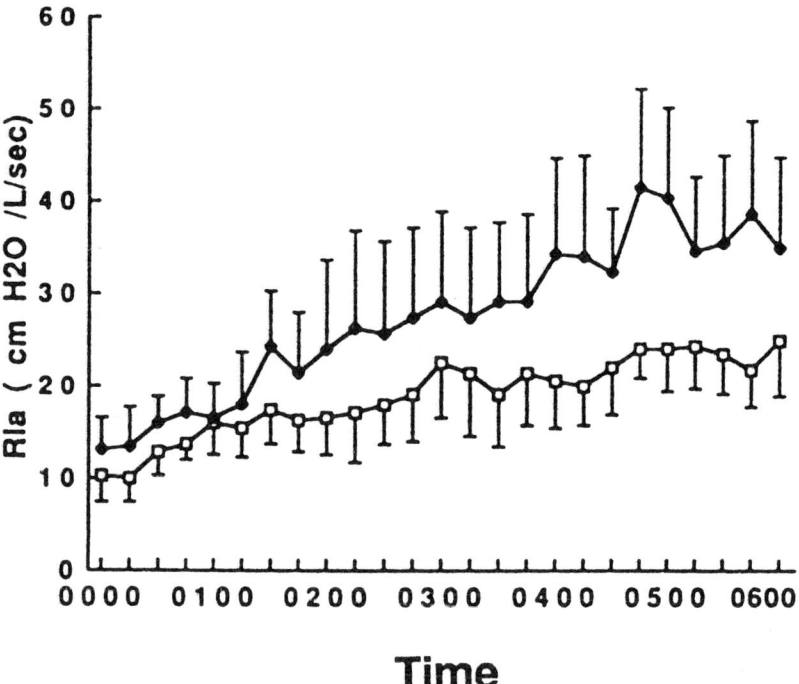

Figure 7. The open squares show the progressive increase in airway resistance throughout the night in the awake state. However, sleep (closed squares) has a more profound effect on increasing airway resistance in asthmatic subjects. From ref. 9, with permission.

Perhaps one should question the use of an inhaled anticholinergic in asthma. Studies using ipratropium bromide in asthma have shown that $beta_2$-agonists have better bronchodilator properties. However, these studies have all been performed during the daytime hours. It is at night that vagal tone is increased in all individuals. At this nocturnal time vagolytic therapy would possibly be most beneficial. In fact, Morrison and colleagues[7] have shown that atropine given at 0400 hours significantly improves the peak flow rates from $260 \pm .04$ to $390 \pm .06$ L/min. At 1600 hours, the same atropine dose has only a minimal effect on the peak flow rates, $400 \pm .04$ to $440 \pm .05$ L/min. This reinforces the principles of chronotherapeutic regimens covered in Chapters 1 to 3. Basically, the body is not a homeostatic mechanism, although virtually all pharmacotherapeutic regimes are tested during the daytime based on this homeostatic hypothesis. All functions of the body, both organ systems and cellular systems, change over a 24-hour period. Thus, therapies need to be based on chronobiological principles.

Speelberg et al. studied 10 nocturnal asthmatic patients in regard to deliberate waking followed by use of both an inhaled $beta_2$-adrenergic agonist and inhaled ipratropium bromide.[32] Two actuations of each medication were used. These patients awoke spontaneously at night at least three times per week with symptoms of wheezing, cough, or dyspnea. For 1 full week, the number and clock time of spontaneous awakenings, peak expiratory flow rates, and symptom scores were recorded. The deliberate waking during the second week was accomplished once a night with the use of an alarm clock set for 1 hour before the prior week's first expected spontaneous awakening.

The mean $\pm$ S.D. clock time the patients spontaneously awoke during the baseline week was 0320 ± 1.3 hours. The frequency of spontaneous awakenings with asthma symptoms the first week was 4.0 ± 0.9. During the treatment intervention week, this frequency of spontaneous awakenings was slightly decreased. By the end of the second week the patients reported improvement in the nocturnal symptom score of a better night's sleep and less disturbance by nocturnal asthma ($P = 0.03$). However, this was not a double-blind placebo controlled study and this subjective effect may have been a "placebo" effect. Objectively, the morning peak flow recording did improve but, again, no controls were used. This study suggests that deliberate awakenings of patients with nocturnal asthma is beneficial to sleep quality, lung function, and subjective symptoms. What is needed is a well-controlled study using objective criteria for sleep architecture and daytime functioning.

A potential drawback to deliberately waking asthmatics from sleep versus attempting to treat the nocturnal asthma by other modalities is the sleep loss that will occur and decreases in the following day's social-school-job functioning.

In a study of 78 asthmatic patients from 18 to 64 years of age with a mean daytime percent predicted FEV_1 of 70% and a mean PC_{20} of 0.2 mg/mL, Ariaansz et al. studied subjective variables of sleep quality and alertness.[33] There were 65 age-matched control subjects. The sleep latency (time to onset of sleep) in the asthmatic population was 30 minutes and for controls was 19 minutes (P = 0.007). The sleep time for the two groups was 7.0 and 7.3 hours, respectively (P = 0.05). The asthmatic patients scored worse on the following variables: ranking of overall sleep quality (P = 0.005), feeling of well-being on morning awakening (P = 0.017), daytime well-being (P = < 0.001), and daytime somnolence (P = 0.008). No significant differences between asthma and control population were found in regard to the subjective feeling of difficulty in initiating sleep, waking up early, or actual sleep time during the day (but as noted above, the asthmatic group felt sleepy during the daytime hours). Thus, it appears that adult asthmatic patients have disturbed sleep which manifests as daytime tiredness. As stated above, what needs to be done are objective studies evaluating the sleep architecture abnormalities and daytime functioning in asthmatic individuals and then determine if a deliberate nocturnal awakening followed by inhaled bronchodilator treatments not only improve the nocturnal aspect, but translate into daytime improvement in mental and physical functioning.

Oxygen Therapy

Severe hypoxia in asthmatic subjects has no significant effect on bronchial responsiveness to eucapnic hypopnea with dry air.[34] However, with the induced severe hypoxia, there was a marked increase in heart rate, suggesting an increase in sympathetic output possibly having an opposing effect to the bronchial response. Alternately, mild isocapnic hypoxia does not alter lung mechanics or produce an increase in plasma catecholamines.[4] With these controls in place, mild hypoxia (oxygen saturation about 87%) was shown to significantly increase bronchial responsiveness to aerosolized methacholine. This was assessed by the heightened FEV_1 and airway resistance responses to methacholine, as well as the steeper slope of the dose-response curve.

The mechanisms involved in the enhancement of nonspecific bronchial responsiveness by mild hypoxia have not been clarified. Hypoxia may have a direct action on the bronchial smooth muscle, be a stimulus for the release of bronchoconstricting mediators and neurotransmitter peptides, or initiate a reflex arc via the peripheral chemoreceptors and vagal efferent to bronchial smooth muscle. Animal data support this latter hypothesis.[3,35] In Denjean et al.'s study[4] in humans, this reflex mechanism may be of importance, but baseline bronchomotor tone was not altered by the hypoxia.

Whatever the mechanism(s) of action of mild hypoxia increasing bronchial responsiveness in asthma, the effect may play a role in nocturnal asthma. It is very uncommon for asthmatic patients to have moderate to severe oxygen desaturation during sleep. Usually there is only mild nonapneic oxygen desaturation intermittently throughout the night, with improvement in oxygenation in association with maneuvers that improve lung function.[36,37]

What would be the effect on the overnight lung function if the mild oxygen desaturation was corrected only by supplemental oxygen? This has not been fully investigated. Martin and Pak,[14] in the above discussed nasal CPAP nonapneic nocturnal asthma study, did place two nocturnal asthmatic individuals on supplemental oxygen at 2 L/min during sleep. These two individuals were the only ones who improved their overnight lung function to any clinically significant degree with nasal CPAP. Table 2 shows the results from the baseline, nasal CPAP, and oxygen nights in these two subjects in regard to sleep architecture, lung function, oxygen saturation, and heart rate. Both the nasal CPAP and oxygen nights improved the overnight change in FEV_1 to the same extent within each patient. The improvement in oxygen saturation was also similar on both intervention nights. Whereas on the nasal CPAP night the sleep architecture was markedly worsened by CPAP, supplemental oxygen alone maintained the baseline night sleep characteristics. Thus, it appears that improvement in the overnight lung function occurred in these two subjects who had the most time spent below an oxygen

Table 2

Supplemental Oxygen Study

Subject		Sleep Efficiency (%)	Awake	Sleep Staging (%) 1	2	3–4	REM	FEV₁(L) Bedtime	% Overnight Change in FEV₁	SaO₂ % < 90%	X̄ HR
#2	Baseline	91	9	5	49	21	16	1.76	−51	26.6	59
	CPAP	68	31	8	44	9	8	1.70	−34	3.5	63
	Oxygen	98	2	5	54	17	22	1.35	−20	0	52
#7	Baseline	96	4	5	36	38	17	2.55	−27	10.9	80
	CPAP	68	33	16	39	14	0	2.67	+4	0.7	75
	Oxygen	94	6	7	35	40	12	2.09	−1	0	65

Reprinted with permission.[14]

saturation of 90%, but still with overall oxygen desaturation in the mild range. Both nasal CPAP and supplemental oxygen corrected the oxygen desaturation, but only the supplemental oxygen did not worsen the sleep characteristics. Of interest, the mean heart rate did not change greatly on nasal CPAP or supplemental oxygen. If anything, on oxygen the mean sleep heart rate fell somewhat. If the heart rate reflects vagal tone, this would suggest that hypoxia and its correction did not alter vagal tone and, thus, the parasympathetic nervous system was not involved in the hypoxic related nocturnal asthma. Caution must be stressed here as to the interpretation of these findings in this very small population of two subjects. Further investigation is needed regarding the potential of mild hypoxia to produce the nocturnal worsening of lung function in asthma.

Inspiratory Muscle Training

In an interesting study on specific inspiratory muscle training in asthmatic patients, Weiner et al. hypothesized inspiratory muscle training would result in an increase in strength and endurance, and this would be associated with improvement in symptoms.[38] Thirty patients with moderate to severe asthma were studied. Fifteen received respiratory muscle training and 15 served as a control population with sham training in this double-blind study. The training was performed with a threshold inspiratory muscle device for 30 minutes per day, 5 days per week for 6 months. Inspiratory muscle strength, as expressed by the maximal inspiratory pressure generated at residual volume significantly increased from 84 ± 4.9 to 107 ± 4.8 cm H_2O (P<0.001). The respiratory muscle endurance measured by the relationship between peak pressure and maximal pressure from residual value increased from $67.5\% \pm 3.6\%$ to $93.1\% \pm 4.6\%$ (P<0.001). These findings were in the treated group; the control group had no significant differences before or during the sham training. Of interest, these objective findings of increased strength and endurance were translated into improvement in nocturnal symptoms (P<0.05), morning tightness (P<0.05), daytime asthma (P<0.01), and cough (P<0.005). Inhaled beta$_2$-adrenergic agonist use, emergency room visits, and sick leave days were all significantly reduced (P<0.05) in the treated group versus the sham control group.

Perhaps by training the inspiratory muscles, the force needed to overcome the increased resistance in asthma is achieved. This would then allow the diaphragm to work at a better mechanical advantage instead of the disadvantageous position of hyperinflation. Also during sleep muscle tone is decreased and, if the inspiratory muscles are strengthened, this decrease in tonicity may play less of a role in the decrement of sleep related lung volume (see Chapter 4). Whatever the mechanism of effect, this interesting finding needs further study.

Secretions

Although asthmatics do not commonly complain of a productive cough as do patients with chronic bronchitis, secretions can still play a role in some patients with nocturnal asthma. At one extreme, autopsies have shown extensive mucus plugging of the airways in asthmatics.[39] In addition, animal studies have shown that the cough reflex mechanisms during sleep are suppressed, particularly during REM sleep.[40] With a suppressed clearance mechanism and potentially irritable airways, secretions may in some manner trigger or be involved in nocturnal asthma.

Not uncommonly, we see patients who develop upper respiratory infections and cough who tolerate the process better during the waking hours than during sleep. Upper respiratory infections can produce reflex bronchoconstriction, but this does not explain the observed awake-to-sleep difference. The exact role of secretions in this situation is not certain but is probably significant.

Asthma is a disease of the airways, and this includes all airways, both intrathoracic and extrathoracic. Chronic sinusitis and/or postnasal drip are frequent problems in asthmatics. Not only will daytime symptoms improve as the sinus and discharge process is cleared, but nocturnal symptoms can dramatically improve. An example of this process is demonstrated by the following history. A 52-year-old female complained of progressive dyspnea and wheezing over a 6-month period. This was associated with postnasal drip and frequent nocturnal awakenings with worsening of breathing. The sinus films showed chronic pansinusitis, FEV_1 was 50% of predicted, and the ratio of FEV_1 to the forced vital capacity was 62%. After a program of inhaled and long-acting oral bronchodilators was initiated, her FEV_1 was 79% of predicted, but she still experienced "poor sleep." The morning FEV_1 was 50% of predicted. She was placed on an intensive program to clear nasal-sinus drainage, which included oral decongestants, saline nasal washes, and nasal steroids. Within several days she slept through the night, and the morning FEV_1 was 75% of predicted.

The exact reason for worsening of symptoms in asthmatics because of sinus problems or extrathoracic secretions is not known. One possibility is a nasal or laryngeal irritation reflex producing bronchoconstriction. Another possibility is the difference between nasal breathing and mouth breathing in asthmatics. Exercise-induced bronchoconstriction is much greater with mouth breathing than nasal breathing.[41] Thus, if nasal congestion is present and the patient is mouth breathing, worsening of nocturnal symptoms may develop. A third possibility is that these secretions are aspirated, which can set off direct or reflex mechanisms worsening the asthma. Brugman et al. have shown in an animal model that induced inflammation in the sinuses does not by itself produce an increase in pulmonary resistance. However, by aspiration of the inflammatory components, pulmonary resistance increases.[42]

Treatment of sinusitis is by extensive nasal-sinus irrigation with normal saline. This procedure involves making a new saline solution for every irrigation. This is done by mixing one-half teaspoon of table salt in an eight ounce glass of warm water with a "pinch" of baking soda. The entire eight ounces should be used per treatment, two to four times a day depending upon the severity of the sinusitis. A simple delivery system is having the patient pour some saline into the palm of the hand and "snuff" the solution into the nose, one nostril at a time. This is best done with the head down, bent over the sink. A better method is the use of a large rubber "bulb" syringe. The syringe is completely filled with the saline solution. Then the syringe is inserted just inside the nostril which is pinched around the tip to prevent the solution from running out of the nose. The bulb is gently squeezed hard enough to make the saline go up and over the palate and out the mouth. This is repeated for the other nostril. While performing this procedure, air should be sniffed through the open nostril to assist the saline in the appropriate direction. For patients with very thick purulent secretions, the use of a Water Pik with a snug fitting nasal adapter is of benefit. The Water Pik is set at the *lowest possible pressure* and the tip inserted just inside the nostril, the head is inclined over the sink with the mouth open so fluid flows out of the mouth.

For very small children, it may be necessary to use a dropper and put 10 to 20 drops of salt solution into each nostril. If necessary, use a bulb syringe to suction mucus. For older children (those who can blow their nose with coaching), use a clean, empty squeeze bottle, such as a nasal spray bottle. Fill with solution, using the recipe given above and gently squeeze the solution into one nostril. Have the child blow his/her nose after each squeeze.

An oral decongestant such as pseudoephedrine 60 mg po t.i.d. to q.i.d. will also be of benefit. Topical nasal steroids are of additional benefit in decreasing the inflammatory process and edema of the nasal mucosa and openings to the sinuses. Two actuations of an aqueous nasal steroid in each nostril following the nasal irrigation is appropriate. It will take several weeks of usage before results from the steroids are seen. The aqueous steroid form is less irritating to the patient. Occasionally antibiotics are needed. Since blood perfusion to the sinus cavities is relatively poor, three weeks of antibiotic usage is needed. Too short a course is one of the most common errors in antibiotic treatment of sinusitis. Rarely, surgical intervention will be needed.

Treatment of the lower airways secretions raises a major problem. Secretions appear to be worse during the sleep related hours. Figure 8A shows a relatively clear endobronchial field at 1600 hours, but in the same individual at 0400 hours secretions are prominent (Fig. 8B). Basically very little is known about what treatments will really improve the depressed clearance mechanisms that are present during sleep. Medications such as theophylline will increase ciliary function during daytime studies and may be of the same benefit during sleep. It is not known if some asthmatics would benefit from postural draining and percussion at bedtime. Indeed, in certain asthma patients this vigorous chest percussion could worsen lung function. A rationale for mucolytic agents such as acetylcysteine or potassium iodide preparations could be made, but again it is unknown if any benefit or worsening would be derived from these agents. If the patient smokes cigarettes, cessation of this ciliostatic agent would be of great benefit.

There is no doubt that intra- and extrathoracic secretions play a major role in asthma and the nocturnal aspect of this syndrome. However, relatively little is known about the problem and, therefore, treatment is limited.

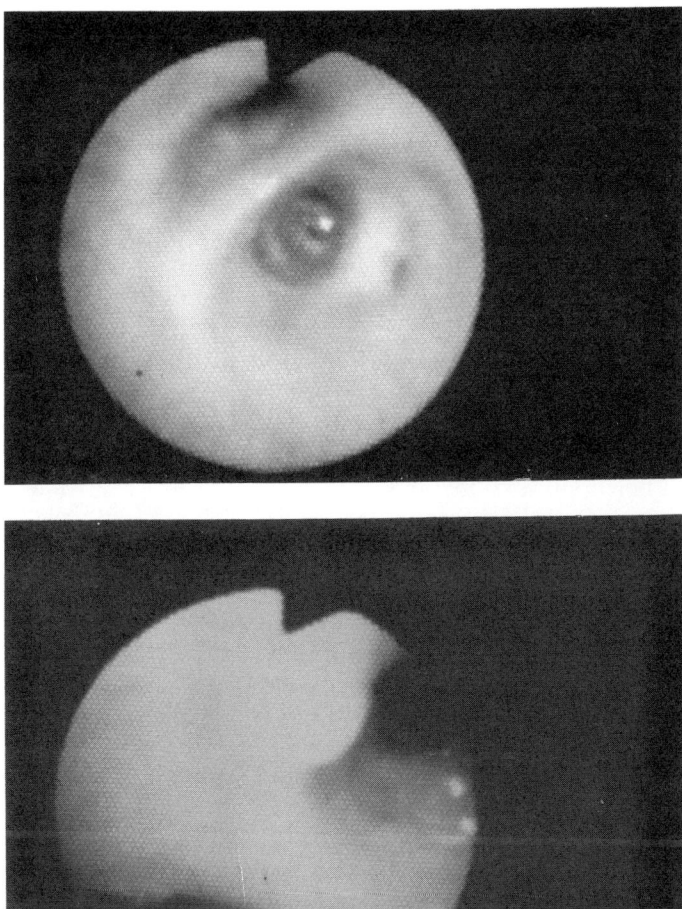

Figure 8. Top: Fiberoptic bronchoscopic view in the bronchus intermedius looking at the right middle (arrow) and lower lobes at 1600 hours in a nocturnal asthmatic individual. At this time, the airways are clear of secretions and the carina is sharp. Bottom: Shows the same individual's airways at 0400 hours with secretions now present and carina is splayed apart by edema. Lung function decreased from bedtime to 0400 hours by greater than 20%.

Sleep Position

Ventilatory function decreases from the upright to the supine position as measured during the daytime in normal subjects by lung compliance, pulmonary resistance, and expiratory reserve volume.[43] Mossberg and Jonssen showed during daytime studies in asthmatic patients that peak expiratory flow rates progressively decreased from upright to supine position over time.[44] Clark and Hetzel, however, demonstrated that the actual overnight decrease in PEFR was not related to the horizontal position. Keeping the asthmatic patient in the horizontal position during the daytime did not prevent the return of peak expiratory flow rates to upright daytime levels. It was only during sleep that the flow rates decreased.

Another study involving a complex protocol had one phase where the subjects were kept awake but supine during the night until a predetermined point where lung function was previously noted to be the lowest during sleep. From that point on they were allowed to fall asleep.[46] Two distinct groups emerged from this study. Six patients sustained virtually all of their nocturnal decrement in peak flow rates while still awake. Five patients showed little or no fall in lung function while kept awake, but their usual decrease ensued while asleep for the short interval before being awakened at 0600 hours. Thus, the supine position itself is not the key factor for all patients. Even in those who had the decrement in lung function awake, it may have been independent of the supine position.

Does sleep position matter for patients with allergic rhinitis? Nasal congestion definitely has a circadian rhythm with worse symptoms occurring during the night. If those patients also have asthma, this increase in congestion could potentiate nocturnal asthma. That is, with mouth breathing the air delivered to the lungs will be cooler and dryer than if it were humidified via the nasal route (see above). It would be of interest to know if in this specific situation, elevation of the head of the bed would be of any benefit. This may be tried empirically in this category of patients.

References

1. Chan CS, Woolcock AJ, Sullivan CE. Nocturnal asthma: role of snoring and obstructive sleep apnea. Am Rev Respir Dis 1988; 137:1502–1504.
2. Nadel JA, Widdicombe JG. Reflex effects of upper airway irritation on

total lung resistance and blood pressure. J Appl Physiol 1962; 17:861–865.

3. Nadel JA, Widdicombe JG. Effect of changes in blood gas and carotid sinus pressure on tracheal volume and total lung resistance to airflow. J Physiol 1962; 163:13–33.

4. Denjean A, Roux C, Herve P, Bunniot JP, Comoy E, Duroux P, Gaultier C. Mild isocapnic hypoxia enhances the bronchial response to methacholine in asthmatic subjects. Am Rev Respir Dis 1988; 138:789–793.

5. Sullivan CE, Issa FG. Obstructive sleep apnea. In: Kryger MH (ed) Symposium on Sleep Disorders, Vol. 6. Clinics in Chest Medicine, WB Saunders, Philadelphia, 1985; 4:646–648.

6. Tilkian AG, Motta J, Guilleminault C. Cardiac arrhythmias in sleep apnea. In: *Sleep Apnea Syndromes,* Guilleminault C, Dement W (eds). Alan R. Liss, Inc. New York, 1978; pp 197–210.

7. Morrison JFJ, Pearson SB, Dean HG. Parasympathetic nervous system in nocturnal asthma. Brit Med J 1988; 296:1427–1429.

8. Guilleminault C, Quera-Salva MA, Powell N, et al. Nocturnal asthma: Snoring, small pharynx and nasal CPAP. Eur Respir J 1988; 1:902–907.

9. Ballard R, Saathoff MC, Patel DK, Kelly PL, Martin RJ. Effect of sleep on nocturnal bronchoconstriction and ventilatory patterns in asthmatics. J Appl Physiol 1989; 67:243–249.

10. Hanly PF, George CF, Millar TW, Kryger MH. Heart rate response to breath-hold, Valsalva and Mueller maneuvers in obstructive sleep apnea. Chest 1989; 95:735–739.

11. Cuccagna G, Mantovani M, Brignani F, Purchi C, Lugaresi E. Continuous recording of the pulmonary and systemic arterial pressure during sleep in syndrome of hypersomnia with periodic breathing. Bull Eur Physiop Thol Respir 1972; 8:1159–1172.

12. Guilleminault C, Tilkian A, Lehrman K, Forno L, Dement WC. Sleep apnea syndrome: states of sleep and autonomic dysfunction. J Neurol Neurosurg Psych 1977; 40:718–725.

13. Guilleminault C, Winkle R, Melvin K, Tilkian A. Cyclical variation of the heart rate in sleep apnea syndrome, mechanisms and usefulness of 24-hour electrocardiography as a screening technique. Lancet 1984; 1:126–136.

14. Martin RJ, Pak J. Nasal CPAP in non-apneic nocturnal asthma. Chest 1991; 100:1024–1027.

15. Chen WY, Horton DJ, Weiser PC. Airway obstruction induced by body cooling in asthmatics. Physiol 1977; 20:16.

16. Chen WY, Horton DJ. Airways obstruction in asthmatics induced by body cooling. Scand J Respir Dis 1978; 59:13–20.

17. Horton DJ, Chen WY. Effects of breathing warm humidified air on bronchoconstriction induced by body cooling and by inhalation of methacholine. Chest 1979; 75:24–28.

18. Petersdorf RA. Disturbance of heat regulation. In: MM Wintrobe, GW Thorn, RD Adams, et al. (eds). Harrison's Principles of Internal Medicine, 7th ed, McGraw-Hill, New York. 1974.

19. Chen WY, Chai H. Airway cooling and nocturnal asthma. Chest 1982; 81:675–680.

20. Overholt RH, Voorhees RJ. Esophageal reflux as a trigger in asthma. Dis Chest 1966; 49:464–466.
21. Urschel HC Jr, Paulson DL. Gastroesophageal reflux and hiatal hernia. Complications and therapy. J Thorac Cardiovasc Surg 1967; 53:32.
22. Davis MV. Evolving concepts regarding hiatus hernia and gastroesophageal reflux. Ann Thorac Surg 1969; 7:120–133.
23. Babb RR, Notarangelo J, Smith VM. Wheezing: a clue to gastroesophageal reflux. Am J Gastroenterol 1970; 53:230–233.
24. Mays EE. Intrinsic asthma in adults: Association with gastroesophageal reflux. J Am Med Assoc 1976; 236:2626–2628.
25. Goodall RJR, Earis JE, Cooper DW, et al. Relationship between asthma and gastro-oesophageal reflux. Thorax 1981; 36:116–121.
26. Mansfield LE, Stein MR. Gastro-oesophageal reflux and asthma: A possible reflex mechanism. Ann Allergy 1978; 41:224–226.
27. Martin ME, Grunstein MM, Larsen GL. The relationship of gastroesophageal reflux to nocturnal wheezing in children with asthma. Ann Allergy 1982; 49:318–322.
28. Spaulding HS Jr, Mansfield LE, Stein MR, Sellner JC, Gremillion DE. Further investigation of the association between gastroesophageal reflux and bronchoconstriction. J Allergy Clin Immunol 1982; 69:516–521.
29. Hughes DM, Spier S, Rivlin J, et al. Gastroesophageal reflux during sleep in asthmatic patients. J Pediatr 1983; 102:666–672.
30. Berquist WE, Rachelefsky GS, Rowshan N, et al. Quantitative gastroesophageal reflux and pulmonary function in asthmatic children and normal adults receiving placebo, theophylline, and metaproterenol sulfate therapy. J Allergy Clin Immunol 1984; 73:253–258.
31. Tan WC, Martin RJ, Pandey R, Ballard RD. Effects of spontaneous and simulated gastroesophageal reflux on sleeping asthmatics. Am Rev Respir Dis 1990; 141:1394–1399.
32. Speelberg B, deMonchy JRG. Is deliberate waking and bronchodilator use a useful therapy in nocturnal asthma? Am Rev Respir Dis 1991; 143:A32.
33. Ariaansz M, Keimpema ARJ, Nauta JJP. Subjective sleep quality and alertness in asthmatics and controls. Am Rev Respir Dis 1991; 143:A31.
34. Tam AK, Goffroy A, Myers DJ, Seltzer J, Sheppard D, Boushey HA. Effect of eucapnic hypoxia on bronchomotor tone and on the bronchomotor response to dry air in asthmatic subjects. Am Rev Respir Dis 1985; 132:690–693.
35. Vidruk EH. Hypoxia potentiates, oxygen attenuates deflation-induced reflex tracheal constriction. J Appl Physiol 1985; 59:941–946.
36. Martin RJ, Cicutto LC, Ballard RD, Goldenheim PD, Cherniack RM. Circadian variations in theophylline concentrations and the treatment of nocturnal asthma. Am Rev Respir Dis 1989; 139:475–478.
37. Zwillich CW, Neagley SR, Cicutto L, White DP, Martin FJ. Nocturnal asthma therapy: Inhaled bitolterol versus sustained release theophylline. Am Rev Respir Dis 1989; 139:470–474.
38. Weiner P, Azgad Y, Ganum R. Specific inspiratory muscle training in patients with bronchial asthma. Chest 1992;102:1357–1361.

39. Hetzel MR, Clark TJH, Branthwaite MA. Asthma: Analysis of sudden deaths and ventilatory arrests in hospital. Br Med J 1977; 1:808–811.
40. Sullivan CE, Murphy E, Kazan LF, et al. Waking and ventilatory responses to laryngeal stimulation in sleeping dogs. J Appl Physiol: Respir Environ Exercise Physiol 1978; 45:681–689.
41. Shturman-Ellstein R, Zeballos RJ, Buckley JM, et al. The beneficial effect of nasal breathing on exercise-induced bronchoconstriction. Am Rev Respir Dis 1978; 118:65–73.
42. Brugman SM, Larsen GL, Henson PM, Irvin CG. Mechanism of the increase in lower airways responsiveness associated with sinusitis in a rabbit model. Am Rev Respir Dis, in press.
43. Behrakis PK, Baydur A, Jaeger MJ, Milic-Emili J. Lung mechanics in sitting and horizontal body positions. Chest 1983; 83:643–646.
44. Mossberg B, Junsson E. Is asthma at night caused by posture? Chest 1985; 87:216–217S.
45. Clark TJH, Hetzel MR. Diurnal variation of asthma. Br J Dis Chest 1977; 71:87–92.
46. Hetzel MR, Clark TJH. Does sleep cause nocturnal asthma? Thorax 1979; 34:749–754.

Index

Abdomen strapping, 130, 132, 133
Acetylcholine, 15
Acid infusion, esophagitis and, 92, 369–370
Adrenaline, chronesthesy of, 39
Adrenal insufficiency, 55
Adrenal suppression, 318–324
Adrenocorticotropic hormone, chronesthesy of, 36, 37
Age, theophylline and, 247–248
Airflow resistance in sleeping asthmatics, 122–127
Airway cooling, 92–93
Airway disease, circadian rhythm and, 13–16, 17
Airway hyperresponsiveness, 333–334
Airway patency, circadian rhythm of, 13, 14
Airway resistance
  lower; *see* Lower airway resistance
  sleep and, 371–372
Airway secretions, 378–381
  lower, 380–381
  lung function and, 90–91
Albuterol, 260–261, 271–272
  theophylline and, 224, 225
Allergen challenge, asthmatic response to, 89–90
Allergic rhinitis, 6, 7
Allergy, 350–351
  circadian rhythm and, 11–13
  lung function and, 88–90
Aminophylline, 34, 228, 229, 230, 249–250
  versus albuterol, 271
Amplitude of rhythm, 3

Angina, 6, 8
Antacids, 103–104
Antiasthma medication, 48–49
Antibiotics, sinusitis and, 349, 380
Anticholinergics
  inhaled, 334–340
  nocturnal inhalation of, 373
  as therapy, 107–108
Antigen challenge, asthmatic response to, 137–138, 178–179
Antihistamines, 340–346
Antioxidants, 175
Arachidonate pathways, 286–288
Armophyllin, 233, 237–241, 250–251
Arousal abnormalities, 74–75
Arthritis, circadian rhythm and, 6–8
Aspiration
  of airway secretions, 91
  of gastric contents, 91, 366, 368–369
Asthma
  circadian rhythm and, 6
  nocturnal; *see* Nocturnal asthma
  sleep and, 119–122, 123
Asthmatic response
  allergen challenge and, 89–90
  antigen challenge and, 137–138, 178–179
Atropine, vagal tone and, 93, 94
Atropine methylnitrate, 337, 339
Awakening
  medication and frequency of, 282
  treatment and, 371–374

BAL; *See* Bronchoalveolar lavage
Bambuterol, 267–270

387

Basophils, 290
Beclomethasone, 260, 314
Beclomethasone dipropionate, 342, 344–345
Beta$_2$-adrenergic agonists
nocturnal inhalation of, 373
as therapy, 104–105
Beta-adrenergic receptors, 289
Beta$_2$-adrenergic receptors, 100–101
Beta$_2$-agonist therapy, 257–272
inhalation, 260–264
sustained-release, 264–272
Betoxycaine, chronesthesy of, 33
Biological rhythm, 3, 283; *See also* Circadian rhythm
chronopharmacology and, 26–42; *See also* Chronopharmacology
chronotherapeutics and, 42–58; *See also* Chronotherapeutics
diagnostic procedures and, 10–16
medication and, 25–69
Biological time structure, 3
Biopsy, endobronchial, 212–213
Blood pooling, intrapulmonary, 131, 139
Blood pressure, circadian rhythm and, 9
Body plethysmograph, horizontal volume-displacement, 127–128
Bronchial hygiene, 103
Bronchial reactivity, 81–83
Bronchial responsiveness
hypoxia and, 375
nocturnal changes in, 136–140
Bronchitis, 103
Bronchoalveolar lavage fluid
corticosteroids and, 311–313
inflammatory cells in, 96–98
lymphocytes and, 213–214
nocturnal asthma and, 176, 177
Bronchoconstriction
nervous system and, 335
toleration of, 143, 146
Bronchodilator
chronotherapy and, 223
nasal continuous positive airway pressure and, 360, 361
peak expiratory flow rate and, 85, 86

Bronchus intermedius, bronchoscopic view in, 381
Bronkodyl-SR, 232
Budesonide, 314, 315
versus terbutaline, 266–267, 268, 269

Candidiasis, inhaled corticosteroids and, 316
Carbon dioxide, expired, 143, 146
Catecholamine
circadian rhythms and, 95
pulmonary function and, 95
CBG; *See* Corticosteroid binding globulin
Cefodizime, 30
Cellular mechanisms, 163–197
eosinophils in, 169–174, 182–186
mast cells in, 167–169, 178–182
neutrophils in, 174–176, 186–188
Cephalosporin, 30
Cerebral hemorrhage, circadian rhythm and, 6
Chemotactic factors, macrophages and, 204–205
Chemotherapy, circadian timing of, 46
Chest wall strapping, 130, 132, 133
Choledyl, 233
Choline theophyllinate, 233
Chronesthesy, 27–39, 40
Chronic obstructive lung disease, 147–151
Chronobiology, 1–23, 76
biological rhythms and, 10–16
circadian rhythms and, 6–10
Chronobiotics, 47
Chronokinetics, 26–27
Chronopharmacology, 26–42, 49–53
of beta$_2$-agonist therapy, 257–280; *See also* Beta$_2$-agonist therapy
of theophylline therapy, 221–257; *See also* Theophylline therapy
Chronotherapeutics, 42–58
Chronotolerance, 39–42
Chronotoxicity, 39–44
Cimetidine, 368

Circadian rhythm; *See also* Biological rhythm
disease and, 6–10
pathophysiology and, 76–102
pulmonary function and, 76–102; *See also* Pulmonary function
Circulating mediators, lung function and, 95–96
Cisplastin, chronotoxicity of, 44
COLD; *See* Chronic obstructive lung disease
Conductance, specific, 128–129, 130
Continuous positive airway pressure, nasal, 358–365; *See also* Nasal continuous positive airway pressure
Corticosteroid binding globulin, 50–51
Corticosteroids, 49–58, 99, 281–331
adrenal suppression and, 318–324
basophils and, 290
eosinophils and, 291–293, 294
inflammatory cells and, 289–290
inhaled, 313–318
lung function and, 98–100
lymphocytes and, 213–215, 295–296, 297, 298
macrophages and, 296–299
mast cells and, 290
monocytes and, 296–299
neutrophils and, 291, 292, 293
plasma cortisol and, 299–300
sensitivity in nocturnal asthma, 300–304
side effects of, 317–318
as therapy, 106–107
withdrawal from, 318–319
Cortisol, 36, 285
circadian variation in, 299–300
corticosteroid withdrawal and, 319–320
Cortisol production, dexamethasone and, 323
Cortisone, 285
CPAP; *See* Continuous positive airway pressure
Cromolyn, 341–343
Cutaneous reactivity, 12
Cystic fibrosis, 152–154

Cytokine, 167–168
Cytometry, lymphocytes and, 210

Death; *See* Mortality
Decongestant, oral, 380
Dexamethasone, 285
cortisol production and, 323
Diaphragm, nonREM sleep and, 131, 134
Dosing schedule
hypoglycemic stress and, 321–322
inhaled corticosteroids and, 314–315
peak expiratory flow rate and, 305, 307
Doxorubicin, chronotoxicity of, 44
Duovent versus albuterol, 261
Dutimelan, 56–57
Dysphonia, 316
Dyspnea, 221–223

ECP; *See* Eosinophil cationic protein
EDN; *See* Eosinophil derived neurotoxin
Elixophyllin, 230
Endobronchial biopsy, lymphocytes and, 212–213
Endogenous hormones, nocturnal wheezing and, 188–190
Endotoxin receptor, macrophages and, 202, 203
Environment as reversible factor, 103
Eosinophil cationic protein, 170
Eosinophil count, 309–311
Eosinophil derived neurotoxin, 170–171
Eosinophil peroxidase, 171
Eosinophils, 169–174, 182–186
asthma and, 166, 171–174
corticosteroids and, 291–293, 294
inflammation and, 96–97
mediators of, 170–171
nocturnal asthma and, 182–186
Epilepsy, circadian rhythm and, 8
Epinephrine
circadian rhythm and, 257
histamine and, 340–341
inhalation of, 257, 259

mast cells and, 180, 181
peak expiratory flow rate and, 188
EPO; *See* Eosinophil peroxidase
Esophagitis, 92, 369
Essential hypertension, circadian rhythm and, 6, 9
Exercise, nocturnal awakening and, 79

$F_{et}CO_2$; *See* Carbon dioxide, expired
Fenoterol, 261
FEV_1; *See* Forced expiratory volume in one second
Flow cytometry, lymphocytes and, 210
Food, drug absorption and, 246–247
Forced expiratory volume in one second, 261, 265, 308, 309, 371
superoxide release and, 187–188
sustained-release theophylline and, 224, 225
FRC; *See* Functional residual capacity
Functional residual capacity, 130, 132, 133, 150, 152
nonREM sleep and, 131, 135
sleep stages and, 128, 129

Gamma globulin, intravenous, 348
Gastroesophageal reflux, 366–371
lung function and, 91–92
therapy for, 103–104
Gastroesophageal sphincter, 366
Gold therapy, 347
Growth factor, platelet-derived, 204–205

Heartburn, 103–104
Heart rate, oxygen and, 377
Hemorrhage, cerebral, 6
Hiatal hernia, 368
Histamine
airway response to, 15
circadian rhythms and, 95–96
epinephrine and, 340–341
mast cells and, 167, 180, 181
nocturnal asthma and, 180, 182
Histamine threshold concentration, 334

Horizontal volume-displacement body plethysmograph, 127–128
House dust, airway response to, 15
H_1-receptor antagonist, 43, 45
Humoral mediator release, sleep apnea and, 362–363
Hydrocortisone infusion, 301–304
Hypercapnia, chronic obstructive lung disease and, 147–150
Hypertension, circadian rhythm and, 6, 9, 10–11
Hypoglycemic stress, corticosteroid dosing schedule and, 321–322
Hypoxemia, 147–150, 362
Hypoxia, 362, 374–377

Immediate asthmatic response, 89–90
Immune complex, macrophages and, 202, 203
Infarction, myocardial, 6, 8
Inflammation
lung function and, 96–98
macrophage and, 201
Inflammatory cells, corticosteroid effect on, 289–290
Inhalation therapy, beta$_2$-agonist, 260–264
Inspiratory muscles, 131, 134, 377–378
Intercostal muscle, nonREM sleep and, 131, 134
Interleukin 8, 204–205
Interleukin-2 receptor, 211
Intrapulmonary blood pooling, 131, 139
Ipratropium bromide, 337, 338–339
nocturnal inhalation of, 373
Irrigation, nasal-sinus, 379–380
Isoproterenol, 260

Ketoprofen, concentration of, 31
Ketotifen, 345–346

LAR; *See* Late asthmatic response
Late asthmatic response, 171–172, 350–351
allergen challenge and, 89–90
antigen challenge and, 178–179

Leukocyte count, 184, 185
Leukotriene B_4, 204–205, 206–207
Leukotrienes, sulfidopeptide, 167, 168
Lidocaine, chronesthesy of, 33
Lower airway resistance, 153, 370
  chronic obstructive lung disease and, 150, 151
  oxygen saturation and, 120–121
  sleep and, 77, 78, 124–126
  ventilation and, 141, 142, 144
Lower airway secretions, 380–381
LTB_4; *See* Leukotriene B_4
Lung function; *See* pulmonary function
Lung volume in sleeping asthmatics, 127–136
Lymphocyte
  asthma and, 209–212
  CD-25 positive, 211
  corticosteroids and, 295–296, 297, 298
  nocturnal asthma and, 212–217

Macrophage
  activation of, 202–203
  asthma and, 200–205
  corticosteroids and, 296–299
  nocturnal asthma and, 205–209
Major basic protein, 170
Mast cells, 167–169, 178–182
  asthma and, 168–169
  biology of, 167–168
  corticosteroids and, 290
  nocturnal asthma and, 178–182
Mast cell stabilizers, 340–346
MBP; *See* Major basic protein
Mediators
  circulating, 95–96
  of eosinophils, 170–171
Medical chronobiology, 1–23; *see also* Chronobiology
Medication
  antiasthma, 48–49
  biological rhythms and, 25–69;
    See also Chronopharmacology;
    Chronotherapeutics
  timing of, 305, 306
Metaproterenol, 35, 260

Methacholine, 143, 147
Methotrexate, 347
Methylprednisolone, 49–53, 285
Monocytes, corticosteroids and, 296–299
Mortality, nocturnal asthma and, 74, 221–223
Mouth breathing, 90–91, 379
Mucolytics, 346
Mucous, corticosteroids and, 289
Müeller maneuver, 87, 363
Muscarinic receptor subtype, 339–340
Muscle
  inspiratory, 131, 134, 377–378
  intercostal, 131, 134
  sternocleidomastoid, 131, 134
Myocardial infarction, circadian rhythm and, 6, 8

Nasal continuous positive airway pressure, 358–365
  oxygen and, 375–377
  sleep apnea and, 85–88
Nasal-sinus irrigation, 379–380
Nasal steroid, aqueous, 380
Nedocromil, 342, 343–344
Nervous system, 334–340
Neutrophils, 174–176, 186–188
  asthma and, 175–176
  biology of, 174
  corticosteroids and, 291, 292, 293
  inflammation and, 96–97
  nocturnal asthma and, 185, 186–188
  reactive oxygen species metabolism and, 174–175
Nocturnal asthma
  frequency of, 72–73
  nonmedication approaches to, 358, 359
  pathophysiology of; *See* Pulmonary function
Nocturnal asthma therapy, 102–109
  anticholinergics in, 107–108
  beta$_2$-adrenergic agonists in, 104–105
  corticosteroids in, 106–107

pharmacological interventions in, 104

reversible factors in, 103

theophylline in, 105–106

Nocturnal awakening

medication and frequency of, 282

treatment and, 371–374

Nonpharmacological interventions, 357–385

inspiratory muscle training as, 377–378

nasal continuous positive airway pressure as, 358–365

nocturnal awakening with treatment in, 371–374

oxygen therapy as, 374–377

reflux and, 366–371

secretions and, 378–381

sleep position and, 382

as therapy, 108–109

warm humidified air as, 365–366, 367

Nuelin, ODSRT250, 238

Nuelin, SR250, 232–233

ODSRT; *See* Once daily sustained-release theophylline

Once daily sustained-release theophylline, 237–245

Oral decongestant, 380

Orciprenaline, circadian rhythm of, 16

Oxitropium bromine, 336–338

Oxygen desaturation

nasal continuous positive airway pressure and, 362

sleep and, 83–85

theophylline and, 224, 226

Oxygen saturation

sleep and, 120–121

ventilation and, 143, 146

Oxygen therapy, 374–377

Parasympathetic nervous system, 335

Peak expiratory flow

Dutimelan and, 56

mast cells and, 180, 181

serum theophylline concentration and, 251

Peak expiratory flow rate

dosing schedule and, 305, 307

epinephrine and, 188

hydrocortisone infusion and, 301

nasal continuous positive airway pressure and, 360, 361, 362

sleep and, 122, 123

PEF; *See* Peak expiratory flow

PEFR; *See* Peak expiratory flow rate

Peptic ulcer disease, circadian rhythm and, 8

Period of rhythm, 3

pH, esophageal, 369–370

Phagocytosis, 202, 203

Phasing of rhythm, 3

Phenotype of macrophages, 201, 205–206

Phyllotemp, 234, 238

Physiology, respiratory, 117–161

Plasma cortisol; *See* Cortisol

Plasma epinephrine concentration, 257

Platelet activation factor, 204–205

Platelet-derived growth factor, 204–205

Plethysmograph, horizontal volume-displacement body, 127–128

Postnasal drip, 378

Posture

lung volume and, 130–131, 132–133

theophylline and, 247

Prednisolone, 285

chronotherapy of, 49–53

Prednisone, 99, 215, 285

macrophages and, 207–208

timing of, 106 107, 305–309

Prinzmetal angina, circadian rhythm and, 6

Propranolol, chronesthesy of, 32

Pseudoephedrine, 380

Pulmidur Forte, 233

Pulmonary function

airway and, 90–94

beta$_2$-adrenergic receptors and, 100–101

bronchial reactivity and, 81–83
circadian rhythm and, 76–102
circulating mediators and, 95–96
corticosteroid levels and, 98–100
histamine threshold concentration and, 334
inflammation and, 96–98
respiratory pattern abnormalities and, 83–85
sleep and, 77–81, 85–88
solar time and, 77–81
vagal cooling and, 93, 94
Pulmonary resistance, 122, 124

R_{la}; *See* Lower airway resistance
Radiotherapy, circadian timing of, 47
Reactive airway disease, circadian rhythm and, 13–16, 17
Reactive oxygen species, metabolism of, 174–175
Receptors, β-adrenergic; *See* Beta-adrenergic receptors
Recumbency, 77–79
Reflux, gastroesophageal, 366–371
lung function and, 91–92
therapy for, 103–104
Reflux asthma, 366
Residual capacity, functional; *see* Functional residual capacity
Resistance
airflow, 122–127
airway, 371–372
lower airway; *See* Lower airway resistance
pulmonary, 122, 124
supraglottic, 126
Respiratory arrest, nocturnal asthma and, 75
Respiratory burst, 175
Respiratory pattern abnormalities, 83–85
Respiratory physiology, sleep and, 117–161; *See also* Sleep
Rhinitis, allergic, 6, 7

Saline infusion, esophagitis and, 92, 369–370
Salmeterol, 261–263

SaO_2; *See* Oxygen saturation
Secretions, 378–381
airway, 90–91
lower airway, 380–381
Serum theophylline concentration
comparison of, 231, 235, 236
peak expiratory flow and, 251
Theo-24 and, 241–245
Theo-dur and, 257, 258
TheoNite and, 257, 258
Uniphyl and, 240–242
Shift work
circadian rhythms and, 5
peak expiratory flow and, 77–79
Sinus irrigation, 379–380
Sinusitis, 349–350, 378
Sleep
airflow resistance and, 122–127
asthma and, 119–122, 123, 127
bronchial responsiveness and, 136–140
chronic obstructive lung disease and, 147–151
cystic fibrosis and, 152–154
lung volume and, 127–136
oxygen desaturation and, 83–85
position, 382
respiratory physiology and, 117–161
ventilation and, 140–147
Sleep alterations, lung function and, 77–81
Sleep apnea
continuous positive airway pressure and, 358–365
lung function and, 85–88
Sleep stages
asthmatic episodes and, 80–81
nasal continuous positive airway pressure and, 363–364
respiratory patterns and, 83–84
Slophyllin, 233
Snoring, sleep apnea and, 88, 358
Solar time, lung function and, 77–81
Somophyllin-CRT, 232, 233
Specific conductance, 128–129, 130
Sphincter, lower esophageal, 366
STC; *See* Serum theophylline concentration

Sternocleidomastoid muscle, 131, 134
Steroid, 99, 380
Steroid hormone-responsive cell, 286, 287
Strapping, lung volume and, 130, 132, 133
Stress, corticosteroid withdrawal and, 319
Stroke, circadian rhythm and, 6
Sulfidopeptide leukotrienes, 167, 168
Superoxide release, 185, 186–188
Supine posture, 130–131, 132
Supraglottic resistance, 126
Synchronizer, 4

Tablet corticosteroids, 49–53
TAO; *See* Troleandomycin
T-cells, asthma and, 164–165
Terbutaline, 264–271
  versus budesonide, 266–267, 268, 269
  versus theophylline, 266
Terfenadine, chronesthesy of, 38
Theo-1, 239
Theo-24, ODSRT, 239, 241–245
Theo-dur, 231–237, 239
  versus TheoNite, 255–256
Theolair, 228, 229, 230
TheoNite, 255–256
Theophylline; *See also* Serum theophylline concentration; Theophylline therapy
  chronesthesy of, 34, 40–41
  chronokinetics of, 227–228
  pharmacokinetics of, 28, 29, 245–248
  sustained-release, 224, 225
  versus terbutaline, 266

as therapy, 105–106, 253, 254
Theophylline chronotherapy, 226–227, 248–258
Theophylline therapy, 224–257
  chronokinetics of, 227–228
  chronotherapy of, 248–257
  immediate-release, 228, 229, 230
  sustained-release, 228–237
  time-dependent dose in, 245–248
  unsustained-release, 237–245
Thromboxane, 206–207
T-lymphocytes
  asthma and, 210
  circadian variations in, 296, 297, 298
Transcortin, 50–51
Triamcinolone, 285, 320, 321
Troleandomycin, 347–348
Tryptase, 167, 168
Typical angina, circadian rhythm and, 6, 8

Uniphyl, 238, 241, 242, 251–253
Uniphyllin, 238, 250, 252

Vagal cooling, lung function and, 93, 94
Vagal tone, 363
  sleep apnea and, 87
Ventilation
  chronic obstructive lung disease and, 150, 151
  cystic fibrosis and, 154
  in sleeping asthmatics, 140–147
  sleep stage and, 117, 118
Ventilatory drive, 143, 145

Warm humidified air, 365–366, 367
Wheezing, 188–190